How the Immune Sys
Self and Nonself

Daisuke Kitamura
Editor

How the Immune System Recognizes Self and Nonself

Immunoreceptors and Their Signaling

Daisuke Kitamura
Professor
Department of Medicinal and Life Science
Faculty of Pharmaceutical Sciences
Division of Molecular Biology
Research Institute for Biological Sciences
Tokyo University of Science
Yamazaki 2669, Noda
Chiba 278-0022, Japan
kitamura@rs.noda.tus.ac.jp

Cover design conceived by Daisuke Kitamura and executed by Chigusa Himuro

ISBN 978-4-431-99831-0 e-ISBN 978-4-431-73884-8

Springer is a part of Springer Science+Business Media
springer.com

Printed in Japan

Preface

How do you discriminate yourself from other people? This question must sound odd to you since you easily recognize others at a glance and, without any effort, would not mistake them for yourself. However, it is not always easy for some people to discriminate themselves from others. For example, patients with schizophrenia often talk with "others" living inside themselves. Thus it is likely that normally your brain actively recognizes and remembers the information belonging to yourself and discriminates it from the information provided by others, although you are not conscious of it. This brain function must have been particularly important for most animals to protect their lives from enemies and for species to survive through evolution. Similarly, higher organisms have also acquired their immune system through evolution that discriminates nonself pathogens and self-body to protect their lives from pathogens such as bacteria or viruses.

The brain system may distinguish integrated images of self and nonself created from many inputs, such as vision, sound, smell, and others. The immune system recognizes and distinguishes a variety of structural features of self and nonself components. The latter actually include almost everything but self: for example, bacteria, viruses, toxins, pollens, chemicals, transplanted organs, and even tumor cells derived from self-tissue. To this end the immune system recruits different kinds of immune cells, such as B and T lymphocytes, natural killer (NK) cells, dendritic cells, and macrophages. These cells have specific functions and are equipped with distinctive sets of receptors to recognize self and/or nonself components. B and T lymphocytes have characteristic antigen receptors whose binding specificities are extremely diverse among individual cells. This diversity is generated through somatic rearrangement of the genes encoding the antigen receptors. The repertoires of the diverse antigen receptors cover a huge variety of self and nonself antigens, which normally results in tolerance to self antigens and the immune responses to the nonself antigens including antibody production, T-cell-mediated cytotoxicity, and inflammation. The antigen recognition signal is critically regulated by activating or inhibitory co-receptors on the same cells. The antigens are also recognized by specific antibodies bound with Fc receptors on various immune cells. NK and NKT cells express collectively what are called NK receptors composed of various pairs of receptors, one recognizing a specific viral or neoplastic antigen and the other self MHC (major histocompatibility complex) antigen on the self cells, and the pair work together to discriminate a target to kill. In addition to the NK receptors, a growing number of the paired immunoglobulin-like receptors have been identified. Finally, Toll-like receptors expressed on the cells such as dendritic cells and macrophages recognize various structural patterns of pathogens to eliminate them as the first defense and also to regulate the immune response by lymphocytes. Single mutations in some of these receptors or their downstream signaling molecules are known to cause autoimmune diseases in mice or humans, suggesting that functions of those immune cells expressing a different set of receptors are highly integrated and cooperate in an immune system to decide what is self and what is nonself.

In contrast to the extreme diversity of the receptors, the molecules involved in the signaling from such receptors is relatively limited and often is shared among different receptor systems. Therefore, the immune cells may use a common strategy for processing the signals from these receptors and for making a decision about their next action. The aim of this book is to try to clarify how the immune cells recognize through receptors a diversity of targets, being either self or nonself, how they translate this recognition into signals and transmit the signals, and how finally they decide to react or not to play a part in the immune response. For this purpose, an extensive and updated review on each receptor system is described in each chapter by an expert in that area. Although our knowledge is still far from complete, this challenge will give you many clues to discover the principle of the signaling machinery and to understand the complexity of cell interaction through receptor signaling in the immune system. Thus this book will help you to imagine the basic strategy of the immune system to distinguish self and nonself. I believe this challenge is particularly important for basic as well as clinical immunologists who are seeking a breakthrough in the regulation of immune diseases such as autoimmunity and allergy. Also, the concept and strategy of self–nonself discrimination in the immune system might be applied to the brain system as briefly mentioned above, or as a security system of a computer network.

I would like to thank all the authors for their invaluable contributions and their patience.

Daisuke Kitamura
Noda, Japan

Contents

List of Authors

Shizuo Akira
Department of Host Defense, Research Institute for Microbial Diseases, Osaka University, Osaka 565-0851, Japan; ERATO, Japan Science and Technology Corporation, Suita, Osaka 565-0871, Japan

Ching-Cheng Chen
Division of Developmental and Clinical Immunology, University of Alabama at Birmingham, Birmingham, AL 35294-3300, USA

Max D. Cooper
Howard Hughes Medical Institute, Birmingham, AL, USA

Yoshimoto Katsura
Division of Cell Regeneration and Transplantation, Advanced Medical Research Center, Nihon University School of Medicine, Tokyo 173-8610, Japan

Hiroshi Kawamoto
Laboratory for Lymphocyte Development, RIKEN Research Center for Allergy and Immunology, Yokohama, Kanagawa 230-0045, Japan

Daisuke Kitamura
Department of Medicinal and Life Science, Faculty of Pharmaceutical Sciences, Division of Molecular Biology, Research Institute for Biological Sciences, Tokyo University of Science, Noda, Chiba 278-0022, Japan

Hiromi Kubagawa
Division of Developmental and Clinical Immunology, University of Alabama at Birmingham, Birmingham, AL 35294-3300, USA

Bernard Malissen
Centre d'Immunologie de Marseille-Luminy, INSERM, U631. CNRS, UMR6102, Université de la Méditerrannée, 13288 Marseille Cedex 9, France

Kyoko Masuda
Laboratory for Lymphocyte Development, RIKEN Research Center for Allergy and Immunology, Yokohama, Kanagawa 230-0045, Japan

Catherine Mazza
Centre d'Immunologie de Marseille-Luminy, INSERM, U631. CNRS, UMR6102, Université de la Méditerrannée, 13288 Marseille Cedex 9, France

Falk Nimmerjahn
Laboratory of Molecular Genetics and Immunology, Rockefeller University, New York, NY 10021, USA; Laboratory of Experimental Immunology and Immunotherapy, University of Erlangen-Nuernberg, Nikolaus-Fiebiger-Center for Molecular Medicine, 91054 Erlangen, Germany

Jeffrey V. Ravetch
Laboratory of Molecular Genetics and Immunology, Rockefeller University, New York, NY 10021, USA

Christopher E. Rudd
Cell Signalling Section, Division of Immunology, Department of Pathology, University of Cambridge, Cambridge, CB2 1QP, UK

Helga Schneider
Cell Signalling Section, Division of Immunology, Department of Pathology, University of Cambridge, Cambridge, CB2 1QP, UK

C. Andrew Stewart
Laboratory of NK cells and Innate Immunity, Centre d'Immunologie de Marseille-Luminy, INSERM, U631. CNRS, UMR6102, Université de la Méditerranée, 13288 Marseille Cedex 9, France

Ikuko Torii
Division of Developmental and Clinical Immunology, University of Alabama at Birmingham, Birmingham, AL 35294-3300, USA

Takeshi Tsubata
Laboratory of Immunology, School of Biomedical Science and Department of Immunology, Medical Research Institute, Tokyo Medical and Dental University, Tokyo 113-8510, Japan

Satoshi Uematsu
Department of Host Defense, Research Institute for Microbial Diseases, Osaka University, Suita, Osaka 565-0851, Japan

Eric Vivier
Laboratory of NK Cells and Innate Immunity, Centre d'Immunologie de Marseille-Luminy, INSERM, U631. CNRS, UMR6102, Université de la Méditerranée, 13288 Marseille Cedex 9, France; Hôpital de la Conception, Assistance Publique-Hôpitaux de Marseille, 13005 Marseille, France

Kozo Watanabe
Laboratory of Immunology, School of Biomedical Science and Department of Immunology, Medical Research Institute, Tokyo Medical and Dental University, Tokyo 113-8510, Japan

Color Plates

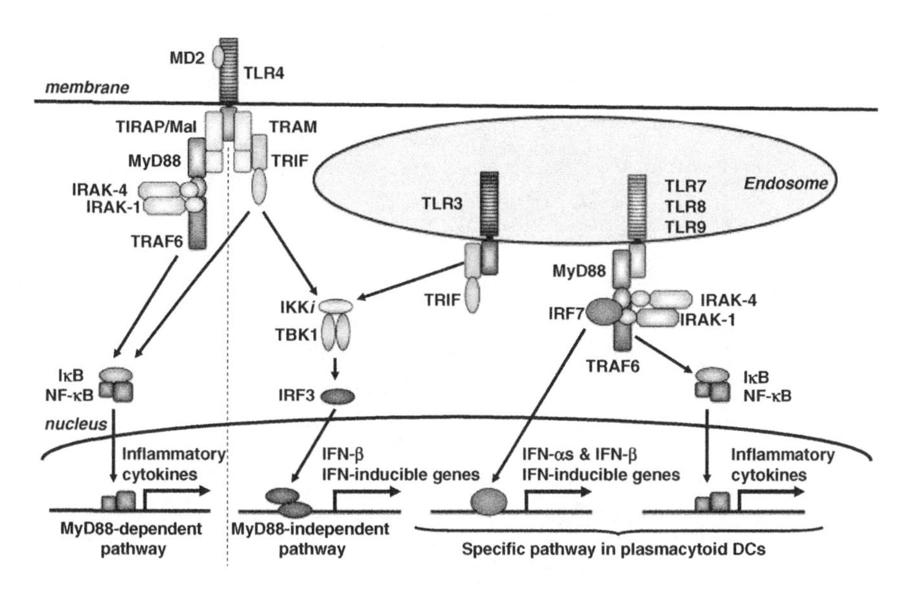

(Fig. 1.2, p. 10)

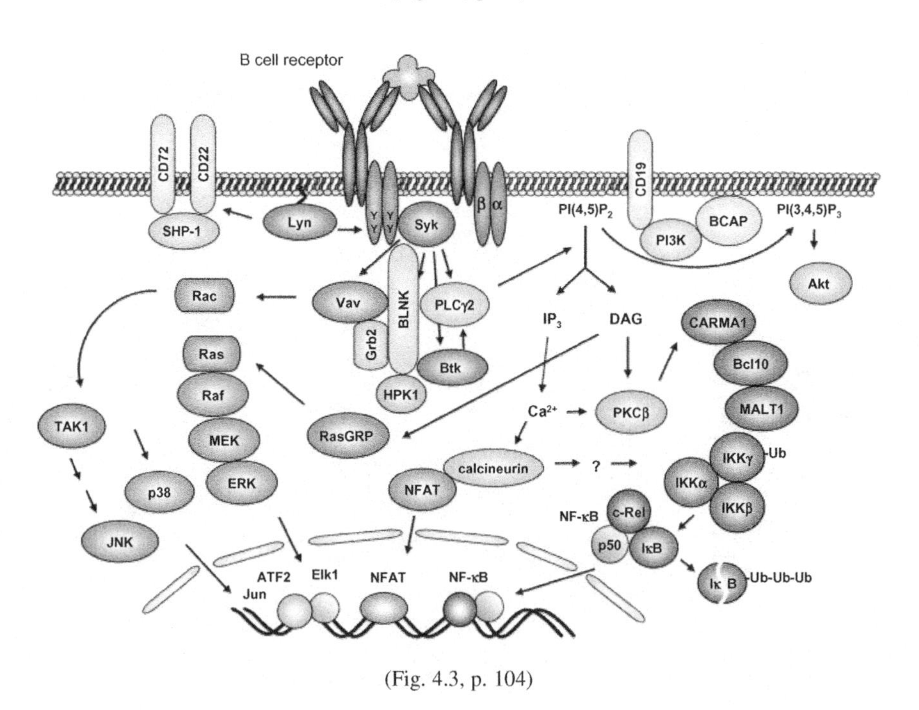

(Fig. 4.3, p. 104)

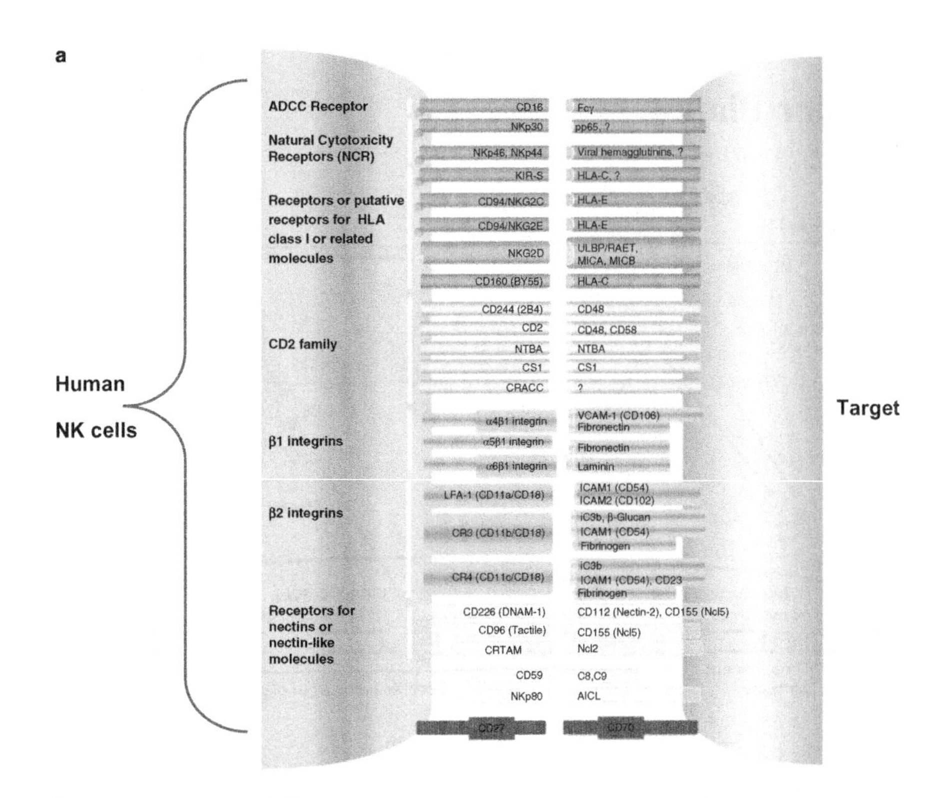
a
ADCC Receptor
CD16
Fcγ
NKp30
pp65, ?
Natural Cytotoxicity Receptors (NCR)
NKp46, NKp44
Viral hemagglutinins, ?
KIR-S
HLA-C, ?
Receptors or putative receptors for HLA class I or related molecules
CD94/NKG2C
HLA-E
CD94/NKG2E
HLA-E
NKG2D
ULBP/RAET, MICA, MICB
CD160 (BY55)
HLA-C
CD2 family
CD244 (2B4)
CD48
CD2
CD48, CD58
NTBA
NTBA
CS1
CS1
CRACC
?
Human
NK cells
Target
β1 integrins
α4β1 integrin
VCAM-1 (CD106) Fibronectin
α5β1 integrin
Fibronectin
α6β1 integrin
Laminin
β2 integrins
LFA-1 (CD11a/CD18)
ICAM1 (CD54) ICAM2 (CD102)
CR3 (CD11b/CD18)
iC3b, β-Glucan ICAM1 (CD54) Fibrinogen
CR4 (CD11c/CD18)
iC3b ICAM1 (CD54), CD23 Fibrinogen
Receptors for nectins or nectin-like molecules
CD226 (DNAM-1)
CD112 (Nectin-2), CD155 (Ncl5)
CD96 (Tactile)
CD155 (Ncl5)
CRTAM
Ncl2
CD59
C8,C9
NKp80
AICL
CD27
CD70

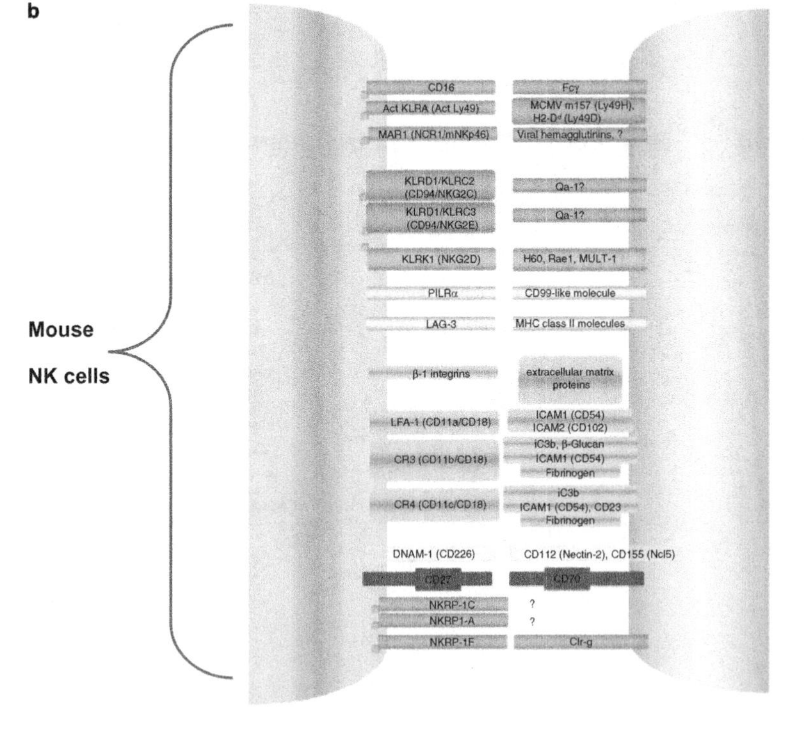
b
CD16
Fcγ
Act KLRA (Act Ly49)
MCMV m157 (Ly49H), H2-D^d (Ly49D)
MAR1 (NCR1/mNKp46)
Viral hemagglutinins, ?
KLRD1/KLRC2 (CD94/NKG2C)
Qa-1?
KLRD1/KLRC3 (CD94/NKG2E)
Qa-1?
KLRK1 (NKG2D)
H60, Rae1, MULT-1
PILRα
CD99-like molecule
LAG-3
MHC class II molecules
Mouse
NK cells
Target
β-1 integrins
extracellular matrix proteins
LFA-1 (CD11a/CD18)
ICAM1 (CD54) ICAM2 (CD102)
CR3 (CD11b/CD18)
iC3b, β-Glucan ICAM1 (CD54) Fibrinogen
CR4 (CD11c/CD18)
iC3b ICAM1 (CD54), CD23 Fibrinogen
DNAM-1 (CD226)
CD112 (Nectin-2), CD155 (Ncl5)
CD27
CD70
NKRP-1C
?
NKRP1-A
?
NKRP-1F
Clr-g

(Fig. 2.4, p. 50)

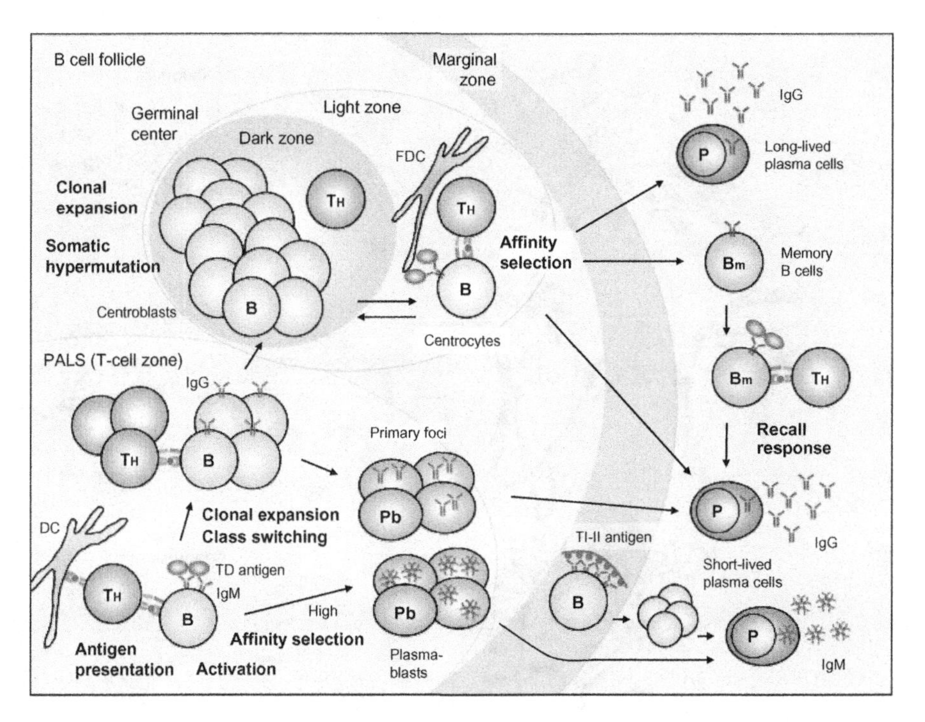
B cell follicle
Marginal zone
Germinal center
Light zone
Dark zone
FDC
IgG
Long-lived plasma cells
Clonal expansion
Somatic hypermutation
Centroblasts
Affinity selection
Memory B cells
Centrocytes
PALS (T-cell zone)
IgG
Recall response
Primary foci
Pb
DC
Clonal expansion
Class switching
TI-II antigen
IgG
TD antigen
IgM
Short-lived plasma cells
High
Affinity selection
Antigen presentation
Activation
Plasma-blasts
IgM

(Fig. 4.4, p. 113)

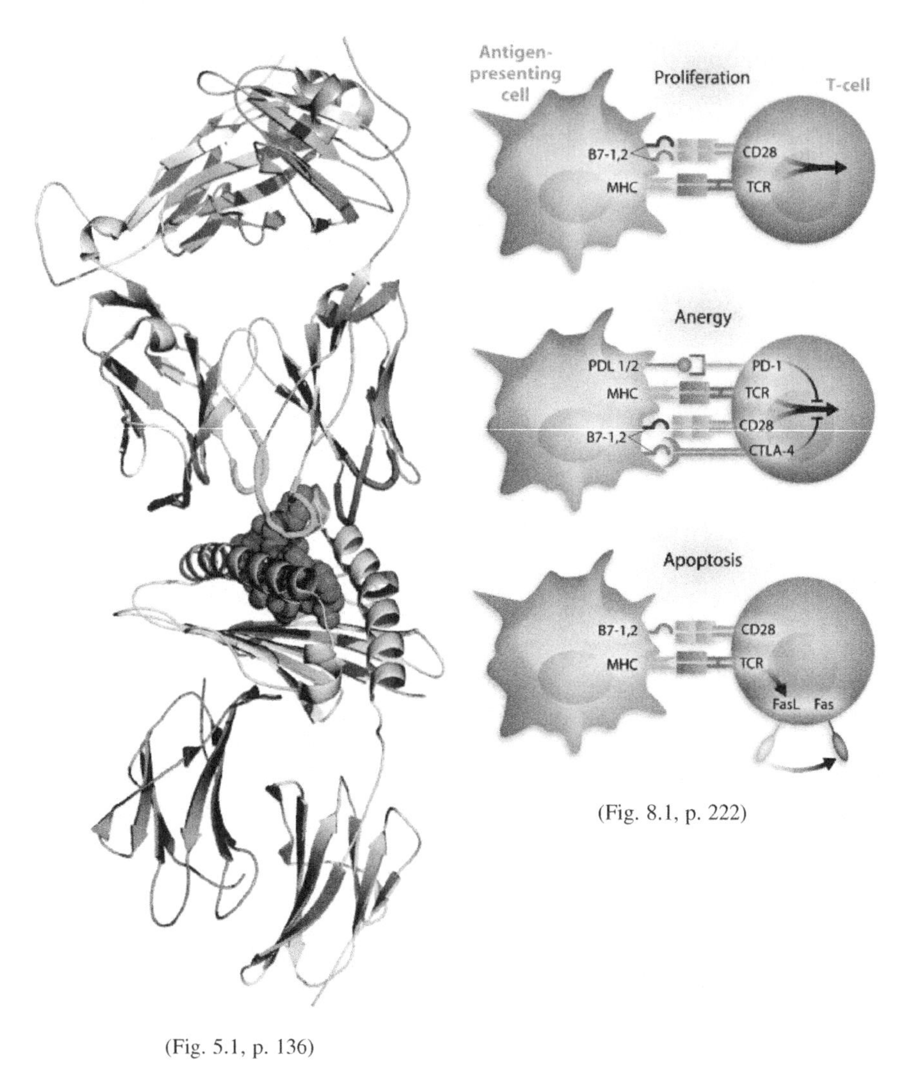

(Fig. 8.1, p. 222)

(Fig. 5.1, p. 136)

1
Recognition of Pathogens: Toll-Like Receptors

Satoshi Uematsu[1] and Shizuo Akira[1,2]

1.1 Introduction

The innate immune response is the first line of defense against microbial infections. Toll-like receptors (TLRs) are evolutionally conserved and recognize specific components of pathogens including bacteria, fungi, protozoa and viruses. Recognition of microbial components by TLRs triggers activation of signal transduction pathways, which then induces dendritic cell maturation and cytokine production, resulting in development of adaptive immunity. Each TLR has its intrinsic signaling pathway and induces specific biological responses against microorganisms. Here, we discuss the rapidly accumulating members of the TLR family, their functions and signaling mechanisms.

1.2 Innate Immunity

The adaptive immune system uses a random and highly diverse repertoire of receptors encoded by rearranging genes, the T- and B-cell receptors, to recognize a variety of antigens. This mechanism allows the host to generate immunological memory. However, it takes time that specific clones expand and differentiate into effector cells before they can serve for host defense. Therefore, the primary adaptive immune system cannot induce immediate responses to invasive pathogens. To induce immediate responses upon encountering with a pathogen, a host is equipped with innate, nonadaptive defenses that form early barriers against infectious diseases. Although

[1]Department of Host Defense, Research Institute for Microbial Diseases, Osaka University, 3-1 Yamada-oka, Suita, Osaka 565-0851, Japan

[2]ERATO, Japan Science and Technology Corporation, 3-1 Yamada-oka, Suita, Osaka 565-0871, Japan

Correspondence to: S. Akira

the innate immune system was first described by Elie Metchnikoff over a century ago, it has long been ignored as just a nonspecific response to simply kill pathogens, and to present antigens to the cells involved in acquired immunity (Brown 2001). In 1996, an epoch-making study showed that the *Drosophila* protein Toll is required for flies to induce effective immune responses to *Aspergillus fumigatus* (Lemaitre et al. 1996). Since then, accumulating evidence has shown that the innate immune system specifically recognizes invading microorganisms. The targets of innate immune recognition are conserved molecular patterns (PAMPs: pathogen-associated molecular patterns) of microorganisms. Therefore, the receptors in innate immunity are called pattern-recognition receptors (PRRs) (Medzhitov and Janeway 1997). PAMPs are generated by microbes and not by the host, suggesting that PAMPs are good targets for innate immunity to discriminate between self and nonself. Furthermore, PAMPs are essential for microbial survival and are conserved structures among a given class, which allows innate immunity to respond to microorganisms with limited numbers of PRRs. There are many PRRs associated with opsonization, phagocytosis, complement and coagulation cascades, proinflammatory signaling pathways, apoptosis and so on. Among them, Toll receptors and the associated signaling pathways represent the most ancient host defense mechanism found in insects, plants and mammals (Akira 2004). Studies conducted by using the fruit fly have shown that the Toll family is one of the most crucial signaling receptors in innate immunity.

1.2.1 Immune Responses in **Drosophila**

Insects do not have counterparts of mammalian B and T cells, and they cannot induce acquired immune responses based on producing antibodies to pathogenic organisms. Nonetheless, insects can recognize the invasion of various microorganisms and induce antimicrobial responses. Recent studies using a model organism, *Drosophila melanogaster*, have shown that the induction of antimicrobial peptides, which are important for survival after infection, depends on Toll and immune deficiency (Imd) signaling pathways (Tanji and Ip 2005). A transmembrane protein, Toll, originally identified as an essential component in dorsal-ventral embryonic development (Wu and Anderson 1997), is also involved in innate immune responses (Lemaitre et al. 1996). Gram-positive bacterial peptidoglycan might bind directly to extracellular peptidoglycan recognition protein (PGRP)-SA (Michel et al. 2001) and SD (Bischoff et al. 2004), which then stimulate the Toll pathway. Another pattern recognition protein, Gram-negative binding protein-1 (GNBP-1) is also involved in the recognition of Gram-positive bacteria (Gobert et al. 2003; Pili-Floury et al. 2004). Not only Gram-positive bacteria but also fungi stimulate the Toll pathway. Fungi are recognized by a serine protease, Persephone, and a protease inhibitor, Necrotic (Levashina et al. 1999; Ligoxygakis et al. 2002). All upstream cascades lead to the cleavage of pro- Spätzle to Spätzle, and the binding of proteolytically processed Spätzle to Toll induces the dimerization of Toll (Hu et al. 2004; Weber et al. 2003). After the activation of Toll, the adapter proteins MyD88 and Tube, and a serine-threonine kinase, Pelle, are recruited to Toll (Sun et al. 2004).

Then, activated Pelle acts on the Cactus, a *Drosophila* IκB. Dif and Dorsal are transcription factors of Rel protein family and are retained in the cytoplasm by Cactus. By the stimulation of the Toll pathway, Cactus is degraded and Dorsal and Dif translocate into the nucleus, leading to the induction of antimicropeptides (Brennan and Anderson 2004; Hoffmann 2003; Hultmark 2003).

The Imd pathway is responsible for the induction of antimicrobial peptides in response to Gram-negative bacteria (Brennan and Anderson 2004; Hoffmann 2003; Hultmark 2003; Lemaitre 2004). Imd is an adapter protein for this pathway (Georgel et al. 2001). Recent reports show that PGRP-LC (Choe et al. 2002; Gottar et al. 2002) and PGRP-LE (Takehana et al. 2004), which have putative transmembrane domains, are the pattern recognition receptors in this pathway. There are at least three branches downstream of Imd (Brennan and Anderson 2004; Hoffmann 2003; Hultmark 2003; Lemaitre 2004). First is Transforming Growth Factor (TGF)-β-activated kinase 1 (TAK1), which induces the proteolytic cleavage of IKK, followed by activation of the transcription factor, Relish (Lu et al. 2001; Rutschmann et al. 2000; Silverman et al. 2000, 2003; Stoven et al. 2003; Vidal et al. 2001). Second is the Fas-associated death domain (FADD)–Dredd pathway that also activates Relish (Balachandran et al. 2004; Chen et al. 1998; Elrod-Erickson et al. 2000; Georgel et al. 2001; Hu and Yang 2000; Leulier et al. 2000, 2002). Two new components, Sickie and Dnr-1 have been identified: whereas Sickie positively regulates the Relish activation by Dredd, Dnr-1 inhibit this pathway (Foley and O'Farrell 2004; Khush et al. 2002). Third is the JNK pathway that is activated through TAK1. The JNK pathway induces immediate early genes after septic shock, which is negatively regulated by Relish (Boutros et al. 2002; Park et al. 2004a).

As stated above, recent genetic and genomic analyses of *D. melanogaster* have shown that insects have an evolutionally primitive recognition and signaling system (Fig. 1). Collectively, these results provide important insights into the mechanism of pathogen recognition and host responses in mammalian systems.

1.2.2 Discovery of TLR in Mammals

A mammalian homolog of Toll receptor (now termed TLR4) was identified through database searches, and was shown to induce expression of the genes involved in inflammatory responses (Medzhitov et al. 1997). Subsequently, a mutation in the *tlr4* gene was identified in C3H/HeJ mice that were hyporesponsive to lipopolysaccharide (LPS) (Poltorak et al. 1998). The TLR family now consists of 13 mammalian members (Akira 2004). As the cytoplasmic portion of TLRs is similar to that of the interleukin (IL)-1 receptor family, it is called the Toll/IL-1 receptor (TIR) domain. However, the extracellular region of TLRs and IL-1R are markedly different; whereas IL-1R possesses an Ig-like domain, TLRs contain leucine-rich repeats (LRR) in the extracellular domain (Akira 2004). Recent genetic studies have revealed that TLRs play an essential role in the recognition of specific components of pathogens. TLRs are capable of sensing organisms ranging from bacteria to fungi, protozoa and viruses (Table 1).

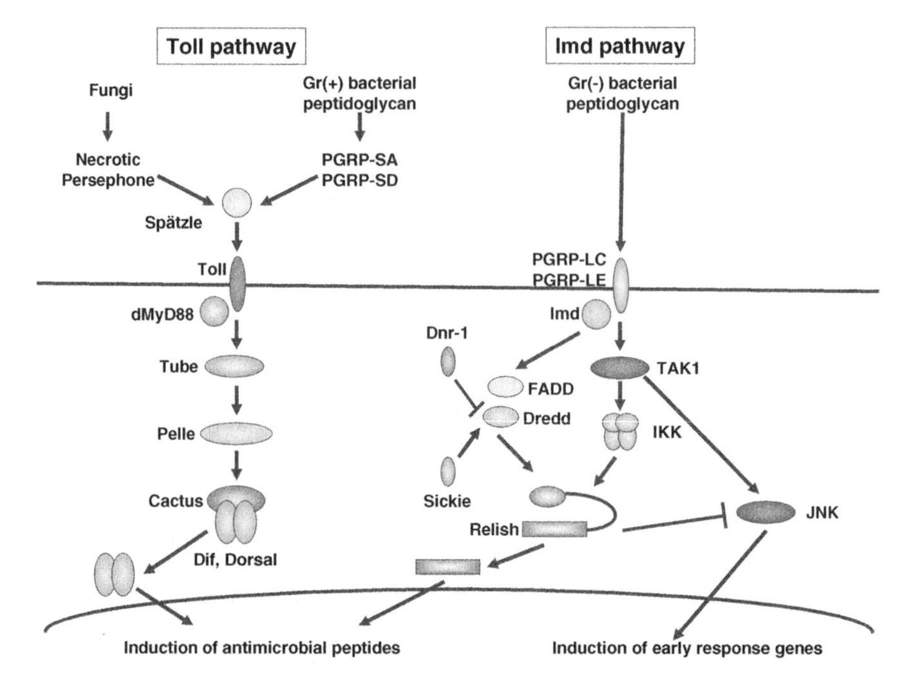

Fig. 1. Toll and immune deficiency (*Imd*) pathways in the *Drosophila* innate immune response. The Toll pathway mediates the response to Fungal and Gr(+) bacterial infection, whereas the Imd pathway mediates the response to Gr(–) bacterial infection. These pathways are similar to the signaling pathway of the mammalian Toll-like receptor, and are essential for *Drosophila* to survive infection. *PGRP*, peptidoglycan recognition protein; *FADD*, Fas-associated death domain; *TAK*, TGF-β-activated kinase; *IKK*, IκB kinase; *JNK*, Jun N-terminal kinase

Table 1. Toll-like receptors and their ligands

TLR	Ligand
TLR1	Triacyl lipopeptides (bacteria)
TLR2	Peptidoglycan, lipoprotein, lipopeptides, atypical LPS (bacteria), zymosan, phospholipomannan (fungi), GPI anchor (protozoa), envelope protein (virus)
TLR3	Poly(I:C), dsRNA (virus)
TLR4	LPS (bacteria), mannan, glucuronyloxylomannan (fungi), glycoinositolphospholipids (protozoa), RSV fusion protein (virus)
TLR5	Flagellin (bacteria)
TLR6	Diacyl lipopeptides (bacteria)
TLR7/TLR8	Synthetic imidazoquinoline-like molecules, ssRNA (virus)
TLR9	CpG DNA (bacteria, protozoa, virus), hemozoin (protozoa)
TLR11	Component of uropathogenic bacteria (bacteria), profolin-like molecule (protozoa)

PLPS, lipopolysaccharide; GPI, glycosyl phosphatidylinositol; RSV, respiratory syncytial virus; dsRNA, double-stranded RNA; ssRNA, single-stranded RNA

1.3 Pathogen Recognition by TLR

1.3.1 Bacteria

Lipopolysaccharide is a cell wall component of Gram-negative bacteria, a strong immunostimulant. As described above, TLR4 is essential for recognition of LPS, which is composed of lipid A (endotoxin), core oligosaccharide and O-antigen. TLR4 recognizes lipid A of LPS; this occurs through the TLR4/MD2/CD14 complex, which is present on various cells such as macrophages and dendritic cells (Shimazu et al. 1999). LPS forms a complex with an accessory protein, LPS-binding protein (LBP) in serum, which converts oligomeric micelles of LPS to a monomer for delivery to CD14, which is a glycosyl phosphatidylinositol (GPI)-anchored, high-affinity membrane protein that can also circulate in a soluble form. CD14 concentrates LPS for binding to the TLR4/MD2 complex (Takeda et al. 2003).

TLR2 recognizes various microbial components, such as lipoproteins/lipopeptides and peptidoglycans from Gram-positive and Gram-negative bacteria, and lipoteichoic acid from Gram-positive bacteria, a phenol-soluble modulin from *Staphylococcus aureus* and glycolipids from *Treponema maltophilum* (Takeda et al. 2003; Takeuchi et al. 1999a). TLR2 is also reported to be involved in the recognition of LPS from nonenterobacteria including *Leptospira interrogans, Porphyromonas gingivalis* and *Helicobacter pylori* (Takeda and Akira 2005). These LPS are atypical LPS, whose structures are different from typical LPS of Gram-negative bacteria (Netea et al. 2002). However, a recent report indicated that lipoproteins contaminated in LPS preparation from *P. gingivalis* stimulated TLR2 and LPS from *P. gingivalis* itself had a poor activity for TLR4 stimulation (Hashimoto et al. 2004). There are also controversial reports regarding peptidoglycan recognition by TLR2. More careful analyses will be needed to exclude any possibility of remaining contaminates.

TLR1 and TLR6 are structurally related to TLR2 (Takeuchi et al. 1999b). TLR2 and TLR1 or TLR6 form a heterodimer, which is involved in the discrimination of subtle changes in lipopeptides from Gram-positive bacteria. TLR6-deficient macrophages do not produce inflammatory cytokines in response to diacyl lipopeptides from mycoplasma; however, they normally produce inflammatory cytokines in response to triacyl lipopeptides derived from a variety of bacteria (Takeuchi et al. 2001). On the contrary, TLR1-deficient macrophages show normal responses to triacyl lipopeptides but not to diacyl lipopeptides (Alexopoulou et al. 2002; Takeuchi et al. 2002). These results suggest that TLR1 and TLR6 functionally associate with TLR2.

CD36 is a member of the class II scavenger family of proteins. A recent report has shown that CD36 serves as a facilitator or co-receptor for diacyl lipopeptide recognition through the TLR2/6 complex (Hoebe et al. 2005).

Bacterial flagellin is a structural protein that forms the major portion of flagella that contribute to virulence through chemotaxis, adhesion to and invasion of host surfaces. TLR5 is responsible for the recognition of flagellin (Hayashi et al. 2001). Gewirtz et al. reported that TLR5 is expressed on the basolateral surface, but not the apical side of intestinal epithelial cells, suggesting that flagellin is detected when bacteria invade across the epithelium (Gewirtz et al. 2001). A common stop codon polymorphism in the ligand-binding domain of TLR5 (TLR5 392STOP SNP) is unable to mediate flagellin signaling and is associated with susceptibility to pneumonia caused by *Legionella pneumophila* (Hawn et al. 2003). However, researchers in Vietnam have reported that the TLR5 392STOP SNP is not associated with susceptibility to typhoid fever (Dunstan et al. 2005). α and ε *Proteobacteria*, including *Helicobacter pylori* and *Campylobacter jejuni*, change the TLR5 recognition site of flagellin without losing flagellar motility (Andersen-Nissen et al. 2005). This modification may contribute to the persistence of these bacteria on mucosal surfaces.

Bacterial DNA, which contains unmethylated CpG motifs, is a potent stimulator of the host immune response. In vertebrates, the frequency of CpG motifs is severely reduced and the cysteine residues of CpG motifs are highly methylated, which leads to abrogation of the immunostimulatory activity. Analysis of TLR9-deficient mice showed that TLR9 is a receptor for CpG DNA (Hemmi et al. 2000).

Mouse TLR11, a relative of TLR5, is expressed abundantly in the kidney and bladder. TLR11-deficient mice are susceptible to uropathogenic bacterial infections, indicating that TLR11 senses the component of uropathogenic bacteria. However, TLR11 is thought to be a pseudogene in human (Zhang et al. 2004).

1.3.2 Fungi

Toll-like receptors have been implicated in the recognition of the fungal pathogens such as *Candida albicans*, *Aspergillus fumigatus*, *Cryptococcus neoformans* and *Pneumocystis carinii* (Netea et al. 2004; Takeda et al. 2003). Several components located in the cell wall or cell surface of fungi have been identified as potential ligands. Yeast zymosan activates TLR2/TLR6 heterodimers, whereas mannan derived from *Saccharomyces cerevisiae* and *C. albicans* is detected by TLR4. Phospholipomannan presented in the cell surface of *C. albicans* is recognized by TLR2, while TLR4 mainly interacts with glucuronoxylomannan, the major capsular polysaccharide of *C. neoformans* (Netea et al. 2004).

Dectin-1 is a lectin family receptor for the fungal cell wall component, β-glucan (Brown et al. 2002). Recently, Dectin-1 has been shown to functionally collaborate with TLR2 and induce a strong immune response to yeast via recruitment of the tyrosine kinase, Syk (Gantner et al. 2003; Rogers et al. 2005; Underhill et al. 2005). TLR2 recognizes a variety of microbial products through functional cooperation with several proteins that are either structurally related or not.

1.3.3 Protozoa

Toll-like receptors also sense the components of protozoa including *Trypanosoma cruzi*, *Trypanosoma brucei*, *Toxoplasma gondii*, *Leishmania major* and *Plasmodium falciparum*. Glycosylphosphatidylinositol anchors, glycoinositolphospholipids and genomic DNA derived from *T. cruzi* are recognized by TLR2, TLR4, and TLR9, respectively (Gazzinelli et al. 2004). Murine TLR11 senses the profilin-like molecule of *T. gondii* (Yarovinsky et al. 2005). Malaria parasites digest host hemoglobin into a hydrophobic heme polymer known as hemozoin. Hemozoin from *P. falciparum* stimulates macrophages and DCs to produce inflammatory cytokines and chemokines in a TLR9-dependent manner (Coban et al. 2005). However, it is unclear how TLR9 recognizes both DNA and non-DNA crystals.

1.3.4 Virus

TLR4 recognizes not only bacterial components but also viral envelope proteins. The fusion (F) protein from respiratory syncytial virus (RSV) is sensed by TLR4 (Kurt-Jones et al. 2000). C3H/HeJ mice were sensitive to RSV infection (Haynes et al. 2001). The envelope protein of mouse mammary tumor virus (MMTV) directly activates B cells via TLR4 (Rassa et al. 2002). TLR2 has also been reported to be involved in the recognition of viral components such as Measles virus, human cytomegalovirus and HSV-1(Bieback et al. 2002; Compton et al. 2003; Kurt-Jones et al. 2004).

Double-stranded (ds) RNA is generated during viral replication. TLR3 is involved in the recognition of a synthetic analog of dsRNA, polyinosine-deoxycytidylic acid (poly I:C), a potent inducer of type I interferons (IFNs) (Alexopoulou et al. 2001; Yamamoto et al. 2003a). Consistent with this result, TLR3-deficient mice were hypersusceptible to mouse cytomegalovirus (Tabeta et al. 2004). Contrarily, TLR3-deficient mice showed more resistance to West Nile virus (WNV) infection. WNV triggers inflammatory responses via TLR3, which results in a disruption of the blood–brain barrier, followed by enhanced brain infection (Wang et al. 2004). These findings suggested that WNV utilizes TLR3 to efficiently enter into the brain.

Mouse splenic DCs are divided into CD11c high B220– and CD11c dull B220+ cells. The latter contain plasmacytoid DCs (pDCs), which induce large amounts of IFN-α during viral infection. CpG DNA motifs are also found in genomes of DNA viruses, such as Herpes simplex virus type 1 (HSV-1), HSV-2 and murine cytomegalovirus (MCMV). Mouse pDCs produce IFN-α by recognizing CpG DNA of HSV-2 via TLR9 (Lund et al. 2003). TLR9-deficient mice were also shown to be susceptible to MCMV infection, suggesting that TLR9 induces anti-viral responses by sensing CpG DNA of DNA virus (Krug et al. 2004a,b; Tabeta et al. 2004). However, in the case of macrophages, HSV-2-induced IFN-α production is not dependent on TLRs. Furthermore, mice lacking TLR9 or the adapter molecule MyD88 can still control HSV-1 infection (Hochrein et al. 2004). Thus, TLR9-mediated IFN-α response to DNA virus is limited to pDCs, and the TLR-independent system also plays an important role in DNA viral infection.

TLR7 and TLR8 are structurally highly conserved proteins (Akira 2004). The synthetic imidazoquinoline-like molecules imiquimod (R-837) and resiquimod (R848) have potent antiviral activities and are used clinically for treatment of viral infections. Analysis of TLR7-deficient mice showed that TLR7 recognizes these synthetic compounds (Hemmi et al. 2002). Human TLR7 and TLR8, but not murine TLR8, recognize imidazoquinoline compounds (Ito et al. 2002). Furthermore, murine TLR7 has also been shown to recognize guanosine analogs such as loxoribine, which has antiviral and antitumor activities (Akira and Hemmi 2003). Since all these compounds are structurally similar to ribonucleic acids, TLR7 and human TLR8 are predicted to recognize a nucleic acid-like structure of a virus. Recently, TLR7 and human TLR8 have been shown to recognize guanosine- or uridine-rich single-stranded RNA (ssRNA) from viruses such as human immunodeficiency virus (HIV), vesicular stomatitis virus (VSV) and influenza virus (Diebold et al. 2004; Heil et al. 2004). Although ssRNA is abundant in hosts, host-derived ssRNA is not usually detected by TLR7 or TLR8. As TLR7 and TLR8 are expressed in the endosome, host-derived ssRNA is not delivered to the endosome and therefore, is not recognized by TLR7 and TLR8.

As well as TLR7 and TLR8, TLR3 and TLR9 are exclusively expressed in endosomal compartments not on cell surfaces (Latz et al. 2004). After phagocytes internalize viruses or virus-infected apoptotic cells, viral nucleic acids are released in phagolysosomes and are recognized by TLRs. However, intracellular localization of TLR9 is not required for ligand recognition but prevents recognition of self DNA. Localization of the nucleic acid-sensing TLRs is critical in discriminating between self and nonself nucleic acids.

1.3.5 Endogenous Ligands

Many reports have suggested that a number of endogenous ligands such as heat shock proteins (hsp) were potent activators of the innate immune system (Tsan and Gao 2004). TLR4 has been shown to be involved in the recognition of endogenous ligands, such as HSPs (HSP60, HSP70 and Gp96) (Asea et al. 2002; Bulut et al. 2002; Dybdahl et al. 2002; Ohashi et al. 2000; Vabulas et al. 2002a,b). Fibrinogens, surfactant protein-A, fibronectin extra domain A, heparan sulfate, β-defensin 2 have since been reported as ligands for TLR4 (Biragyn et al. 2002; Guillot et al. 2002; Johnson et al. 2002; Okamura et al. 2001; Smiley et al. 2001; Termeer et al. 2002). It has also been shown that HSP60, HSP70, gp96 and HMGB1 protein are endogenous ligands for TLR2 and TLR4 (Asea et al. 2002; Park et al. 2004b; Vabulas et al. 2001, 2002b); and that mRNA is an endogenous ligand for TLR3 (Kariko et al. 2004). The extracellular matrix hyaluronan is produced after tissue injury.

A recent study showed that hyaluronan is recognized by both TLR2 and TLR4, and that this interaction regulates both innate inflammatory responses and epithelial cell integrity, which are both crucial for recovery from acute lung injury (Jiang et al. 2005). However, most of these endogenous ligands require very high concentrations to activate TLRs. In addition, it has been shown that the cytokine effect of HSP70 was a result of the contaminating LPS (Wallin et al. 2002). The endogenous ligands used in the previous studies were recombinant products, purified native molecules or purified fragments of macromolecules. As recombinant products are produced by genetically engineered *Escherichia coli*, the final preparations may have been contaminated with bacterial products; similarly, purified preparations are also frequently contaminated with bacterial cell wall products such as LPS and lipoproteins (Tsan and Gao 2004). More careful experiments will be needed to conclude that TLRs recognize these endogenous ligands.

1.4 Toll-Like Receptor Signaling Pathways (Fig. 2)

The engagement of TLRs by microbial components triggers the induction of specific gene profiles that are suited to the removal invading *pathogens*. After ligand binding, TLRs dimerize and undergo the conformational changes required for the recruitment of downstream signaling molecules (Akira 2004). All TLR signals originate from the TIR domain. A crucial role for the TIR domain was first identified in the C3H/HeJ mouse strain, which has a point mutation that results in an amino acid change of the cytoplasmic proline residue at position 712 to histidine (Poltorak et al. 1998). This proline residue in the TIR domain is conserved among all TLRs, except for TLR3, and its substitution to histidine causes a dominant negative effect on TLR-mediated signaling (Akira and Takeda 2004). There are four adapter proteins containing TIR domains, including myeloid differentiation factor 88 (MyD88) (Adachi et al. 1998) (Medzhitov et al. 1998), TIR-associated protein (TIRAP)/MyD88-adapter-like (MAL) (Horng et al. 2001; Fitzgerald et al. 2001; Horng et al. 2002; Yamamoto et al. 2002a), TIR-domain-containing adapter protein inducing IFN-β (TRIF)/TIR-domain containing molecule 1 (TICAM1) (Hoebe et al. 2003; Oshiumi et al. 2003a; Yamamoto et al. 2003a,b), and TRIF-related adapter molecule (TRAM)/TICAM2 (Bin et al. 2003; Oshiumi et al. 2003b; Yamamoto et al. 2003b). They are recruited to the TIR domains of TLRs upon ligand stimulation and transduce signals from TIR domains, activating protein kinases and then transcription factors that induce inflammatory responses. Individual TLRs mediate distinctive responses by association with different combinations of these adapters. Nuclear factor (NF)-κB is the most crucial transcription factor and is universally used by all TLRs. It is also involved in the induction of various genes including proinflammatory cytokines. Recent studies have revealed that specific members of the interferon (IFN)-regulatory factor (IRF) family also play pivotal roles in the induction of TLR-responsive genes such as type I IFNs and interferon inducible genes. Moreover, in TLR7 and TLR9 signaling pathways, a

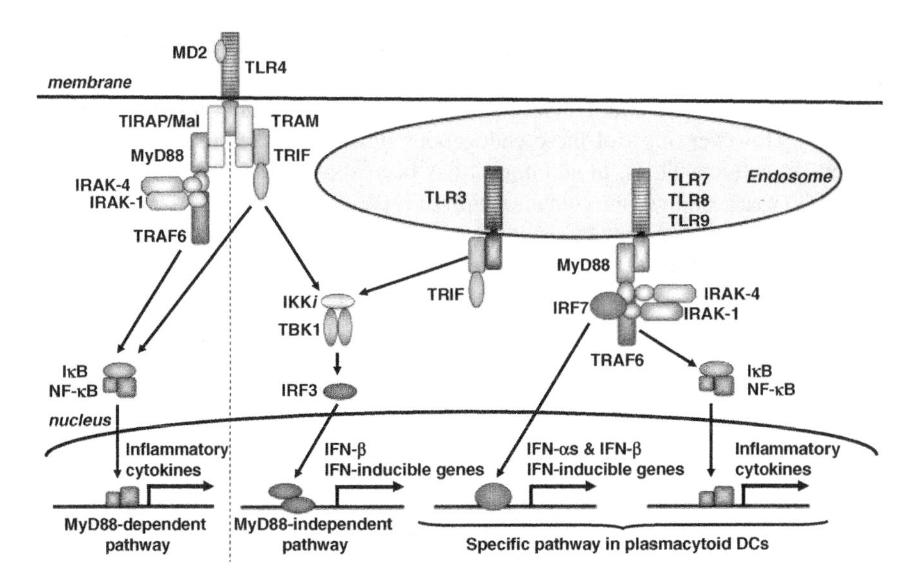

Fig. 2. Toll-like receptor (*TLR*) signaling pathways. All TLR signals originate from the Toll/interleukin-1 receptor (*TIR*) domain. There are four adapter proteins containing the TIR domain: MyD88, TIR-associated protein (*TIRAP*), TIR-domain-containing adapter protein inducing IFN-β *TRIF*), and TRIF-related adapter molecule (*TRAM*). They are recruited to the TIR domains of TLRs upon ligand stimulation and transduce signals from the TIR domains, activating protein kinases and then transcription factors that induce inflammatory responses. Individual TLRs mediate distinctive responses by association with different combinations of these adapters. Nuclear factor κB (*NF-κB*) is the most crucial transcription factor and is universally used by all TLRs, and is also involved in the induction of various genes including proinflammatory cytokines. Specific members of the interferon (*IFN*)-regulatory factor (*IRF*) family also play pivotal roles in the induction of TLR-responsive genes such as type I IFNs and interferon-inducible genes. Moreover, in TLR7 and TLR9 signaling pathways, a special subset of dendritic cells (*DCs*), plasmacytoid DCs (*pDCs*), have an intrinsic signaling pathway to induce robust IFN-αs. (See Color Plates)

special subset of dendritic cells (DCs), plasmacytoid DCs (pDCs) have an intrinsic signaling pathway to induce robust IFN-αs (Akira and Takeda 2004). In the next sections, we focus on signaling molecules used by TLRs and give an overview of the most recent findings regarding TLR signaling pathways.

1.4.1 Proinflammatory Cytokine Pathways

Toll-like receptor family members use many of the same signaling components as IL-1R since both have a conserved cytoplasmic domain, the TIR domain. The TIR domain-containing adapter MyD88 is utilized by all TLRs, with the exception of TLR3. MyD88 was originally isolated as a gene induced rapidly during IL-6-

stimulated differentiation of M1 cells into macrophages (Lord et al. 1990). MyD88 possesses a TIR domain in its C-terminus and a death domain in its N-terminus. Upon stimulation, MyD88 recruits IL-1 receptor kinases (IRAKs) to TLRs/IL-1Rs through interaction of the death domains (DDs) of the molecules (Akira and Takeda 2004). MyD88 functions as an adapter linking TLRs/IL-1Rs with downstream signaling molecules that have DDs (Adachi et al. 1998) (Medzhitov et al. 1998).

The association of TLRs and MyD88 recruits members of IRAKs (Janssens and Beyaert 2003; Muzio et al. 1997; Wesche et al. 1997). IRAK was originally identified as a serine/threonine kinase associated with the IL-1 receptor, which also harbors a death domain (Li et al. 1999). Four members have so far been identified: IRAK-1, IRAK-2, IRAK-M, and IRAK-4. IRAKs contain an N-terminal DD and a central serine/threonine-kinase domain. IRAK-1 and IRAK-4 have intrinsic kinase activity, whereas IRAK-2 and IRAK-M have none that is detectable. The kinase activity of IRAK-1 strongly increases following TLRs/IL-1Rs stimulation and its kinase domain but not the kinase activity itself is essential for the activation of NF-κB (Akira and Takeda 2004). However, IRAK-1-deficient mice have a partially impaired response to LPS (Kanakaraj et al. 1998; Swantek et al. 2000; Thomas et al. 1999). Contrarily, overexpression of a kinase-inactive mutant of IRAK-4 inhibits IL-1-mediated NF-κB activation. Furthermore, a biochemical study indicated that IRAK-4 acts upstream of IRAK-1 and phosphorylates IRAK-1 upon stimulation, suggesting that IRAK-1 is a direct substrate of IRAK-4 but not vice versa (Li et al. 2002). IRAK-4-deficient mice show virtually no response to IL-1, LPS or other bacterial components (Suzuki et al. 2002). Recently, it has been reported that patients with autosomal recessive amorphic mutations in *IRAK-4* have recurrent bacterial infections (Picard et al. 2003). There have now been 10 patients reported with genetically proven IRAK-4 deficiency and three others with probable IRAK-4 deficiency. Taken together, these results show that IRAK-4 is a central molecule in TLR-mediated MyD88-dependent pathways and acts upstream of IRAK-1.

TRAF6 is a member of the tumor necrosis factor receptor (TNFR)-associated factor (TRAF) family, which comprises two C-terminal TRAF domains (TRAF-N and TRAF-C), an N-terminal RING finger, and a zinc finger domain. The N-terminal domain is essential for the activation of downstream signaling cascades, and the TRAF domain permits self-association and interactions with receptors and other signaling proteins (Akira and Takeda 2004). Upon stimulation of TLRs, TRAF6 is recruited to the receptor complex and activated by IRAK-1. Then the IRAK-1/TRAF6 complex dissociates from the receptor to mediate signals. TRAF-6-deficient mice showed an impaired response to various TLR ligands.

After dissociation, IRAK-1/TRAF6 complex associates with TGF-β-activated kinase 1 (TAK1) and TAK1-binding proteins, TAB1 and TAB2, at the membrane portion. TAK1 is a member of mitogen-activated protein kinase kinase kinase (MAPKKK) family, which was originally identified as a kinase involved in TGF-β signaling (Yamaguchi et al. 1995). TAK1 has been shown to function as an 'upstream' signaling molecule of NF-κB and MAPKs in IL-1R signaling pathways (Takaesu et al. 2003). Furthermore, TAK1 is activated by TNF-α, LPS and latent membrane protein 1 from Epstein-Barr virus (Wan et al. 2004). Recently, conditional TAK1-deficient mice were generated, showing that TAK1 is essential for TLR, IL-1R, TNFR, and BCR cellular responses and signaling pathways leading to the activation of JNK and/or NF-κB (Sato et al. 2005). TAB1 is thought to be an activator of TAK1 because it enhances the kinase activity of TAK1 when co-expressed ectopically (Shibuya et al. 1996). TAB2 was identified as a novel adaptor protein that mediates activation of TAK1 by linking TAK1 to TRAF6 in the IL-1 signal transduction pathway (Takaesu et al. 2000). However, TAB2-deficient embryonic fibroblasts showed no impairment in either IL-1/LPS- or TNF-induced activation of NF-κB (Sanjo et al. 2003). One possible explanation for this phenotype is redundancy: another molecule can compensate for the loss of TAB2. A mammalian TAB2 homolog, termed TAB3, has been identified recently, and RNAi of both TAB2 and TAB3 have been shown to inhibit both IL-1- and TNF-induced activation of TAK1 and NF-κB (Ishitani et al. 2003). Thus, TAB2 and TAB3 function redundantly as mediators of TAK1 activation.

It has been shown that ubiquitination has an important role in TAK1 activation and that TRAF6 functions as an E3 ubiquitin ligase (Deng et al. 2000). The complex of TRAF6, TAB1, TAB2 and TAB3 forms a larger complex with other proteins such as the E2 ligase Ubiquitin-conjugating enzyme 13 (Ubc13) and Ubiquitin-conjugating enzyme E2 variant 1 (Uev1A). The Ubc13 and Uev1A complex catalyzes the synthesis of a Lys63-linked polyubiquitin chain of TRAF6 and thereby induces TRAF6-mediated activation of TAK1, which in turn activates the transcription factors nuclear factor NF-κB and activator protein-1 through the canonical IκB kinase (IKK) complex and the mitogen-activated protein kinase pathway, respectively (Wang et al. 2001).

NF-κB proteins are usually sequestered in the cytoplasm in an inactive form by a family of inhibitors of NF-κB (IκB). Upon stimulation of the cells by TLR ligands, IκB is phosphorylated on two serine residues, which triggers its ubiquitination and degradation by the 26S proteasome. NF-κB is released to enter the nucleus and to activate the transcription of target genes. Activation of NF-κB through IκB phosphorylation and degradation depends on the activation of IκB kinases (IKKs). The IKK complex comprises two catalytic subunits, IKK-α and IKK-β, and the regulatory subunit IKK-γ/NF-κB essential modulator (NEMO) (Karin and Ben-Neriah 2000). This signaling pathway, mentioned above, is called the "MyD88-dependent pathway" and is essential for the expression of inflammatory cytokine genes, including TNF-α, IL-6, IL-12, and IL-1β, and co-stimulatory molecules (Akira and Takeda, 2004).

1.4.2 TRIF-Dependent Pathway

In addition to proinflammatory signals, recent studies have shown that some members of the TLR family trigger the induction of type I IFNs. In particular, TLR3 and TLR4 have the ability to induce IFN-β and IFN-inducible genes in MyD88-

deficient cells (Kawai et al. 1999, 2001). Moreover, TLR7, 8 and 9 can also induce type I IFNs in certain dendritic cell subtypes, plasmacytoid DCs (Akira and Hemmi 2003; Hemmi et al. 2003).

In accordance with the loss of inflammatory cytokine production, activation of NF-κB and JNK by TLR2, TLR7, and TLR9 ligands was abolished in MyD88-deficient mice. However, activation of NF-κB and JNK in response to LPS was observed with delayed kinetics in MyD88-deficient cells (Kawai et al. 1999). This finding indicated the presence of a MyD88-independent pathway for LPS stimulation. To elucidate this signaling pathway, subtraction analysis was performed using mRNA extracted from unstimulated and LPS-stimulated MyD88-deficient macrophages. This analysis revealed that several IFN-inducible genes, including glucocorticoid-attenuated response gene 16 (GARG-16), immunoresponsive gene 1 (IRG-1), and CXC-chemokine ligand 10 (CXCL10)/IFN-g-induced 10-kDa protein (IP-10), were induced by LPS in the MyD88-deficient cells. Furthermore, LPS stimulation led to activation of the transcription factor IRF-3 that induced IFN-β in the MyD88-independent pathway. In turn, IFN-β activated Stat1 through IFN-α/βR, leading to the induction of several IFN-inducible genes (Hoshino et al. 2002; Kawai et al. 2001).

In addition to the TLR4 ligand, TLR3 ligand dsRNA has been shown to activate NF-κB in MyD88-deficient cells. dsRNA also activates IRF-3 and IFN-β as well as IFN-inducible genes. These data indicate that TLR3 also utilizes a MyD88-independent pathway (Akira and Takeda 2004).

A second TIR domain-containing adaptor, TIR-domain-containing adaptor protein (TIRAP)/MyD88-adaptor-like protein (Mal), was identified from a database search (Fitzgerald et al. 2001; Horng et al. 2001). As TIRAP/Mal does not have a death domain, it was initially thought to mediate the MyD88-independent pathway of TLR4-signaling. However, TIRAP/Mal-deficient mice showed the impaired production of inflammatory cytokines in response to LPS, and the LPS-induced MyD88-independent pathway was not impaired. Interestingly, cytokine production in response to TLR2 ligands was also impaired in TIRAP/Mal-deficient mice, suggesting that TIRAP/Mal acts as a bridging adapter binding MyD88 with TLR2 and TLR4 (Yamamoto et al. 2002a) (Horng et al. 2002).

Subsequent studies identified a third adapter TIR-domain-containing adapter protein inducing IFN-β (TRIF)/TIR-domain-containing molecule 1 (TICAM-1) (Oshiumi et al. 2003a; Yamamoto et al. 2002b). TRIF/TICAM-1-deficient mice and *Lps2*-mice with a frame shift mutation of *Trif* gene by the alkylating agent *N*-ethyl-*N*-nitrosourea (ENU) failed to show MyD88-independent activation of NF-κB and are nonresponsive to TLR4 or TLR3 ligand with respect to the expression of IFN-inducible genes (Hoebe et al. 2003; Yamamoto et al. 2003a). Therefore, TRIF is essential for the TLR3- and TLR4-mediated MyD88-independent pathway. Additionally, TRIF-deficient mice displayed defective TLR4-mediated inflammatory cytokine production, although activation of the MyD88-dependent pathway, such as IRAK-1 phosphorylation and early phase NF-κB activation, was not impaired. Thus, TLR4 requires both MyD88-dependent and MyD88-independent signals to induce the expression of inflammatory cytokines.

TRAF6 is reported to bind TRIF and cooperatively activate NF-κB (Sato et al. 2003). However, cells doubly deficient in TRAF6 and MyD88 still partially activated NF-κB in response to LPS, suggesting that TRIF activates NF-κB through both TRAF6-dependent and -independent pathways in TLR4 signaling (Gohda et al. 2004; Kawai et al. 2001).

Recently, receptor-interacting protein-1 (RIP1) has been implicated in the TLR3-mediated NF-κB response to dsRNA. RIP1 binds the C-terminus of TRIF/TICAM-1 via a Rip homotypic interaction motif (RHIM). In cells lacking RIP1, TLR3-mediated NF-κB activation and the subsequent induction of Icam-1 was severely impaired, whereas IFNβ induction was intact. By contrast, responses to LPS were normal in the absence of RIP1. However, LPS failed to stimulate NF-κB activation in $rip^{-/-}MyD88^{-/-}$ cells, revealing that RIP1 is also required for the TRIF-dependent TLR4-induced NF-κB pathway. In the TNF-α pathway, RIP1 interacts with the E3 ubiquitin ligase TRAF2 and is modified by polyubiquitin chain. Upon TLR3 activation, RIP1 is also modified by polyubiquitin chains and is recruited to TLR3 along with TRAF6 and TAK1, suggesting that RIP1 uses a similar ubiquitin-dependent mechanism to activate IKK-β in response to TNF-α and TLR3 ligands (Meylan et al. 2004).

TRIF also binds to RIP3 through its RHIM domain, and RIP3 is likely to negatively regulate the TRIF-RIP1-induced NF-κB pathway. Because RIP3 deficient mice have yet to be analyzed, the physiological role of RIP3 remains unknown (Meylan et al. 2004).

A fourth TIR-domain-containing adapter, TRAM (TRIF-related adaptor molecule)/TIRP (TIR domain-containing protein)/TICAM-2 has been identified (Bin et al. 2003; Oshiumi et al. 2003b; Yamamoto et al. 2003b). The analysis of TRAM/TICAM-2-deficient mice demonstrated that TRAM is essential for the TLR4-mediated MyD88-independent/TRIF-dependent pathway (Yamamoto et al. 2003b). The N terminus of TRAM has a myristoylation site, the mutation of which alters its normal membrane localization. TRAM/TICAM-2 acts as a bridging adapter between TLR4 and TRIF/TICAM-1 (Oshiumi et al. 2003b).

There is another TIR-domain-containing adapter, SARM (sterile α and armadillo-motif-containing protein). While an ortholog of mammalian SARM, the *Caenorhabditis elegans* TIR-domain-containing protein 1 (TIR1), has recently been shown to mediate the expression of genes that encodes antimicrobial peptides (Couillault et al. 2004), the function of SARM in mammalian innate immunity remains unknown. All analyses of these adaptor molecules have indicated that TIR domain-containing adaptors regulate the TLR signaling pathways by providing the specificity for each TLR cascade.

Recently, two noncanonical IKKs, inducible IKK (IKK*i*)/IKKε and TRAF family member-associated NF-κB activator (TANK)-binding kinase 1 (TBK1)/NF-κB-activating kinase (NAK)/TRAF2-associated kinase (T2K), have been shown responsible for dsRNA-induced IRF-3 activation (Sharma et al. 2003). These two IKKs are structurally related to IKKα and IKKβ, the catalytic subunits of the IKK complex that are essential for phosphorylation of two serine residues in IkBα. IKK*i*/IKKε and TBK1/NAK/T2K only induce phosphorylation of one serine residue in IκBα, and produce NF-κB activation via different mechanisms from IKKα and IKKβ. Overexpression of TBK1/NAK/T2K activates the promoter of IFN-β and IFN-inducible genes (Fitzgerald et al. 2003; McWhirter et al. 2004; Sharma et al. 2003). TBK1/NAK/T2K and IKK*i*/IKKε were shown to interact with IRF-3 and phosphorylate it using an in vitro kinase assay (Sharma et al. 2003). More recently, IKK*i*/IKKε-deficient and TBK1/NAK/T2K-deficient mice were generated and analyzed (Hemmi et al. 2004; Perry et al. 2004). TBK1/NAK/T2K and IKK*i*/IKKε were essential for the induction of IFN-β and IFN-inducible genes via activation of IRF-3 in both TLR-stimulated and virus-infected cells.

1.4.3 TLR7- and TLR9-Mediated Type I IFN Production

pDCs are specialized for producing large amounts of type I IFNs during viral infection (Liu 2005). pDCs highly express TLR7 and TLR9 and produce high levels of IFN-α in response to TLR7 and TLR9 ligands (Akira and Hemmi 2003). It appears that IFN-α induction by TLR7 and TLR9 depends entirely on MyD88 (Hemmi et al. 2003). IRF7 is a transcriptional factor, structurally related to IRF3, which is expressed constitutively in pDCs. Overexpression of IRF7 activates IFN-α- and IFN-β-dependent promoters. IRF7 forms a signaling complex with MyD88 and TRAF6 in the cytoplasm (Honda et al. 2004; Kawai et al. 2004). After ligand stimulation, IRF7 translocates into the nucleus to induce IFN-αs (Honda et al. 2004; Kawai et al. 2004). Similar to IRF3, IRF7 is activated by its phosphorylation. However, in TBK1-deficient cells, TLR9-mediated IFN-α production was still observed (Kawai et al. 2004). Mouse pDCs lacking IRAK-4 failed to produce both inflammatory cytokines and IFN-αs (Honda et al. 2004). Human TLR7-, TLR8- and TLR9-mediated induction of IFN-α/β and -γ was also IRAK-4 dependent (Yang et al. 2005). Recently, IRAK-1 has been shown to serve as an IRF7 kinase. In IRAK-1-deficient mice, TLR7- and TLR9-induced IFN-α production was completely abolished, although inflammatory cytokines were normally produced. Furthermore, IRF7 activation by TLR9 ligand was impaired in IRAK-1-deficient mice, in spite of normal NF-κB activation. IRAK-1 but not IRAK-4 could directly bind and phosphorylate IRF7; thus, IRAK-1 specifically mediates IFN-α induction downstream of MyD88 and IRAK-4 (Uematsu et al. 2005).

IRF8 is involved in TLR9-mediated responses because pDCs lacking IRF8 failed to produce proinflammatory cytokines and IFN-α in response to TLR9 ligand; in IRF8-deficient cells, TLR9-stimulated NF-κB activation was unexpectedly impaired, suggesting that IRF8 mediates NF-κB activation in TLR9 signaling (Tsujimura et al. 2004).

1.4.4 Other Molecules Involved in TLR Signaling

Several other molecules have been implicated in the TLR-mediated signaling pathway. Toll-interacting protein (Tollip) was originally identified through a yeast-two-hybrid screen using the IL-1R accessory protein (Burns et al. 2000). Tollip is present in a complex with IRAK-1. Upon stimulation with IL-1, the Tollip-IRAK-1 complex is recruited to the IL-1 receptor complex. IRAK-1 is then phosphorylated, which leads to the rapid dissociation of IRAK-1 from Tollip, thereby inducing activation of TRAF6. Tollip has subsequently been shown to negatively regulate the TLR-mediated signaling pathway (Zhang and Ghosh 2002). Overexpression of Tollip inhibited activation of NF-κB in response to IL-1, the TLR2 and TLR4 ligands. In addition, Tollip expression is elevated in intestinal epithelial cells, which are hyporesponsive to TLR2 ligands (Melmed et al. 2003). A unique C2-like domain in the N-terminus of Tollip has recently identified (Li et al. 2004). C2 domains in other proteins are involved in binding various phospholipids. It has also been shown that Tollip preferentially binds to phosphatidylinositol-3-phosphate and phosphatidylinositol-3,4,5-phosphate. Mutation of a vital lysine residue (K150) to glutamic acid within the C2 domain of Tollip inhibits LPS-induced NF-κB activation, indicating that its lipid-binding capability is somehow connected with the inhibitory role of Tollip (Liew et al. 2005). However, it remains unclear how Tollip is physiologically involved in TLR signaling.

Pellino was originally identified in *Drosophila* as a molecule that associates with Pelle, a *Drosophila* homolog of IRAK (Grosshans et al. 1999). Three mammalian homologs have so far been identified (Jensen and Whitehead 2003b; Jiang et al. 2003; Yu et al. 2002). Mammalian Pellino1 interacts with IRAK-1, IRAK-4, TRAF6, and TAK1, and is required for the activation of NF-κB but is not involved in the activation of the MAP kinase pathway (Jiang et al. 2003; Yu et al. 2002). Mammalian Pellino2 also interacts with these same proteins (Yu et al. 2002). There are controversial reports on the involvement of Pellino2 in NF-κB activation (Liu et al. 2004; Strelow et al. 2003). However, it has recently been shown that Pellino2 promotes activation of ERK1/2 and JNK (Jensen and Whitehead, 2003a). Human Pellino3 interacts with IRAK-1, TRAF6 and TAK1 but is incapable of activating NF-κB. However, like Pellino2, Pellino3 activates ERK1/2 and JNK (Jensen and Whitehead 2003b). A more recent report has shown that Pellino3 is an upstream regulator of p38 MAPK and activates CREB in a p38-dependent manner (Butler et al. 2005).

ECSIT (evolutionarily conserved signaling intermediate in Toll pathways) has no homology with any known protein and was cloned as a TRAF6-interacting protein by yeast two-hybrid screening (Kopp et al. 1999). ECSIT interacts with the conserved TRAF domain of TRAF6. A *Drosophila* homolog of ECSIT has been identified, and the interaction between TRAF6 and ECSIT is also conserved in *Drosophila*. ECSIT also interacts with MEKK1 (MAPK/ERK (extracellular signal-regulated kinase) kinase kinase 1), which can phosphorylate and activate the IKK complex. Expression of a dominant-negative mutant of ECSIT blocks signaling through TLR4, indicating that ECSIT might transduce TLR signals by bridging TRAF6 and the IKK complex. Furthermore, the inhibition of ECSIT expression, using siRNA in a macrophage cell line, resulted in impaired LPS-induced, but not TNF-induced, NF-κB activation (Xiao et al. 2003). The physiological function of ECSIT was studied by generating ECSIT-deficient mice, which were found to die on about embryonic day 7.5 (Xiao et al. 2003). Further characterization showed that ECSIT is an obligatory intermediate in bone morphogenetic protein (BMP) signaling, and therefore ECSIT is an essential component in both the TLR- and BMP-signaling pathways.

Members of the MAPKKK family are implicated in IKK/NF-κB and MAPK activation. Among them, MEKK3 has been shown to be involved in signaling through TLR4 but not TLR9 (Huang et al. 2004). MEKK3-deficient EFs showed impaired IL-6 production and defective activation of NF-κB, JNK and p38 MAPK in response to the TLR4 ligand. Stimulation of TLR4 also induced association of MEKK3 with TRAF6. Another member of the MKKK family, TPL2 (tumor-progression locus 2; also known as cancer Osaka thyroid, COT), has been shown to be involved in the TLR4-mediated activation of ERK (Dumitru et al. 2000). In response to TLR4 ligand, TPL2-deficient mice showed impaired TNF-α production and defective activation of ERK.

Recently, IRF5 has been shown to be involved in a MyD88/TRAF6 complex (Schoenemeyer et al. 2005; Takaoka et al. 2005). In IRF5-deficient macrophages and DCs, induction of proinflammatory cytokines in response to various TLR ligands (TLR3, 4, 5, 7 and 9) was severely impaired (Takaoka et al. 2005). Upon TLR ligand stimulation, IRF5 was shown to translocate to the nucleus from the cytoplasm and bind the potential IFN-stimulated response element (ISRE) motifs present in the promoter region of proinflammatory cytokine genes. IRF5 associates with both MyD88 and TRAF6. TLR4 and TLR9 ligands promote nuclear translocation of IRF5 in a MyD88-dependent manner. However, downstream signals of IRF5 are still unclear. Although IRF5 is phosphorylated by TBK1 and IKK-*i*/IKKε, this phosphorylation does not induce nuclear translocation of IRF5. The mechanism of TLR3-mediated IRF5 activation is also yet to be clarified as TLR3 does not use MyD88. Furthermore, Schoenemeyer and colleagues showed that IRF5 is not a target of the TLR3 signaling pathway but is activated by TLR7 or TLR8 signaling (Schoenemeyer et al. 2005). Further study is necessary to better characterize the IRF5 pathway.

TRAF3 was identified as a molecule that is required for type I IFN production in response to both TLR activation and viral infection (Hacker et al. 2006; Oganesyan et al. 2006). TRAF3 associates with the TLR adapter TRIF and IRAK-1, as well as downstream IRF3/7 kinases TBK1 and IKK-*i*/IKKε, and serves as a critical link between TLR adaptors and downstream regulatory kinases for IRF activation. TRAF3 is also essential for the induction of the anti-inflammatory cytokine, IL-10 (Hacker et al. 2006).

1.5 Negative Regulation of TLR Signaling

Toll-like receptors sense microbial products and induce the expression of immune and proinflammatory genes to eliminate invading pathogens. However, excessive activation of TLRs is harmful for a host since it induces immune-mediated or

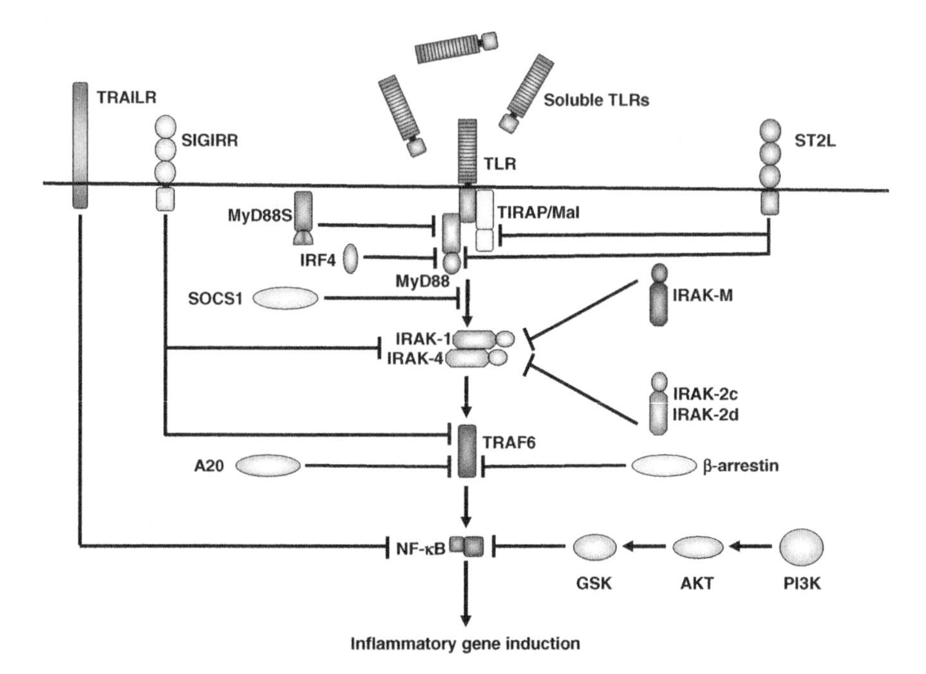

Fig. 3. Negative regulation of TLR signaling. TLR signaling pathways are negatively regulated by several molecules that are induced by TLR stimulations. Soluble forms of TLRs function as the first line of blockade of TLR signaling. Single Ig interleukin-1R-related molecule (*SIGIRR*) and ST2L have been shown to negatively modulate TLR signaling. MyD88S antagonizes MyD88. Interferon regulatory factor 4 (*IRF4*) also participates in the negative feedback regulation of TLR signaling. Inhibitory proteins such as suppressor of cytokine signaling-1 (*SOCS1*), interleukin-1 receptor kinase (IRAK)-M, IRAK-2c, and IRAK-2d selectively suppress IRAK functions. Phosphatidylinositol-3-kinase (*PI3K*) negatively regulates some TLR responses through glycogen synthase kinase-3 (*GSK*). A20 and β-arrestin deubiquitylate tumor necrosis factor receptor-associated factor 6 (*TRAF6*). Tumor necrosis factor-related apoptosis-inducing ligand receptor (*TRAILR*) suppresses nuclear factor kappa-B (*NF-κB*) activation during later TLR signaling events

inflammatory disorders. Living organisms have evolved multiple mechanisms of negative regulation to modulate TIR signaling, maintaining the balance between defending pathogens and preventing chronic inflammation and autoimmune diseases (Liew et al. 2005). Here, we discuss the molecules that are thought to negatively regulate TLR signaling (Fig. 3).

There are soluble TLRs (sTLRs) which function as decoys to effectively block TLR signaling. sTLR4 and sTLR2 have been identified in mammals. sTLR4 consists of 122 amino acids, of which 86 are identical to those of the extracellular domain of TLR4, and the remaining 36 amino acids share 70% homology with the N-terminal end of mouse phosphatidylinositol 3-kinase. sTLR4 is thought to block the interaction of LPS with TLR4 and co-receptor complexes, such as MD2 and CD14 (Iwami et al. 2000). sTLR2 was found to be constantly released from blood monocytes (LeBouder et al. 2003). There are six sTLR2 isoforms, which are generated by the post-translational modification of the transmembrane receptor protein. sTLR2 is present in human milk and plasma, where sTLR co-expresses and physically interacts with sCD14 (Iwaki et al. 2002). sTLRs work as antagonists of TLR2 in monocytes.

MyD88 is composed of three domains that are encoded by five exons (Hardiman et al. 1997). Exon1 encodes the death domain, Exon2 encodes the interdomain and the last three exons encode the TIR domain. MyD88s is an alternative spliced variant of MyD88, which lacks the interdomain (Burns et al. 2003; Janssens et al. 2002). MyD88s is only detected in the spleen but is upregulated in monocytes by LPS stimulation. Unlike MyD88, MyD88s does not interact with IRAK-4, but with IRAK-1. Overexpression of MyD88s does not induce IRAK-1 phosphorylation. MyD88s inhibits LPS-induced NF-κB activation by blocking the recruitment of IRAK-4 to the receptor complex and the following IRAK-1 phosphorylation (Burns et al. 2003).

Unlike other IRAKs, which are expressed predominantly by peripheral blood leukocytes, the expression of IRAK-M is restricted to monocytes and macrophages and is upregulated following stimulation with TLR ligands (Wesche et al. 1999). IRAK-M-deficient mice showed increased inflammatory responses to various TLR ligands and bacterial infections, and also impaired LPS tolerance (Kobayashi et al., 2002). Biochemical analysis has revealed that IRAK-M prevents the dissociation of the IRAK-1-IRAK-4 complex from MyD88, thereby preventing formation of the IRAK-1-TRAF6 complex. These findings indicated that IRAK-M functions as a negative regulator of TLR signaling.

Four splice variants of the mouse *Irak2* gene, *Irak2a, Irak2b, Irak2c* and *Irak2d*, have been identified, and are generated by alternative splicing at the 5′ end of the gene (Hardy and O'Neill 2004). Overexpression of IRAK-2c and IRAK-2d was shown to inhibit LPS-induced NF-κB activation. Both IRAK-2c and IRAK-2d lack the death domain which IRAK-2a, IRAK-2b, and full length IRAK-2 contain. Furthermore, LPS induced IRAK-2c but not IRAK-2a in a mouse macrophage cell line RAW264.7, indicating that IRAK-2c and IRAK-2d work as negative regulators in TLR signaling pathways.

SOCS1 is a member of the SOCS (suppressors of cytokine signaling) family, which is known to be important in suppressing cytokine signaling (Yasukawa et al. 2000). Lipopolysaccharide and CpG DNA have been shown to induce the expression of SOCS1 in macrophages (Dalpke et al. 2001; Stoiber et al. 1999). SOCS1-deficient mice overproduced nitric oxide and pro-inflammatory cytokines in response to LPS and CpG DNA, and were hypersensitive to LPS-induced endotoxin shock (Kinjyo et al. 2002; Nakagawa et al. 2002). Furthermore, LPS tolerance was not induced in SOCS1-deficient mice and the ectopic expression of SOCS1 in macrophages inhibited LPS-induced NF-κB activation, suggesting that SOCS1 directly inhibits TLR signaling pathways. This inhibition seems to occur by targeting IRAK-1, but the precise mechanism of its inhibition remains unclear.

Class IA PI3K (phosphatidylinositol-3-kinase) consists of a p85 regulatory subunit and a p110 catalytic chain. PI3Ks are activated by various TLR ligand stimulations in DCs (Katso et al. 2001). P85-deficient mice show enhanced TLR signaling and a dominant Th1-cell response. PI3K-deficient dendritic cells overproduce IL-12 in response to the TLR2, TLR4 and TLR9 ligands (Fukao et al. 2002); furthermore, PI3K-deficient mice on a BALB/c background were resistant to *Leishmania major* infection, unlike wild-type mice, because of the skewed Th1 response in these mice. These results indicated that PI3K is an effective negative regulator of TLR signaling. Activation of PI3K can mediate the recruitment and subsequent activation of signaling molecules with pleckstrin homology domains, including serine-threonine kinase Akt (Franke et al. 1997; Lawlor and Alessi 2001; Stokoe et al. 1997). After recruitment, Akt is activated by phosphorylation, subsequently phosphorylating several downstream targets of the PI3K pathway, including the constitutively active serine-threonine kinase glycogen synthase kinase (GSK3)-β. Phosphorylation of GSK3-β results in its inhibition (Cross et al. 1995). Recently, GSK3-β has been shown to differentially regulate TLR-mediated production of pro- and anti-inflammatory cytokines (Martin et al. 2005). Inhibition of GSK3 resulted in profound increases in IL-10 production after TLR2, TLR4, TLR5 or TLR 9 stimulations, whereas the concurrent production of proinflammatory cytokines, including IL-1β, IL-6, TNF-α, IL-12, and IFN-γ, by human monocytes and peripheral blood mononuclear cells, was substantially reduced. Administration of GSK3 inhibitor protected mice from endotoxin shock. cAMP response element-binding protein (CREB) is an important transcription factor for IL-10 production in human monocytes, whereas NF-κB regulates many diverse cellular process, including proinflammatory cytokine responses. Both CREB and NF-κB p65 utilize a nuclear coactivator CREB-binding protein (CBP) to induce their target genes. As nuclear amounts of CBP are limiting, CREB and NF-κB p65 compete for CBP binding. Without stimulation, GSK3 negatively regulates the activation and DNA-binding activity of CREB. However, when stimulated with TLR ligands, GSK3 is inhibited and increased association of CREB and CBP suppress NF-κB activity. Thus, GSK3 may regulate inflammatory responses by differentially affecting the nuclear amounts of the transcription factors NF-κB subunit p65 and CREB interacting with CBP (Martin et al. 2005).

A20 was initially identified as a zinc-finger protein induced by TNF-α, which suppresses TNF-mediated NF-κB activation (Krikos et al. 1992; Opipari et al. 1990). A20 was reported to be induced by LPS, and macrophages in A20-deficient mice show increased production of inflammatory cytokines in response to various TLR ligands. Mice reconstituted with fetal liver hematopoietic stem cells from A20-deficient mice are hypersensitive to LPS-induced shock but LPS tolerance was not impaired in A20-deficient mice. In vitro study also showed evidence of a role for A20 as a negative regulator in TLR signaling. Overexpression of A20 inhibited TLR-mediated NF-κB activation, and A20 was shown to be a cysteine protease deubiquitylating enzyme that cleaves the ubiquitin chain of TRAF6 to block TLR signaling (Boone et al. 2004).

ST2 (also known as T1, Fit-1 or DER4), which has been an orphan member of the IL-1 receptor family for a long time, was recently shown to be the receptor of the cytokine, IL-33. IL-33 promoted Th2-type responses via ST2 (Schmitz et al. 2005). ST2 has also been described as a negative regulator of Toll-like receptor–IL-1 receptor signaling (Brint et al. 2004). Two variants of the *St2* gene (ST2L and sST2) have been identified, and are generated by mRNA splicing (Bergers et al. 1994; Klemenz et al. 1989; Tominaga 1989). ST2L is a type I membrane protein and contains three extracellular immunoglobulin-like domains and an intracellular TIR domain. sST2 is the soluble protein, which consists of the extracellular domains of ST2L with an extra 9 amino acids in the C-terminal (Bergers et al. 1994; Klemenz et al. 1989; Tominaga 1989). ST2-deficient mice (lacking both ST2L and sST2) showed increased production of inflammatory cytokines in response to LPS, CpG DNA and IL-1 but not to the TLR3 ligand, polyI:C. ST2L co-precipitated with MyD88 and TIRAP/MAL through a conserved proline residue in TIR domain, but not with IRAK or TRIF. Taken together, ST2L blocks TLR signaling by sequestering MyD88 and TIRAP/MAL with its TIR domain from TLRs. Although ST2-deficient mice showed impairment in LPS tolerance, ST2-deficient mice were not susceptible to LPS shock. ST2L normally stays in cytoplasm and expresses on the cell surface after at least 4 h of LPS stimulation. The discrepancy of phenotype between LPS shock and LPS tolerance in ST2-deficient mice may due to this time lag (Brint et al. 2004). sST2 is also thought to be a negative regulator of TLR signaling. sST2 expression is upregulated in pro-inflammatory cytokines and LPS (Kumar et al. 1997; Saccani et al. 1998). It is also present in normal human serum and its levels are increased in various inflammatory diseases. sST2 is thought to enhance cytokine production by binding macrophages through a putative ST2 receptor. There is also another report that sST2 downregulates the expression of TLR4 and TLR1 in LPS-stimulated macrophages. However, the precise mechanism of sST2 inhibition is currently unknown (Sweet et al. 2001).

SIGIRR (single Ig IL-1R-related molecule) is also an orphan receptor, and contains a single extracellular immunoglobulin domain and a cytoplasmic TIR domain (Thomassen et al. 1999). SIGIRR-deficient bone marrow DCs exaggerated responses to LPS and CpG DNA but not to the PolyI:C (Garlanda et al. 2004). Furthermore, SIGIRR-deficient mice were highly susceptible to LPS shock. Following TLR stimulation, SIGIRR has been shown to interact transiently with TLR4, IRAK-1 and TRAF6, thereby negatively regulating TLR signaling (Wald et al. 2003).

TRAILR (tumor necrosis factor-related apoptosis-inducing ligand receptor) belongs to the TNF superfamily and does not have a TIR domain (Wu et al. 1999). TRAIL (Tumor necrosis factor related apoptosis inducing ligand) expression is upregulated by TLR2, TRL3 and TLR4 but not TLR9 ligands. TRAILR-deficient mice showed increased cytokine production in response to TLR ligands. TRAILR seems to inhibit TLR signaling by stabilizing IκBα, which results in the decreased nuclear translocation of NF-κB (Diehl et al. 2004).

RP105 is a TLR homolog originally cloned as a B-cell specific molecule (Miyake et al. 1994; Miyake et al. 1995). RP105 has a conserved extracellular leucine-rich repeat domain but lacks a signaling domain. Divanovic et al. recently showed that RP105 is widely expressed and mirrors the expression of TLR4 (Divanovic et al. 2005). RP105 and its helper molecule, MD1, interacted directly with the TLR4/MD2 complex, inhibiting its ability to bind LPS. RP105 has been shown to be a physiological regulator of TLR4 signaling in primary DCs, and of responses to LPS in vivo.

A member of the IRF family, IRF4 has recently been reported as a negative regulator of TLR signaling (Honma et al. 2005; Negishi et al. 2005). IRF4 mRNA is induced by TLR activation and IRF4 competes with IRF5 for interaction with MyD88. IRF4-deficent mice were sensitive to LPS-induced shock, and their macrophages produced high levels of proinflammatory cytokines. Activation of NF-κB and JNK was enhanced after LPS stimulation in IRF4-deficient macrophages. IRF4 may participate in the negative feedback regulation of TLR signaling.

The arrestin family consists of ubiquitously expressed β-arrestin 1, β-arrestin 2 and two more arrestins expressed exclusively in retinae (Lefkowitz and Whalen 2004). β-arrestins have been shown to directly interact with TRAF6 after TLR-IL-1R activation. Formation of the β-arrestin-TRAF6 complex prevented auto-ubiquitination of TRAF6 and activation of NF-κB and AP-1. Endotoxin-treated β-arrestin 2-deficient mice had higher expressions of proinflammatory cytokines and were more susceptible to endotoxic shock, suggesting that β-arrestins are essential negative regulators of the innate immune activation via TLR-IL-1R signaling (Wang et al. 2006).

1.6 Conclusion and Future Prospects

Following the discovery of Toll, the molecular mechanisms of pathogen recognition by TLRs have been elucidated over the past few years. Toll-like receptors have now been identified as important components of the innate immune system. However, it is reported that viruses and intracellular bacteria also induce innate immune responses through a TLR-independent mechanism. For example, intracellular administration of polyI:C induces type I IFNs from DCs lacking TLR3 (Hemmi et al. 2004; Yamamoto et al. 2003a). The DExD/H box RNA helicase RIG-I (retinoic acid inducible gene I) has recently been identified as a cytoplasmic sensor for dsRNA (Kato et al. 2005; Yoneyama et al. 2004). Another DExD/H box RNA helicase, Mda-5 (Melanoma differentiation associated gene 5), which is closely related to RIG-I, has been suggested to be receptor-like molecule for viral RNA (Andrejeva et al. 2004; Kang et al. 2004). Recently, the RIG-I/Mda5-associated adapter IPS-1 (interferon-beta promoter stimulator-1)/MAVS (mitochondrial antiviral signaling)/VISA (virus-induced signaling adaptor)/CARDIF (CARD adaptor inducing interferon-beta) was identified, and is required for IFN-β and NF-κB activation (Kawai et al. 2005; Meylan et al. 2005; Seth et al. 2005; Xu et al. 2005). Furthermore, DNA, including host DNA, can be recognized independently of TLR9 (Boule et al. 2004; Decker et al. 2005). DNaseII in macrophages is important for cleaving DNA of engulfed apoptotic cells or debris (Kawane et al. 2003). Without cleaning off DNA in phagosomes, macrophages produced IFN-β in a TLR-independent manner, indicating that there is a TLR-independent, endogenous DNA-sensing mechanism to activate innate immunity (Okabe et al. 2005; Yoshida et al. 2005). Recently, Ishii et al. reported that intracellular administration of double stranded B-form DNA (B-DNA) but not Z-DNA triggered antiviral responses including production of type I IFNs and chemokines independently of TLR. B-DNA-mediated signaling might utilize a similar pathway of RIG-I or Mda-5 (Ishii et al. 2006).

The newly discovered CATERPILLER gene family regulates inflammatory and apoptotic responses, and some act as sensors to detect pathogen products (Ting and Davis 2005). Among them, NOD1 (nucleotide-binding oligomerization domain) and NOD2, which contain N-terminal CARD (caspase recruiting domain) domains, a central NACHT (neuronal apoptosis inhibitory protein (NAIP), MHC class II transcription activator (CIITA), incompatibility locus protein from *Podospora anserina* (HET-E), and telomerase-associated protein (TP1)) domain and C-terminal LRRs, recognize γ-D-glutamyl-*meso*-diaminopimelic acid (ie-DAP), a cell wall derivative from Gram-negative bacteria, and muramyl dipeptide (MDP), found in both Gram-positive and Gram-negative bacteria, respectively (Chamaillard et al. 2003; Girardin et al. 2003). Mutation in the human *nod2* gene is reportedly associated with the pathogenesis of Crohn's disease and Blau syndrome (Hugot et al. 2001; Ogura et al. 2001; Philpott and Girardin 2004). NAIP (neuronal apoptosis inhibitor protein) is a member of the inhibitor of apoptosis (IAP) family, which contain a BIR (baculovirus inhibitor of apoptosis protein repeat) domain in their N-termini. NAIP5 has also been implicated in host responses to *Legionella* infection (Diez et al. 2003).

As stated above, recent reports have brought rapid clarification of TLR-independent pathogen recognition. As TLRs recognize pathogens at cell surfaces or in phagosomes, they cannot sense pathogens such as intracellular bacteria and viruses that replicate in the cytoplasm. Hence, hosts have developed other mechanisms to sense and exclude them. To more comprehensively understand innate immune responses, further studies of both TLR-dependent and TLR-independent mechanism will be needed.

Acknowledgments We thank our colleagues in our laboratory for helpful discussion and M. Hashimoto for secretarial assistance. This work is supported in part by grants from the Special Coordination Funds of the Japanese Ministry of Education, Culture, Sports, Science and Technology.

References

Adachi O, Kawai T, Takeda K, Matsumoto M, Tsutsui H, Sakagami M, Nakanishi K, Akira S (1998) Targeted disruption of the MyD88 gene results in loss of IL-1- and IL-18-mediated function. Immunity 9:143–150

Akira S (2004) Toll receptor families: structure and function. Semin Immunol 16:1–2

Akira S, Hemmi H (2003) Recognition of pathogen-associated molecular patterns by TLR family. Immunol Lett 85:85–95

Akira S, Takeda K (2004) Toll-like receptor signalling. Nat Rev Immunol 4:499–511

Alexopoulou L, Holt AC, Medzhitov R, Flavell RA (2001) Recognition of double-stranded RNA and activation of NF-kappaB by Toll-like receptor 3. Nature 413:732–738

Alexopoulou L, Thomas V, Schnare M, Lobet Y, Anguita J, Schoen RT, Medzhitov R, Fikrig E, Flavell RA (2002) Hyporesponsiveness to vaccination with *Borrelia burgdorferi* OspA in humans and in TLR1- and TLR2-deficient mice. Nat Med 8:878–884

Andersen-Nissen E, Smith KD, Strobe KL, Barrett SL, Cookson BT, Logan SM, Aderem A (2005) Evasion of Toll-like receptor 5 by flagellated bacteria. Proc Natl Acad Sci USA 102: 9247–9252

Andrejeva J, Childs KS, Young DF, Carlos TS, Stock N, Goodbourn S, Randall RE (2004) The V proteins of paramyxoviruses bind the IFN-inducible RNA helicase, mda-5, and inhibit its activation of the IFN-beta promoter. Proc Natl Acad Sci USA 101:17264–17269

Asea A, Rehli M, Kabingu E, Boch JA, Bare O, Auron PE, Stevenson MA, Calderwood SK (2002) Novel signal transduction pathway utilized by extracellular HSP70: role of toll-like receptor (TLR) 2 and TLR4. J Biol Chem 277:15028–15034

Balachandran S, Thomas E, Barber GN (2004) A FADD-dependent innate immune mechanism in mammalian cells. Nature 432:401–405

Bergers G, Reikerstorfer A, Braselmann S, Graninger P, Busslinger M (1994) Alternative promoter usage of the Fos-responsive gene Fit-1 generates mRNA isoforms coding for either secreted or membrane-bound proteins related to the IL-1 receptor. EMBO J 13:1176–1188

Bieback K, Lien E, Klagge IM, Avota E, Schneider-Schaulies J, Duprex WP, Wagner H, Kirschning CJ, Ter Meulen V, Schneider-Schaulies S (2002) Hemagglutinin protein of wild-type measles virus activates toll-like receptor 2 signaling. J Virol 76:8729–8736

Bin LH, Xu LG, Shu HB (2003) TIRP, a novel Toll/interleukin-1 receptor (TIR) domain-containing adapter protein involved in TIR signaling. J Biol Chem 278:24526–24532

Biragyn A, Ruffini PA, Leifer CA, Klyushnenkova E, Shakhov A, Chertov O, Shirakawa AK, Farber JM, Segal DM, Oppenheim JJ, Kwak LW (2002) Toll-like receptor 4-dependent activation of dendritic cells by beta-defensin 2. Science 298:1025–1029

Bischoff V, Vignal C, Boneca IG, Michel T, Hoffmann JA, Royet J (2004) Function of the drosophila pattern-recognition receptor PGRP-SD in the detection of Gram-positive bacteria. Nat Immunol 5:1175–1180

Boone DL, Turer EE, Lee EG, Ahmad RC, Wheeler MT, Tsui C, Hurley P, Chien M, Chai S, Hitotsumatsu O, et al. (2004) The ubiquitin-modifying enzyme A20 is required for termination of Toll-like receptor responses. Nat Immunol 5:1052–1060

Boule MW, Broughton C, Mackay F, Akira S, Marshak-Rothstein A, Rifkin IR (2004) Toll-like receptor 9-dependent and -independent dendritic cell activation by chromatin-immunoglobulin G complexes. J Exp Med 199:1631–1640

Boutros M, Agaisse H, Perrimon N (2002) Sequential activation of signaling pathways during innate immune responses in *Drosophila*. Dev Cell 3:711–722

Brennan CA, Anderson KV (2004) *Drosophila*: the genetics of innate immune recognition and response. Annu Rev Immunol 22:457–483

Brint EK, Xu D, Liu H, Dunne A, McKenzie AN, O'Neill LA, Liew FY (2004) ST2 is an inhibitor of interleukin 1 receptor and Toll-like receptor 4 signaling and maintains endotoxin tolerance. Nat Immunol 5:373–379

Brown P (2001) Cinderella goes to the ball. Nature 410:1018–1020

Brown GD, Taylor PR, Reid DM, Willment JA, Williams DL, Martinez-Pomares L, Wong SY, Gordon S (2002) Dectin-1 is a major beta-glucan receptor on macrophages. J Exp Med 196:407–412

Bulut Y, Faure E, Thomas L, Karahashi H, Michelsen KS, Equils O, Morrison SG, Morrison RP, Arditi M (2002) Chlamydial heat shock protein 60 activates macrophages and endothelial cells through Toll-like receptor 4 and MD2 in a MyD88-dependent pathway. J Immunol 168:1435–1440

Burns K, Clatworthy J, Martin L, Martinon F, Plumpton C, Maschera B, Lewis A, Ray K, Tschopp J, Volpe F (2000) Tollip, a new component of the IL-1RI pathway, links IRAK to the IL-1 receptor. Nat Cell Biol 2:346–351

Burns K, Janssens S, Brissoni B, Olivos N, Beyaert R, Tschopp J (2003) Inhibition of interleukin 1 receptor/Toll-like receptor signaling through the alternatively spliced, short form of MyD88 is due to its failure to recruit IRAK-4. J Exp Med 197:263–268

Butler MP, Hanly JA, Moynagh PN (2005) Pellino3 is a novel upstream regulator of p38 MAPK and activates CREB in a p38-dependent manner. J Biol Chem 280:27759–27768

Chamaillard M, Hashimoto M, Horie Y, Masumoto J, Qiu S, Saab L, Ogura Y, Kawasaki A, Fukase K, Kusumoto S, et al. (2003) An essential role for NOD1 in host recognition of bacterial peptidoglycan containing diaminopimelic acid. Nat Immunol 4:702–707

Chen P, Rodriguez A, Erskine R, Thach T, Abrams JM (1998) Dredd, a novel effector of the apoptosis activators reaper, grim, and hid in *Drosophila*. Dev Biol 201:202–216

Choe KM, Werner T, Stoven S, Hultmark D, Anderson KV (2002) Requirement for a peptidoglycan recognition protein (PGRP) in Relish activation and antibacterial immune responses in *Drosophila*. Science 296:359–362

Coban C, Ishii KJ, Kawai T, Hemmi H, Sato S, Uematsu S, Yamamoto M, Takeuchi O, Itagaki S, Kumar N, et al (2005) Toll-like receptor 9 mediates innate immune activation by the malaria pigment hemozoin. J Exp Med 201:19–25

Compton T, Kurt-Jones EA, Boehme KW, Belko J, Latz E, Golenbock DT, Finberg RW (2003) Human cytomegalovirus activates inflammatory cytokine responses via CD14 and Toll-like receptor 2. J Virol 77:4588–4596

Couillault C, Pujol N, Reboul J, Sabatier L, Guichou JF, Kohara Y, Ewbank JJ (2004) TLR-independent control of innate immunity in Caenorhabditis elegans by the TIR domain adaptor protein TIR-1, an ortholog of human SARM. Nat Immunol 5:488–494

Cross DA, Alessi DR, Cohen P, Andjelkovich M, Hemmings BA (1995) Inhibition of glycogen synthase kinase-3 by insulin mediated by protein kinase B. Nature 378:785–789

Dalpke AH, Opper S, Zimmermann S, Heeg K (2001) Suppressors of cytokine signaling (SOCS)-1 and SOCS-3 are induced by CpG-DNA and modulate cytokine responses in APCs. J Immunol 166:7082–7089

Decker P, Singh-Jasuja H, Haager S, Kotter I, Rammensee HG (2005) Nucleosome, the main autoantigen in systemic lupus erythematosus, induces direct dendritic cell activation via a MyD88-independent pathway: consequences on inflammation. J Immunol 174:3326–3334

Deng L, Wang C, Spencer E, Yang L, Braun A, You J, Slaughter C, Pickart C, Chen ZJ (2000) Activation of the IkappaB kinase complex by TRAF6 requires a dimeric ubiquitin-conjugating enzyme complex and a unique polyubiquitin chain. Cell 103:351–361

Diebold SS, Kaisho T, Hemmi H, Akira S, Reis E, Sousa C (2004) Innate antiviral responses by means of TLR7-mediated recognition of single-stranded RNA. Science 303:1529–1531

Diehl GE, Yue HH, Hsieh K, Kuang AA, Ho M, Morici LA, Lenz LL, Cado D, Riley LW, Winoto A (2004) TRAIL-R as a negative regulator of innate immune cell responses. Immunity 21:877–889

Diez E, Lee SH, Gauthier S, Yaraghi Z, Tremblay M, Vidal S, Gros P (2003) Birc1e is the gene within the Lgn1 locus associated with resistance to *Legionella pneumophila*. Nat Genet 33:55–60

Divanovic S, Trompette A, Atabani SF, Madan R, Golenbock DT, Visintin A, Finberg RW, Tarakhovsky A, Vogel SN, Belkaid Y, et al. (2005) Negative regulation of Toll-like receptor 4 signaling by the Toll-like receptor homolog RP105. Nat Immunol 6:571–578

Dumitru CD, Ceci JD, Tsatsanis C, Kontoyiannis D, Stamatakis K, Lin JH, Patriotis C, Jenkins NA, Copeland NG, Kollias G, Tsichlis PN (2000) TNF-alpha induction by LPS is regulated posttranscriptionally via a Tpl2/ERK-dependent pathway. Cell 103:1071–1083

Dunstan SJ, Hawn TR, Hue NT, Parry CP, Ho VA, Vinh H, Diep TS, House D, Wain J, Aderem A, et al. (2005) Host susceptibility and clinical outcomes in toll-like receptor 5-deficient patients with typhoid fever in Vietnam. J Infect Dis 191:1068–1071

Dybdahl B, Wahba A, Lien E, Flo TH, Waage A, Qureshi N, Sellevold OF, Espevik T, Sundan A (2002) Inflammatory response after open heart surgery: release of heat-shock protein 70 and signaling through toll-like receptor-4. Circulation 105:685–690

Elrod-Erickson M, Mishra S, Schneider D (2000) Interactions between the cellular and humoral immune responses in *Drosophila*. Curr Biol 10:781–784

Fitzgerald KA, Palsson-McDermott EM, Bowie AG, Jefferies CA, Mansell AS, Brady G, Brint E, Dunne A, Gray P, Harte MT, et al. (2001) Mal (MyD88-adapter-like) is required for Toll-like receptor-4 signal transduction. Nature 413:78–83

Fitzgerald KA, McWhirter SM, Faia KL, Rowe DC, Latz E, Golenbock DT, Coyle AJ, Liao SM, Maniatis T (2003) IKKepsilon and TBK1 are essential components of the IRF3 signaling pathway. Nat Immunol 4:491–496

Foley E, O'Farrell PH (2004) Functional dissection of an innate immune response by a genome-wide RNAi screen. PLoS Biol 2, E203

Franke TF, Kaplan DR, Cantley LC, Toker A (1997) Direct regulation of the Akt proto-oncogene product by phosphatidylinositol-3,4-bisphosphate. Science 275:665–668

Fukao T, Tanabe M, Terauchi Y, Ota T, Matsuda S, Asano T, Kadowaki T, Takeuchi T, Koyasu S (2002) PI3K-mediated negative feedback regulation of IL-12 production in DCs. Nat Immunol 3:875–881

Gantner BN, Simmons RM, Canavera SJ, Akira S, Underhill DM (2003) Collaborative induction of inflammatory responses by dectin-1 and Toll-like receptor 2. J Exp Med 197:1107–1117

Garlanda C, Riva F, Polentarutti N, Buracchi C, Sironi M, De Bortoli M, Muzio M, Bergottini R, Scanziani E, Vecchi A, et al. (2004) Intestinal inflammation in mice deficient in Tir8, an inhibitory member of the IL-1 receptor family. Proc Natl Acad Sci USA 101:3522–3526

Gazzinelli RT, Ropert C, Campos MA (2004) Role of the Toll/interleukin-1 receptor signaling pathway in host resistance and pathogenesis during infection with protozoan parasites. Immunol Rev 201:9–25

Georgel P, Naitza S, Kappler C, Ferrandon D, Zachary D, Swimmer C, Kopczynski C, Duyk G, Reichhart JM, Hoffmann JA (2001) *Drosophila* immune deficiency (IMD) is a death domain protein that activates antibacterial defense and can promote apoptosis. Dev Cell 1:503–514

Gewirtz AT, Navas TA, Lyons S, Godowski PJ, Madara JL (2001) Cutting edge: bacterial flagellin activates basolaterally expressed TLR5 to induce epithelial proinflammatory gene expression. J Immunol 167:1882–1885

Girardin SE, Boneca IG, Viala J, Chamaillard M, Labigne A, Thomas G, Philpott DJ, Sansonetti PJ (2003) Nod2 is a general sensor of peptidoglycan through muramyl dipeptide (MDP) detection. J Biol Chem 278:8869–8872

Gobert V, Gottar M, Matskevich AA, Rutschmann S, Royet J, Belvin M, Hoffmann JA, Ferrandon D (2003) Dual activation of the *Drosophila* toll pathway by two pattern recognition receptors. Science 302:2126–2130

Gohda J, Matsumura T, Inoue J (2004) Cutting edge: TNFR-associated factor (TRAF) 6 is essential for MyD88-dependent pathway but not toll/IL-1 receptor domain-containing adaptor-inducing IFN-beta (TRIF)-dependent pathway in TLR signaling. J Immunol 173:2913–2917

Gottar M, Gobert V, Michel T, Belvin M, Duyk G, Hoffmann JA, Ferrandon D, Royet J (2002) The *Drosophila* immune response against Gram-negative bacteria is mediated by a peptidoglycan recognition protein. Nature 416:640–644

Grosshans J, Schnorrer F, Nusslein-Volhard C (1999) Oligomerisation of Tube and Pelle leads to nuclear localisation of dorsal. Mech Dev 81:127–138

Guillot L, Balloy V, McCormack FX, Golenbock DT, Chignard M, Si-Tahar M (2002) Cutting edge: the immunostimulatory activity of the lung surfactant protein-A involves Toll-like receptor 4. J Immunol 168:5989–5992

Hacker H, Redecke V, Blagoev B, Kratchmarova I, Hsu LC, Wang GG, Kamps MP, Raz E, Wagner H, Hacker G, et al. (2006) Specificity in Toll-like receptor signalling through distinct effector functions of TRAF3 and TRAF6. Nature 439:204–207

Hardiman G, Jenkins NA, Copeland NG, Gilbert DJ, Garcia DK, Naylor SL, Kastelein RA, Bazan JF (1997) Genetic structure and chromosomal mapping of MyD88. Genomics 45:332–339

Hardy MP, O'Neill LA (2004) The murine IRAK2 gene encodes four alternatively spliced isoforms, two of which are inhibitory. J Biol Chem 279:27699–27708
Hashimoto M, Asai Y, Ogawa T (2004) Separation and structural analysis of lipoprotein in a lipopolysaccharide preparation from Porphyromonas gingivalis. Int Immunol 16:1431–1437
Hawn TR, Verbon A, Lettinga KD, Zhao LP, Li SS, Laws RJ, Skerrett SJ, Beutler B, Schroeder L, Nachman A, et al. (2003) A common dominant TLR5 stop codon polymorphism abolishes flagellin signaling and is associated with susceptibility to legionnaires' disease. J Exp Med 198:1563–1572
Hayashi F, Smith KD, Ozinsky A, Hawn TR, Yi EC, Goodlett DR, Eng JK, Akira S, Underhill DM, Aderem A (2001) The innate immune response to bacterial flagellin is mediated by Toll-like receptor 5. Nature 410:1099–1103
Haynes LM, Moore DD, Kurt-Jones EA, Finberg RW, Anderson LJ, Tripp RA (2001) Involvement of toll-like receptor 4 in innate immunity to respiratory syncytial virus. J Virol 75:10730–10737
Heil F, Hemmi H, Hochrein H, Ampenberger F, Kirschning C, Akira S, Lipford G, Wagner H, Bauer S (2004) Species-specific recognition of single-stranded RNA via toll-like receptor 7 and 8. Science 303:1526–1529
Hemmi H, Takeuchi O, Kawai T, Kaisho T, Sato S, Sanjo H, Matsumoto M, Hoshino K, Wagner H, Takeda K, Akira S (2000) A Toll-like receptor recognizes bacterial DNA. Nature 408:740–745
Hemmi H, Kaisho T, Takeuchi O, Sato S, Sanjo H, Hoshino K, Horiuchi T, Tomizawa H, Takeda K, Akira S (2002) Small anti-viral compounds activate immune cells via the TLR7 MyD88-dependent signaling pathway. Nat Immunol 3:196–200
Hemmi H, Kaisho T, Takeda K, Akira S (2003) The roles of Toll-like receptor 9, MyD88, and DNA-dependent protein kinase catalytic subunit in the effects of two distinct CpG DNAs on dendritic cell subsets. J Immunol 170:3059–3064
Hemmi H, Takeuchi O, Sato S, Yamamoto M, Kaisho T, Sanjo H, Kawai T, Hoshino K, Takeda K, Akira S (2004) The roles of two IkappaB Kinase-related kinases in lipopolysaccharide and double stranded RNA signaling and viral infection. J Exp Med 199:1641–1650
Hochrein H, Schlatter B, O'Keeffe M, Wagner C, Schmitz F, Schiemann M, Bauer S, Suter M, Wagner H (2004) Herpes simplex virus type-1 induces IFN-alpha production via Toll-like receptor 9-dependent and -independent pathways. Proc Natl Acad Sci USA 101: 11416–11421
Hoebe K, Du X, Georgel P, Janssen E, Tabeta K, Kim SO, Goode J, Lin P, Mann N, Mudd S, et al. (2003) Identification of Lps2 as a key transducer of MyD88-independent TIR signalling. Nature 424:743–748
Hoebe K, Georgel P, Rutschmann S, Du X, Mudd S, Crozat K, Sovath S, Shamel L, Hartung T, Zahringer U, Beutler B (2005) CD36 is a sensor of diacylglycerides. Nature 433:523–527
Hoffmann JA (2003) The immune response of *Drosophila*. Nature 426:33–38
Honda K, Yanai H, Mizutani T, Negishi H, Shimada N, Suzuki N, Ohba Y, Takaoka A, Yeh WC, Taniguchi T (2004) Role of a transductional-transcriptional processor complex involving MyD88 and IRF-7 in Toll-like receptor signaling. Proc Natl Acad Sci USA 101:15416–15421
Honma K, Udono H, Kohno T, Yamamoto K, Ogawa A, Takemori T, Kumatori A, Suzuki S, Matsuyama T, Yui K (2005) Interferon regulatory factor 4 negatively regulates the production of proinflammatory cytokines by macrophages in response to LPS. Proc Natl Acad Sci USA 102:16001–16006
Horng T, Barton GM, Medzhitov R (2001) TIRAP: an adapter molecule in the Toll signaling pathway. Nat Immunol 2:835–841
Horng T, Barton GM, Flavell RA, Medzhitov R (2002) The adaptor molecule TIRAP provides signalling specificity for Toll-like receptors. Nature 420:329–333
Hoshino K, Kaisho T, Iwabe T, Takeuchi O, Akira S (2002) Differential involvement of IFN-beta in Toll-like receptor-stimulated dendritic cell activation. Int Immunol 14:1225–1231

Hu S, Yang X (2000) dFADD, a novel death domain-containing adapter protein for the *Drosophila* caspase DREDD. J Biol Chem 275:30761–30764
Hu X, Yagi Y, Tanji T, Zhou S, Ip YT (2004) Multimerization and interaction of Toll and Spatzle in *Drosophila*. Proc Natl Acad Sci USA 101:9369–9374
Huang Q, Yang J, Lin Y, Walker C, Cheng J, Liu ZG, Su B (2004) Differential regulation of interleukin 1 receptor and Toll-like receptor signaling by MEKK3. Nat Immunol 5:98–103
Hugot JP, Chamaillard M, Zouali H, Lesage S, Cezard JP, Belaiche J, Almer S, Tysk C, O'Morain CA, Gassull M, et al (2001) Association of NOD2 leucine-rich repeat variants with susceptibility to Crohn's disease. Nature 411:599–603
Hultmark D (2003) *Drosophila* immunity: paths and patterns. Curr Opin Immunol 15:12–19
Ishii KJ, Coban C, Kato H, Takahashi K, Torii Y, Takeshita F, Ludwig H, Sutter G, Suzuki K, Hemmi H, et al (2006) A Toll-like receptor-independent antiviral response induced by double-stranded B-form DNA. Nat Immunol 7:40–48
Ishitani T, Takaesu G, Ninomiya-Tsuji J, Shibuya H, Gaynor RB, Matsumoto K (2003) Role of the TAB2-related protein TAB3 in IL-1 and TNF signaling. EMBO J 22:6277–6288
Ito T, Amakawa R, Kaisho T, Hemmi H, Tajima K, Uehira K, Ozaki Y, Tomizawa H, Akira S, Fukuhara S (2002) Interferon-alpha and interleukin-12 are induced differentially by Toll-like receptor 7 ligands in human blood dendritic cell subsets. J Exp Med 195:1507–1512
Iwaki D, Mitsuzawa H, Murakami S, Sano H, Konishi M, Akino T, Kuroki Y (2002) The extracellular toll-like receptor 2 domain directly binds peptidoglycan derived from *Staphylococcus aureus*. J Biol Chem 277:24315–24320
Iwami KI, Matsuguchi T, Masuda A, Kikuchi T, Musikacharoen T, Yoshikai Y (2000) Cutting edge: naturally occurring soluble form of mouse Toll-like receptor 4 inhibits lipopolysaccharide signaling. J Immunol 165:6682–6686
Janssens S, Beyaert R (2003) Functional diversity and regulation of different interleukin-1 receptor-associated kinase (IRAK) family members. Mol Cell 11:293–302
Janssens S, Burns K, Tschopp J, Beyaert R (2002) Regulation of interleukin-1- and lipopolysaccharide-induced NF-kappaB activation by alternative splicing of MyD88. Curr Biol 12:467–471
Jensen LE, Whitehead AS (2003a) Pellino2 activates the mitogen activated protein kinase pathway. FEBS Lett 545:199–202
Jensen LE, Whitehead AS (2003b) Pellino3, a novel member of the Pellino protein family, promotes activation of c-Jun and Elk-1 and may act as a scaffolding protein. J Immunol 171:1500–1506
Jiang Z, Johnson HJ, Nie H, Qin J, Bird TA, Li X (2003) Pellino 1 is required for interleukin-1 (IL-1)-mediated signaling through its interaction with the IL-1 receptor-associated kinase 4 (IRAK4)-IRAK-tumor necrosis factor receptor-associated factor 6 (TRAF6) complex. J Biol Chem 278:10952–10956
Jiang D, Liang J, Fan J, Yu S, Chen S, Luo Y, Prestwich GD, Mascarenhas MM, Garg HG, Quinn DA, et al (2005) Regulation of lung injury and repair by Toll-like receptors and hyaluronan. Nat Med 11:1173–1179
Johnson GB, Brunn GJ, Kodaira Y, Platt JL (2002) Receptor-mediated monitoring of tissue well-being via detection of soluble heparan sulfate by Toll-like receptor 4. J Immunol 168: 5233–5239
Kanakaraj P, Schafer PH, Cavender DE, Wu Y, Ngo K, Grealish PF, Wadsworth SA, Peterson PA, Siekierka JJ, Harris CA, Fung-Leung WP (1998) Interleukin (IL)-1 receptor-associated kinase (IRAK) requirement for optimal induction of multiple IL-1 signaling pathways and IL-6 production. J Exp Med 187:2073–2079
Kang DC, Gopalkrishnan RV, Lin L, Randolph A, Valerie K, Pestka S, Fisher PB (2004) Expression analysis and genomic characterization of human melanoma differentiation associated

gene-5, mda-5: a novel type I interferon-responsive apoptosis-inducing gene. Oncogene 23:1789–1800

Kariko K, Ni H, Capodici J, Lamphier M, Weissman D (2004) mRNA is an endogenous ligand for Toll-like receptor 3. J Biol Chem 279:12542–12550

Karin M, Ben-Neriah Y (2000) Phosphorylation meets ubiquitination: the control of NF-[kappa]B activity. Annu Rev Immunol 18:621–663

Kato H, Sato S, Yoneyama M, Yamamoto M, Uematsu S, Matsui K, Tsujimura T, Takeda K, Fujita T, Takeuchi O, Akira S (2005) Cell type-specific involvement of RIG-I in antiviral response. Immunity 23:19–28

Katso R, Okkenhaug K, Ahmadi K, White S, Timms J, Waterfield MD (2001) Cellular function of phosphoinositide 3-kinases: implications for development, homeostasis, and cancer. Annu Rev Cell Dev Biol 17:615–675

Kawai T, Adachi O, Ogawa T, Takeda K, Akira S (1999) Unresponsiveness of MyD88-deficient mice to endotoxin. Immunity 11:115–122

Kawai T, Takeuchi O, Fujita T, Inoue J, Muhlradt PF, Sato S, Hoshino K, Akira S (2001) Lipopolysaccharide stimulates the MyD88-independent pathway and results in activation of IFN-regulatory factor 3 and the expression of a subset of lipopolysaccharide-inducible genes. J Immunol 167:5887–5894

Kawai T, Sato S, Ishii KJ, Coban C, Hemmi H, Yamamoto M, Terai K, Matsuda M, Inoue J, Uematsu S, et al. (2004) Interferon-alpha induction through Toll-like receptors involves a direct interaction of IRF7 with MyD88 and TRAF6. Nat Immunol 5:1061–1068

Kawai T, Takahashi K, Sato S, Coban C, Kumar H, Kato H, Ishii KJ, Takeuchi O, Akira S (2005) IPS-1, an adaptor triggering RIG-I- and Mda5-mediated type I interferon induction. Nat Immunol 6:981–988

Kawane K, Fukuyama H, Yoshida H, Nagase H, Ohsawa Y, Uchiyama Y, Okada K, Iida T, Nagata S (2003) Impaired thymic development in mouse embryos deficient in apoptotic DNA degradation. Nat Immunol 4:138–144

Khush RS, Cornwell WD, Uram JN, Lemaitre B (2002) A ubiquitin-proteasome pathway represses the *Drosophila* immune deficiency signaling cascade. Curr Biol 12:1728–1737

Kinjyo I, Hanada T, Inagaki-Ohara K, Mori H, Aki D, Ohishi M, Yoshida H, Kubo M, Yoshimura A (2002) SOCS1/JAB is a negative regulator of LPS-induced macrophage activation. Immunity 17:583–591

Klemenz R, Hoffmann S, Werenskiold AK (1989) Serum- and oncoprotein-mediated induction of a gene with sequence similarity to the gene encoding carcinoembryonic antigen. Proc Natl Acad Sci USA 86:5708–5712

Kobayashi K, Hernandez LD, Galan JE, Janeway CAJ, Medzhitov R, Flavell R (2002) IRAK-M is a negative regulator of Toll-like receptor signaling. Cell 110:191–202

Kopp E, Medzhitov R, Carothers J, Xiao C, Douglas I, Janeway CA, Ghosh S (1999) ECSIT is an evolutionarily conserved intermediate in the Toll/IL-1 signal transduction pathway. Genes Dev 13:2059–2071

Krikos A, Laherty CD, Dixit VM (1992) Transcriptional activation of the tumor necrosis factor alpha-inducible zinc finger protein, A20, is mediated by kappa B elements. J Biol Chem 267:17971–17976

Krug A, French AR, Barchet W, Fischer JA, Dzionek A, Pingel JT, Orihuela MM, Akira S, Yokoyama WM, Colonna M (2004a) TLR9-dependent recognition of MCMV by IPC and DC generates coordinated cytokine responses that activate antiviral NK cell function. Immunity 21:107–119

Krug A, Luker GD, Barchet W, Leib DA, Akira S, Colonna M (2004b) Herpes simplex virus type 1 activates murine natural interferon-producing cells through toll-like receptor 9. Blood 103:1433–1437

Kumar S, Tzimas MN, Griswold DE, Young PR (1997) Expression of ST2, an interleukin-1 receptor homologue, is induced by proinflammatory stimuli. Biochem Biophys Res Commun 235:474–478

Kurt-Jones EA, Popova L, Kwinn L, Haynes LM, Jones LP, Tripp RA, Walsh EE, Freeman MW, Golenbock DT, Anderson LJ, Finberg RW (2000) Pattern recognition receptors TLR4 and CD14 mediate response to respiratory syncytial virus. Nat Immunol 1:398–401

Kurt-Jones EA, Chan M, Zhou S, Wang J, Reed G, Bronson R, Arnold MM, Knipe DM, Finberg RW (2004) Herpes simplex virus 1 interaction with Toll-like receptor 2 contributes to lethal encephalitis. Proc Natl Acad Sci USA 101:1315–1320

Latz E, Schoenemeyer A, Visintin A, Fitzgerald KA, Monks BG, Knetter CF, Lien E, Nilsen NJ, Espevik T, Golenbock DT (2004) TLR9 signals after translocating from the ER to CpG DNA in the lysosome. Nat Immunol 5:190–198

Lawlor MA, Alessi DR (2001) PKB/Akt: a key mediator of cell proliferation, survival and insulin responses? J Cell Sci 114:2903–2910

LeBouder E, Rey-Nores JE, Rushmere NK, Grigorov M, Lawn SD, Affolter M, Griffin GE, Ferrara P, Schiffrin EJ, Morgan BP, Labeta MO (2003) Soluble forms of Toll-like receptor (TLR)2 capable of modulating TLR2 signaling are present in human plasma and breast milk. J Immunol 171:6680–6689

Lefkowitz RJ, Whalen EJ (2004) beta-arrestins: traffic cops of cell signaling. Curr Opin Cell Biol 16:162–168

Lemaitre B (2004) The road to Toll. Nat Rev Immunol 4:521–527

Lemaitre B, Nicolas E, Michaut L, Reichhart JM, Hoffmann JA (1996) The dorsoventral regulatory gene cassette spatzle/Toll/cactus controls the potent antifungal response in *Drosophila* adults. Cell 86:973–983

Leulier F, Rodriguez A, Khush RS, Abrams JM, Lemaitre B (2000) The *Drosophila* caspase Dredd is required to resist gram-negative bacterial infection. EMBO Rep 1:353–358

Leulier F, Vidal S, Saigo K, Ueda R, Lemaitre B (2002) Inducible expression of double-stranded RNA reveals a role for dFADD in the regulation of the antibacterial response in *Drosophila* adults. Curr Biol 12:996–1000

Levashina EA, Langley E, Green C, Gubb D, Ashburner M, Hoffmann JA, Reichhart JM (1999) Constitutive activation of toll-mediated antifungal defense in serpin-deficient *Drosophila*. Science 285:1917–1919

Li X, Commane M, Burns C, Vithalani K, Cao Z, Stark GR (1999) Mutant cells that do not respond to interleukin-1 (IL-1) reveal a novel role for IL-1 receptor-associated kinase. Mol Cell Biol 19:4643–4652

Li S, Strelow A, Fontana EJ, Wesche H (2002) IRAK-4: a novel member of the IRAK family with the properties of an IRAK-kinase. Proc Natl Acad Sci USA 99:5567–5572

Li T, Hu J, Li L (2004) Characterization of Tollip protein upon Lipopolysaccharide challenge. Mol Immunol 41:85–92

Liew FY, Xu D, Brint EK, O'Neill LA (2005) Negative regulation of toll-like receptor-mediated immune responses. Nat Rev Immunol 5:446–458

Ligoxygakis P, Pelte N, Hoffmann JA, Reichhart JM (2002) Activation of *Drosophila* Toll during fungal infection by a blood serine protease. Science 297:114–116

Liu YJ (2005) IPC: professional type 1 interferon-producing cells and plasmacytoid dendritic cell precursors. Annu Rev Immunol 23:275–306

Liu Y, Dong W, Chen L, Xiang R, Xiao H, De G, Wang Z, Qi Y (2004) BCL10 mediates lipopolysaccharide/toll-like receptor-4 signaling through interaction with Pellino2. J Biol Chem 279:37436–37444

Lord KA, Hoffman-Liebermann B, Liebermann DA (1990) Nucleotide sequence and expression of a cDNA encoding MyD88, a novel myeloid differentiation primary response gene induced by IL6. Oncogene 5:1095–1097

Lu Y, Wu LP, Anderson KV (2001) The antibacterial arm of the drosophila innate immune response requires an IkappaB kinase. Genes Dev 15:104–110

Lund J, Sato A, Akira S, Medzhitov R, Iwasaki A (2003) Toll-like receptor 9-mediated recognition of Herpes simplex virus-2 by plasmacytoid dendritic cells. J Exp Med 198:513–520

Martin M, Rehani K, Jope RS, Michalek SM (2005) Toll-like receptor-mediated cytokine production is differentially regulated by glycogen synthase kinase 3. Nat Immunol 6:777–784
McWhirter SM, Fitzgerald KA, Rosains J, Rowe DC, Golenbock DT, Maniatis T (2004) IFN-regulatory factor 3-dependent gene expression is defective in Tbk1-deficient mouse embryonic fibroblasts. Proc Natl Acad Sci USA 101:233–238
Medzhitov R, Janeway CJ (1997) Innate immunity: the virtues of a nonclonal system of recognition. Cell 91:295–298
Medzhitov R, Preston-Hurlburt P, Janeway CJ (1997) A human homologue of the *Drosophila* Toll protein signals activation of adaptive immunity. Nature 388:394–397
Medzhitov R, Preston-Hurlburt P, Kopp E, Stadlen A, Chen C, Ghosh S, Janeway CJ (1998) MyD88 is an adaptor protein in the hToll/IL-1 receptor family signaling pathways. Mol Cell 2:253–258
Melmed G, Thomas LS, Lee N, Tesfay SY, Lukasek K, Michelsen KS, Zhou Y, Hu B, Arditi M, Abreu MT (2003) Human intestinal epithelial cells are broadly unresponsive to Toll-like receptor 2-dependent bacterial ligands: implications for host-microbial interactions in the gut. J Immunol 170:1406–1415
Meylan E, Burns K, Hofmann K, Blancheteau V, Martinon F, Kelliher M, Tschopp J (2004) RIP1 is an essential mediator of Toll-like receptor 3-induced NF-kappaB activation. Nat Immunol 5:503–507
Meylan E, Curran J, Hofmann K, Moradpour D, Binder M, Bartenschlager R, Tschopp J (2005) Cardif is an adaptor protein in the RIG-I antiviral pathway and is targeted by hepatitis C virus. Nature 437:1167–1172
Michel T, Reichhart JM, Hoffmann JA, Royet J (2001) *Drosophila* Toll is activated by Gram-positive bacteria through a circulating peptidoglycan recognition protein. Nature 414:756–759
Miyake K, Yamashita Y, Hitoshi Y, Takatsu K, Kimoto M (1994) Murine B-cell proliferation and protection from apoptosis with an antibody against a 105-kD molecule: unresponsiveness of X-linked immunodeficient B cells. J Exp Med 180:1217–1224
Miyake K, Yamashita Y, Ogata M, Sudo T, Kimoto M (1995) RP105, a novel B-cell surface molecule implicated in B-cell activation, is a member of the leucine-rich repeat protein family. J Immunol 154:3333–3340
Muzio M, Ni J, Feng P, Dixit VM (1997) IRAK (Pelle) family member IRAK-2 and MyD88 as proximal mediators of IL-1 signaling. Science 278:1612–1615
Nakagawa R, Naka T, Tsutsui H, Fujimoto M, Kimura A, Abe T, Seki E, Sato S, Takeuchi O, Takeda K, et al. (2002) SOCS-1 participates in negative regulation of LPS responses. Immunity 17:677–687
Negishi H, Ohba Y, Yanai H, Takaoka A, Honma K, Yui K, Matsuyama T, Taniguchi T, Honda K (2005) Negative regulation of Toll-like-receptor signaling by IRF-4. Proc Natl Acad Sci USA 102:15989–15994
Netea MG, van Deuren M, Kullberg BJ, Cavaillon JM, Van der Meer JW (2002) Does the shape of lipid A determine the interaction of LPS with Toll-like receptors? Trends Immunol 23:135–139
Netea MG, Van der Graaf C, Van der Meer JW, Kullberg BJ (2004) Recognition of fungal pathogens by Toll-like receptors. Eur J Clin Microbiol Infect Dis 23:672–676
Oganesyan G, Saha SK, Guo B, He JQ, Shahangian A, Zarnegar B, Perry A, Cheng G (2006) Critical role of TRAF3 in the Toll-like receptor-dependent and -independent antiviral response. Nature 439:208–211
Ogura Y, Bonen DK, Inohara N, Nicolae DL, Chen FF, Ramos R, Britton H, Moran T, Karaliuskas R, Duerr RH, et al. (2001) A frameshift mutation in NOD2 associated with susceptibility to Crohn's disease. Nature 411:603–606
Ohashi K, Burkart V, Flohe S, Kolb H (2000) Cutting edge: heat shock protein 60 is a putative endogenous ligand of the toll-like receptor-4 complex. J Immunol 164:558–561

Okabe Y, Kawane K, Akira S, Taniguchi T, Nagata S (2005) Toll-like receptor-independent gene induction program activated by mammalian DNA escaped from apoptotic DNA degradation. J Exp Med 202:1333–1339

Okamura Y, Watari M, Jerud ES, Young DW, Ishizaka ST, Rose J, Chow JC, Strauss JF, 3rd (2001). The extra domain A of fibronectin activates Toll-like receptor 4. J Biol Chem 276:10229–10233

Opipari AW, Jr., Boguski MS, Dixit VM (1990) The A20 cDNA induced by tumor necrosis factor alpha encodes a novel type of zinc finger protein. J Biol Chem 265:14705–14708

Oshiumi H, Matsumoto M, Funami K, Akazawa T, Seya T (2003a) TICAM-1, an adaptor molecule that participates in Toll-like receptor 3-mediated interferon-beta induction. Nat Immunol 4:161–167

Oshiumi H, Sasai M, Shida K, Fujita T, Matsumoto M, Seya T (2003b) TIR-containing adapter molecule (TICAM)-2, a bridging adapter recruiting to toll-like receptor 4 TICAM-1 that induces interferon-beta. J Biol Chem 278:49751–49762

Park JM, Brady H, Ruocco MG, Sun H, Williams D, Lee SJ, Kato T, Jr., Richards N, Chan K, Mercurio F, et al. (2004a). Targeting of TAK1 by the NF-kappa B protein Relish regulates the JNK-mediated immune response in *Drosophila*. Genes Dev 18:584–594

Park JS, Svetkauskaite D, He Q, Kim JY, Strassheim D, Ishizaka A, Abraham E (2004b). Involvement of toll-like receptors 2 and 4 in cellular activation by high mobility group box 1 protein. J Biol Chem 279:7370–7377

Perry AK, Chow EK, Goodnough JB, Yeh WC, Cheng G (2004) Differential requirement for TANK-binding kinase-1 in type I interferon responses to toll-like receptor activation and viral infection. J Exp Med 199:1651–1658

Philpott DJ, Girardin SE (2004) The role of Toll-like receptors and Nod proteins in bacterial infection. Mol Immunol 41:1099–1108

Picard C, Puel A, Bonnet M, Ku CL, Bustamante J, Yang K, Soudais C, Dupuis S, Feinberg J, Fieschi C, et al. (2003) Pyogenic bacterial infections in humans with IRAK-4 deficiency. Science 299:2076–2079

Pili-Floury S, Leulier F, Takahashi K, Saigo K, Samain E, Ueda R, Lemaitre B (2004) In vivo RNA interference analysis reveals an unexpected role for GNBP1 in the defense against Gram-positive bacterial infection in *Drosophila* adults. J Biol Chem 279:12848–12853

Poltorak A, He X, Smirnova I, Liu MY, Van Huffel C, Du X, Birdwell D, Alejos E, Silva M, Galanos C, et al. (1998) Defective LPS signaling in C3H/HeJ and C57BL/10ScCr mice: mutations in Tlr4 gene. Science 282:2085–2088

Rassa JC, Meyers JL, Zhang Y, Kudaravalli R, Ross SR (2002) Murine retroviruses activate B cells via interaction with toll-like receptor 4. Proc Natl Acad Sci USA 99:2281–2286

Rogers NC, Slack EC, Edwards AD, Nolte MA, Schulz O, Schweighoffer E, Williams DL, Gordon S, Tybulewicz VL, Brown GD, Reis ESC (2005) Syk-dependent cytokine induction by Dectin-1 reveals a novel pattern recognition pathway for C type lectins. Immunity 22:507–517

Rutschmann S, Jung AC, Zhou R, Silverman N, Hoffmann JA, Ferrandon D (2000) Role of *Drosophila* IKK gamma in a toll-independent antibacterial immune response. Nat Immunol 1:342–347

Saccani S, Polentarutti N, Penton-Rol G, Sims JE, Mantovani A (1998) Divergent effects of LPS on expression of IL-1 receptor family members in mononuclear phagocytes in vitro and in vivo. Cytokine 10:773–780

Sanjo H, Takeda K, Tsujimura T, Ninomiya-Tsuji J, Matsumoto K, Akira S (2003) TAB2 is essential for prevention of apoptosis in fetal liver but not for interleukin-1 signaling. Mol Cell Biol 23:1231–1238

Sato S, Sugiyama M, Yamamoto M, Watanabe Y, Kawai T, Takeda K, Akira S (2003) Toll/IL-1 receptor domain-containing adaptor inducing IFN-beta (TRIF) associates with TNF receptor-associated factor 6 and TANK-binding kinase 1, and activates two distinct transcription factors, NF-kappa B and IFN-regulatory factor-3, in the Toll-like receptor signaling. J Immunol 171:4304–4310

Sato S, Sanjo H, Takeda K, Ninomiya-Tsuji J, Yamamoto M, Kawai T, Matsumoto K, Takeuchi O, Akira S (2005) Essential function for the kinase TAK1 in innate and adaptive immune responses. Nat Immunol 6:1087–1095

Schmitz J, Owyang A, Oldham E, Song Y, Murphy E, McClanahan TK, Zurawski G, Moshrefi M, Qin J, Li X, et al. (2005) IL-33, an interleukin-1-like cytokine that signals via the IL-1 receptor-related protein ST2 and induces T helper type 2-associated cytokines. Immunity 23:479–490

Schoenemeyer A, Barnes BJ, Mancl ME, Latz E, Goutagny N, Pitha PM, Fitzgerald KA, Golenbock DT (2005) The interferon regulatory factor, IRF5, is a central mediator of toll-like receptor 7 signaling. J Biol Chem 280:17005–17012

Seth RB, Sun L, Ea CK, Chen ZJ (2005) Identification and characterization of MAVS, a mitochondrial antiviral signaling protein that activates NF-kappaB and IRF 3. Cell 122:669–682

Sharma S, tenOever BR, Grandvaux N, Zhou GP, Lin R, Hiscott J (2003) Triggering the interferon antiviral response through an IKK-related pathway. Science 300:1148–1151

Shibuya H, Yamaguchi K, Shirakabe K, Tonegawa A, Gotoh Y, Ueno N, Irie K, Nishida E, Matsumoto K (1996) TAB1: an activator of the TAK1 MAPKKK in TGF-beta signal transduction. Science 272:1179–1182

Shimazu R, Akashi S, Ogata H, Nagai Y, Fukudome K, Miyake K, Kimoto M (1999) MD-2, a molecule that confers lipopolysaccharide responsiveness on Toll-like receptor 4. J Exp Med 189:1777–1782

Silverman N, Zhou R, Stoven S, Pandey N, Hultmark D, Maniatis T (2000) A *Drosophila* IkappaB kinase complex required for Relish cleavage and antibacterial immunity. Genes Dev 14:2461–2471

Silverman N, Zhou R, Erlich RL, Hunter M, Bernstein E, Schneider D, Maniatis T (2003) Immune activation of NF-kappaB and JNK requires *Drosophila* TAK1. J Biol Chem 278:48928–48934

Smiley ST, King JA, Hancock WW (2001) Fibrinogen stimulates macrophage chemokine secretion through toll-like receptor 4. J Immunol 167:2887–2894

Stoiber D, Kovarik P, Cohney S, Johnston JA, Steinlein P, Decker T (1999) Lipopolysaccharide induces in macrophages the synthesis of the suppressor of cytokine signaling 3 and suppresses signal transduction in response to the activating factor IFN-gamma. J Immunol 163:2640–2647

Stokoe D, Stephens LR, Copeland T, Gaffney PR, Reese CB, Painter GF, Holmes AB, McCormick F, Hawkins PT (1997) Dual role of phosphatidylinositol-3,4,5-trisphosphate in the activation of protein kinase B. Science 277:567–570

Stoven S, Silverman N, Junell A, Hedengren-Olcott M, Erturk D, Engstrom Y, Maniatis T, Hultmark D (2003) Caspase-mediated processing of the *Drosophila* NF-kappaB factor Relish. Proc Natl Acad Sci USA 100:5991–5996

Strelow A, Kollewe C, Wesche H (2003) Characterization of Pellino2, a substrate of IRAK1 and IRAK4. FEBS Lett 547:157–161

Sun H, Towb P, Chiem DN, Foster BA, Wasserman SA (2004) Regulated assembly of the Toll signaling complex drives *Drosophila* dorsoventral patterning. EMBO J 23:100–110

Suzuki N, Suzuki S, Duncan GS, Millar DG, Wada T, Mirtsos C, Takada H, Wakeham A, Itie A, Li S, et al. (2002) Severe impairment of interleukin-1 and Toll-like receptor signalling in mice lacking IRAK-4. Nature 416:750–756

Swantek JL, Tsen MF, Cobb MH, Thomas JA (2000) IL-1 receptor-associated kinase modulates host responsiveness to endotoxin. J Immunol 164:4301–4306

Sweet MJ, Leung BP, Kang D, Sogaard M, Schulz K, Trajkovic V, Campbell CC, Xu D, Liew FY (2001) A novel pathway regulating lipopolysaccharide-induced shock by ST2/T1 via inhibition of Toll-like receptor 4 expression. J Immunol 166:6633–6639

Tabeta K, Georgel P, Janssen E, Du X, Hoebe K, Crozat K, Mudd S, Shamel L, Sovath S, Goode J, et al. (2004) Toll-like receptors 9 and 3 as essential components of innate immune defense against mouse cytomegalovirus infection. Proc Natl Acad Sci USA 101:3516–3521

Takaesu G, Kishida S, Hiyama A, Yamaguchi K, Shibuya H, Irie K, Ninomiya-Tsuji J, Matsumoto K (2000) TAB2, a novel adaptor protein, mediates activation of TAK1 MAPKKK by linking TAK1 to TRAF6 in the IL-1 signal transduction pathway. Mol Cell 5:649–658
Takaesu G, Surabhi RM, Park KJ, Ninomiya-Tsuji J, Matsumoto K, Gaynor RB (2003) TAK1 is critical for IkappaB kinase-mediated activation of the NF-kappaB pathway. J Mol Biol 326:105–115
Takaoka A, Yanai H, Kondo S, Duncan G, Negishi H, Mizutani T, Kano S, Honda K, Ohba Y, Mak TW, Taniguchi T (2005) Integral role of IRF-5 in the gene induction programme activated by Toll-like receptors. Nature 434:243–249
Takeda K, Akira S (2005) Toll-like receptors in innate immunity. Int Immunol 17:1–14
Takeda K, Kaisho T, Akira S (2003) Toll-like receptors. Annu Rev Immunol 21:335–376
Takehana A, Yano T, Mita S, Kotani A, Oshima Y, Kurata S (2004) Peptidoglycan recognition protein (PGRP)-LE and PGRP-LC act synergistically in *Drosophila* immunity. EMBO J 23:4690–4700
Takeuchi O, Hoshino K, Kawai T, Sanjo H, Takada H, Ogawa T, Takeda K, Akira S (1999a) Differential roles of TLR2 and TLR4 in recognition of gram-negative and gram-positive bacterial cell wall components. Immunity 11:443–451
Takeuchi O, Kawai T, Sanjo H, Copeland NG, Gilbert DJ, Jenkins NA, Takeda K, Akira S (1999b) TLR6: a novel member of an expanding toll-like receptor family. Gene 231:59–65
Takeuchi O, Kawai T, Muhlradt PF, Morr M, Radolf JD, Zychlinsky A, Takeda K, Akira S (2001) Discrimination of bacterial lipoproteins by Toll-like receptor 6. Int Immunol 13:933–940
Takeuchi O, Sato S, Horiuchi T, Hoshino K, Takeda K, Dong Z, Modlin RL, Akira S (2002) Cutting edge: role of Toll-like receptor 1 in mediating immune response to microbial lipoproteins. J Immunol 169:10–14
Tanji T, Ip YT (2005) Regulators of the Toll and Imd pathways in the *Drosophila* innate immune response. Trends Immunol 26:193–198
Termeer C, Benedix F, Sleeman J, Fieber C, Voith U, Ahrens T, Miyake K, Freudenberg M, Galanos C, Simon JC (2002) Oligosaccharides of Hyaluronan activate dendritic cells via toll-like receptor 4. J Exp Med 195:99–111
Thomas JA, Allen JL, Tsen M, Dubnicoff T, Danao J, Liao XC, Cao Z, Wasserman SA (1999) Impaired cytokine signaling in mice lacking the IL-1 receptor-associated kinase. J Immunol 163:978–984
Thomassen E, Renshaw BR, Sims JE (1999) Identification and characterization of SIGIRR, a molecule representing a novel subtype of the IL-1R superfamily. Cytokine 11:389–399
Ting JP, Davis BK (2005) CATERPILLER: a novel gene family important in immunity, cell death, and diseases. Annu Rev Immunol 23:387–414
Tominaga S (1989) A putative protein of a growth specific cDNA from BALB/c-3T3 cells is highly similar to the extracellular portion of mouse interleukin 1 receptor. FEBS Lett 258:301–304
Tsan MF, Gao B (2004) Endogenous ligands of Toll-like receptors. J Leukoc Biol 76:514–519
Tsujimura H, Tamura T, Kong HJ, Nishiyama A, Ishii KJ, Klinman DM, Ozato K (2004) Toll-like receptor 9 signaling activates NF-kappaB through IFN regulatory factor-8/IFN consensus sequence binding protein in dendritic cells. J Immunol 172:6820–6827
Uematsu S, Sato S, Yamamoto M, Hirotani T, Kato H, Takeshita F, Matsuda M, Coban C, Ishii KJ, Kawai T, et al (2005) Interleukin-1 receptor-associated kinase-1 plays an essential role for Toll-like receptor (TLR)7- and TLR9-mediated interferon-{alpha} induction. J Exp Med 201:915–923
Underhill DM, Rossnagle E, Lowell CA, Simmons RM (2005) Dectin-1 activates Syk tyrosine kinase in a dynamic subset of macrophages for reactive oxygen production. Blood 106:2543–2550
Vabulas RM, Ahmad-Nejad P, da Costa C, Miethke T, Kirschning CJ, Hacker H, Wagner H (2001) Endocytosed HSP60s use toll-like receptor 2 (TLR2) and TLR4 to activate the toll/interleukin-1 receptor signaling pathway in innate immune cells. J Biol Chem 276:31332–31339

Vabulas RM, Ahmad-Nejad P, Ghose S, Kirschning CJ, Issels RD, Wagner H (2002a) HSP70 as endogenous stimulus of the Toll/interleukin-1 receptor signal pathway. J Biol Chem 277:15107–15112

Vabulas RM, Braedel S, Hilf N, Singh-Jasuja H, Herter S, Ahmad-Nejad P, Kirschning CJ, Da Costa C, Rammensee HG, Wagner H, Schild H (2002b) The endoplasmic reticulum-resident heat shock protein Gp96 activates dendritic cells via the Toll-like receptor 2/4 pathway. J Biol Chem 277:20847–20853

Vidal S, Khush RS, Leulier F, Tzou P, Nakamura M, Lemaitre B (2001) Mutations in the *Drosophila* dTAK1 gene reveal a conserved function for MAPKKKs in the control of rel/NF-kappaB-dependent innate immune responses. Genes Dev 15:1900–1912

Wald D, Qin J, Zhao Z, Qian Y, Naramura M, Tian L, Towne J, Sims JE, Stark GR, Li X (2003) SIGIRR, a negative regulator of Toll-like receptor-interleukin 1 receptor signaling. Nat Immunol 4:920–927

Wallin RP, Lundqvist A, More SH, von Bonin A, Kiessling R, Ljunggren HG (2002) Heat-shock proteins as activators of the innate immune system. Trends Immunol 23:130–135

Wan J, Sun L, Mendoza JW, Chui YL, Huang DP, Chen ZJ, Suzuki N, Suzuki S, Yeh WC, Akira S, et al. (2004) Elucidation of the c-Jun N-terminal kinase pathway mediated by Estein-Barr virus-encoded latent membrane protein 1. Mol Cell Biol 24:192–199

Wang C, Deng L, Hong M, Akkaraju GR, Inoue J, Chen ZJ (2001) TAK1 is a ubiquitin-dependent kinase of MKK and IKK. Nature 412:346–351

Wang T, Town T, Alexopoulou L, Anderson JF, Fikrig E, Flavell RA (2004) Toll-like receptor 3 mediates West Nile virus entry into the brain causing lethal encephalitis. Nat Med 10:1366–1373

Wang Y, Tang Y, Teng L, Wu Y, Zhao X, Pei G (2006) Association of beta-arrestin and TRAF6 negatively regulates Toll-like receptor-interleukin 1 receptor signaling. Nat Immunol 7:139–147

Weber AN, Tauszig-Delamasure S, Hoffmann JA, Lelievre E, Gascan H, Ray KP, Morse MA, Imler JL, Gay NJ (2003) Binding of the *Drosophila* cytokine Spatzle to Toll is direct and establishes signaling. Nat Immunol 4:794–800

Wesche H, Henzel WJ, Shillinglaw W, Li S, Cao Z (1997) MyD88: an adapter that recruits IRAK to the IL-1 receptor complex. Immunity 7:837–847

Wesche H, Gao X, Li X, Kirschning CJ, Stark GR, Cao Z (1999) IRAK-M is a novel member of the Pelle/interleukin-1 receptor-associated kinase (IRAK) family. J Biol Chem 274:19403–19410

Wu LP, Anderson KV (1997) Related signaling networks in *Drosophila* that control dorsoventral patterning in the embryo and the immune response. Cold Spring Harb Symp Quant Biol 62:97–103

Wu GS, Burns TF, Zhan Y, Alnemri ES, El-Deiry WS (1999) Molecular cloning and functional analysis of the mouse homologue of the KILLER/DR5 tumor necrosis factor-related apoptosis-inducing ligand (TRAIL) death receptor. Cancer Res 59:2770–2775

Xiao C, Shim JH, Kluppel M, Zhang SS, Dong C, Flavell RA, Fu XY, Wrana JL, Hogan BL, Ghosh S (2003) Ecsit is required for Bmp signaling and mesoderm formation during mouse embryogenesis. Genes Dev 17:2933–2949

Xu LG, Wang YY, Han KJ, Li LY, Zhai Z, Shu HB (2005) VISA is an adapter protein required for virus-triggered IFN-beta signaling. Mol Cell 19:727–740

Yamaguchi K, Shirakabe K, Shibuya H, Irie K, Oishi I, Ueno N, Taniguchi T, Nishida E, Matsumoto K (1995) Identification of a member of the MAPKKK family as a potential mediator of TGF-beta signal transduction. Science 270:2008–2011

Yamamoto M, Sato S, Hemmi H, Sanjo H, Uematsu S, Kaisho T, Hoshino K, Takeuchi O, Kobayashi M, Fujita T, et al (2002a) Essential role for TIRAP in activation of the signalling cascade shared by TLR2 and TLR4. Nature 420:324–329

Yamamoto M, Sato S, Mori K, Hoshino K, Takeuchi O, Takeda K, Akira S (2002b) Cutting edge: a novel Toll/IL-1 receptor domain-containing adapter that preferentially activates the IFN-beta promoter in the Toll-like receptor signaling. J Immunol 169:6668–6672

Yamamoto M, Sato S, Hemmi H, Hoshino K, Kaisho T, Sanjo H, Takeuchi O, Sugiyama M, Okabe M, Takeda K, Akira S (2003a) Role of adaptor TRIF in the MyD88-independent toll-like receptor signaling pathway. Science 301:640–643

Yamamoto M, Sato S, Hemmi H, Uematsu S, Hoshino K, Kaisho T, Takeuchi O, Takeda K, Akira S (2003b) TRAM is specifically involved in the Toll-like receptor 4-mediated MyD88-independent signaling pathway. Nat Immunol 4:1144–1150

Yang K, Puel A, Zhang S, Eidenschenk C, Ku CL, Casrouge A, Picard C, von Bernuth H, Senechal B, Plancoulaine S, et al (2005) Human TLR-7-, -8-, and -9-mediated induction of IFN-alpha/beta and -lambda Is IRAK-4 dependent and redundant for protective immunity to viruses. Immunity 23:465–478

Yarovinsky F, Zhang D, Andersen JF, Bannenberg GL, Serhan CN, Hayden MS, Hieny S, Sutterwala FS, Flavell RA, Ghosh S, Sher A (2005) TLR11 activation of dendritic cells by a protozoan profilin-like protein. Science 308:1626–1629

Yasukawa H, Sasaki A, Yoshimura A (2000) Negative regulation of cytokine signaling pathways. Annu Rev Immunol 18:143–164

Yoneyama M, Kikuchi M, Natsukawa T, Shinobu N, Imaizumi T, Miyagishi M, Taira K, Akira S, Fujita T (2004) The RNA helicase RIG-I has an essential function in double-stranded RNA-induced innate antiviral responses. Nat Immunol 5:730–737

Yoshida H, Okabe Y, Kawane K, Fukuyama H, Nagata S (2005) Lethal anemia caused by interferon-beta produced in mouse embryos carrying undigested DNA. Nat Immunol 6:49–56

Yu KY, Kwon HJ, Norman DA, Vig E, Goebl MG, Harrington MA (2002) Cutting edge: mouse pellino-2 modulates IL-1 and lipopolysaccharide signaling. J Immunol 169:4075–4078

Zhang G, Ghosh S (2002) Negative regulation of toll-like receptor-mediated signaling by Tollip. J Biol Chem 277:7059–7065

Zhang D, Zhang G, Hayden MS, Greenblatt MB, Bussey C, Flavell RA, Ghosh S (2004) A toll-like receptor that prevents infection by uropathogenic bacteria. Science 303:1522–1526

2 Strategies of Natural Killer (NK) Cell Recognition and Their Roles in Tumor Immunosurveillance

C. Andrew Stewart[1,2,3] and Eric Vivier[1,2,3,4]

2.1 Introduction: An Interesting Epistemological Case

Natural Killer cells (NK cells) represent an interesting epistemological example in Immunology. First considered as "background noise" in T-cell cytolytic assays, Natural Killer (NK) cells were characterized more than 30 years ago as cytotoxic effectors of the innate immune system (Kiessling et al. 1975). Later, NK cells were recognized as a peculiar type of large granular lymphocytes that are widespread throughout the body (Lanier et al. 1986), being present in both lymphoid organs and non-lymphoid peripheral tissues (Cooper et al. 2004; Ferlazzo and Munz 2004). Their specificity for a variety of tumor cells, virus-infected cells or allogeneic cells along with their lack of antigen-specific receptors, have puzzled immunologists for many years. Since this time, a series of discoveries have shed light on the mechanisms of NK cell effector function and have simultaneously broadened our views on immune detection strategies (Carayannopoulos and Yokoyama 2004; Lanier 2005; Moretta et al. 2002; Stewart et al. 2006; Vivier and Biron 2002). Such discoveries include "missing-self recognition" (via major histocompatibility complex [MHC] class I) (Kärre et al. 1986), the identification of inhibitory cell surface receptors that modulate NK cell activation (via Immunoreceptor Tyrosine-based Inhibition Motifs: ITIM) (Vély and Vivier 1997) or the "stress-induced self recognition" (via NKG2D) (Raulet 2003) (Fig. 1). The involvement of NK cells in the control of viral and parasitic infections, in auto-immunity, in reproduction as well

[1]Laboratory of NK cells and Innate Immunity, Centre d'Immunologie de Marseille-Luminy, INSERM, U631, 13288 Marseille, France

[2]Université de la Méditerranée, Case 906, 13288 Marseille Cedex 9, France

[3]CNRS, UMR6102, 13288 Marseille, France

[4]Hôpital de la Conception, Assistance Publique-Hôpitaux de Marseille, 13005 Marseille, France

Correspondence to: C.A. Stewart, National Cancer Institute, NCI-Frederick, MD, USA

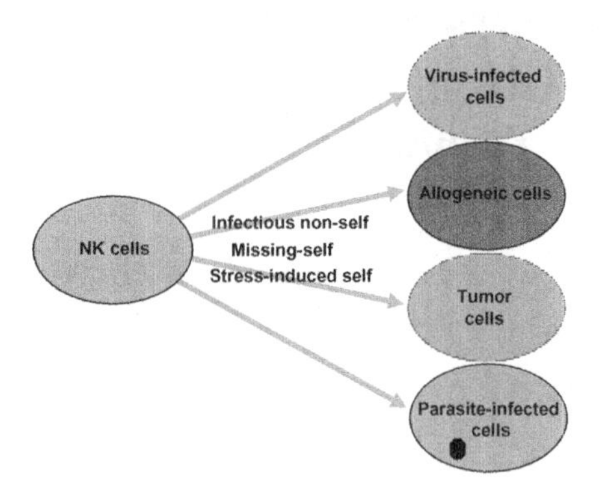

Fig. 1. Natural killer cell recognition strategies. Schematic representation of the mode of NK cell interaction with partner cells (see text for details)

as in the clinical outcome of hematopoietic transplants has been reviewed recently (Carayannopoulos and Yokoyama 2004; Johansson et al. 2005; Korbel et al. 2004; Lodoen and Lanier 2005; Orange and Ballas 2006; Parham 2005; Ruggeri et al. 2005; Zhang et al. 2005). Similarly, extensive reviews on NK cell signaling and developmental pathways and the interaction between NK cells and other innate immune sentinels such as dendritic cells (DC) have been published (Degli-Esposti and Smyth 2005; Di Santo 2006; MacFarlane and Campbell 2005; Moretta et al. 2005; Vivier et al. 2004; Walzer et al. 2005). Herein, we focus on reviewing the molecular devices by which NK cells discriminate "resistant" cells from "target" cells, and address how these recognition strategies may govern the involvement of NK cells in cancer immunosurveillance processes.

2.2 Natural Killer Cells in Innate Immunity

The appearance of multicellular organisms is an outstanding step in the evolution of life. Multicellularity was accompanied by specialization of cells which allowed the acquisition of novel biological functions. All cells have an autonomous capacity to respond to various forms of stress, such as physical stress (UV, heat shock, etc.), chemical stress (pH, osmolarity, etc.) or microbial stress. Multicellularity permitted the emergence of cells specialized in recognition of pathological stress. These cells form the immune system. They form an extrinsic system that complements autonomous stress responses and shares the burden of stress-recognition and stress-resolution with affected cells. Cells of the immune system act at the organism-level to facilitate controlled, measured, appropriate, and memorable responses (Fig. 2).

The immune response is organized in two complementary systems: the innate immune system and the adaptive immune system (Vivier and Malissen 2005). Adaptive immunity in higher vertebrates is due to T and B lymphocytes, which recognize a colossal antigenic repertoire via specific receptors that are the products of combinatorial gene segment rearrangements. Alternative forms of adaptive

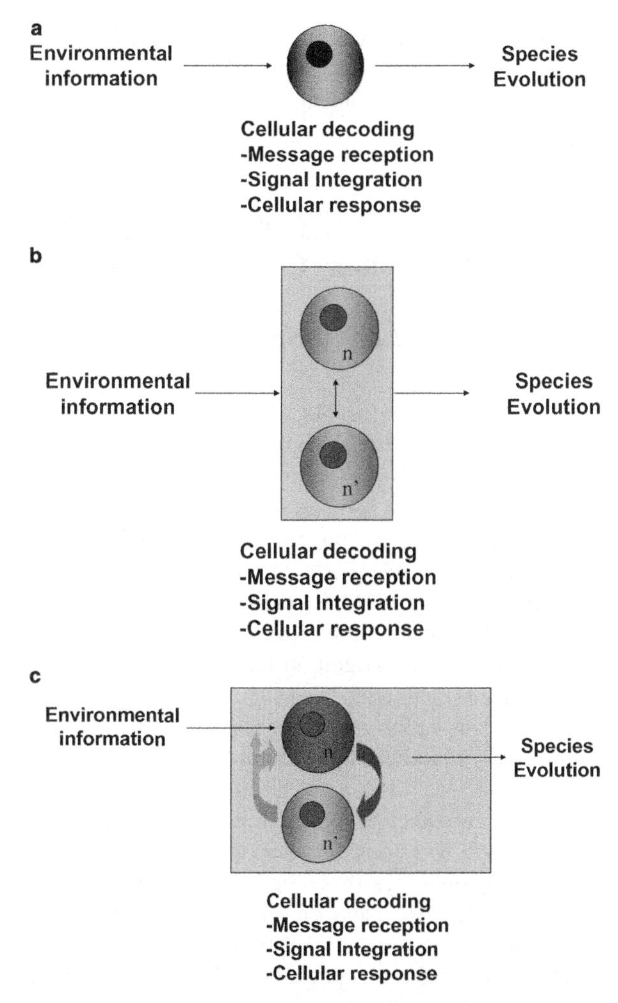

Fig. 2a–c. The immune system in multicellular organisms. (**a**) Driving forces in the evolution of unicellular organisms. (**b**) Complexity of multicellular organisms. (**c**) Schematic representation of the immune system, as a group of cells (*n*′) that are alerted by cells in distress (*n*), without having been exposed to the stress themselves. This mode of operation distinguishes a "stress response" that every cell can exert in every organism from an immune response

immunity have also recently been described in lower vertebrates (Pancer and Cooper 2006). The innate immune system is present in all metazoans and is characterized by a functional duality: On one hand, innate immune cells can directly exert their effector function against the "aggressors" of the organism (e.g. via the production of type I interferons, via cell cytotoxicity), and on the other hand, they also initiate and orient the adaptive immune response (e.g. via the antigenic presentation, via the production of cytokines and chemokines).

The innate immune system is fundamental in antimicrobial responses. Indeed, a variety of receptors have been identified, which confer to innate immune cells the capacity to recognize microbial products (e.g. Toll-like receptors). These receptors are characterized by their phylogenetic conservation, which bears witness to their evolutionary selection (Beutler 2004; Janeway and Medzhitov 2002). However, the roles of innate immunity in the defense mechanisms against tumors, in allogeneic reactions and in other processes are still poorly understood. Amongst cells of the innate immune system, NK cells represent an interesting and important case as these lymphocytes were initially characterized for their antitumoral and anti-allogeneic cytolytic functions.

2.3 Natural Killer Cell Recognition of "Missing-Self": An Array of Inhibitory Cell Surface Receptors

2.3.1 MHC Class I-Specific Inhibitory Receptors

The demonstration that NK cell cytolytic activity inversely correlates with the level of MHC class I expression on target cells provided the first evidence for missing-self recognition by NK cells (Kärre et al. 1986). Part of the intellectual appeal of the missing-self hypothesis in MHC class I recognition is how it complements the adaptive immune system, providing a means of detecting cells that might otherwise escape immune recognition through loss of MHC class I (Kärre 1997). The widespread presence of MHC class I on normal tissues supports the argument that missing-self recognition prevents NK cell auto-immune reactions. This is also true for missing-self recognition mediated by other processes including for example, complement activation. The alternative pathway of complement activation involves spontaneous hydrolysis of the labile complement component C3. The complement regulatory proteins CD46 (membrane cofactor protein, MCP), CD55 (decay accelerating factor, DAF), and CD35 (complement receptor type 1, CR1) are expressed by most normal cells and coordinately reduce the creation and accelerate the decay of C3 and C5 convertases (Kim and Song 2006). Without adaptation to mimic host cells, microbes lack these regulators and are susceptible to complement attack.

Multiple MHC class I receptors have been identified that inhibit cell activation through recruitment of the protein tyrosine phosphatases (PTPs) SHP-1 and/or SHP-2 via their cytoplasmic immunoreceptor tyrosine-based inhibitory motifs (ITIMs; I/V/S/LxYxxL/V, where x can be any amino-acid) (Burshtyn et al. 1996; Long 1999; Olcese et al. 1996; Vivier and Daëron 1997). The receptors can be classified into the C-type lectin CD94/NKG2 heterodimers, the C-type lectin Ly49 family and the immunoglobulin (Ig)-like receptors including the killer Ig-like receptors (KIR) and Ig-like transcripts (ILT) (Fig. 3a,b).

The CD94/NKG2 receptors are composed of C-type lectin type II transmembrane proteins. These recognize the nonclassical MHC class I molecules human leukocyte antigen (HLA)-E (in human) and Qa-1^b (in mouse) (Braud et al. 1998; Borrego et al. 1998; Lee et al. 1998; Vance et al. 1998). Of the NKG2 family members that associate with CD94 (including NKG2A, C, and E), NKG2A contains ITIMs in its cytoplasmic domain conferring inhibitory function. Recognition of HLA-E or Qa-1^b by CD94/NKG2 requires the presence of peptides in their peptide binding grooves. These peptides are frequently derived from the leader sequences

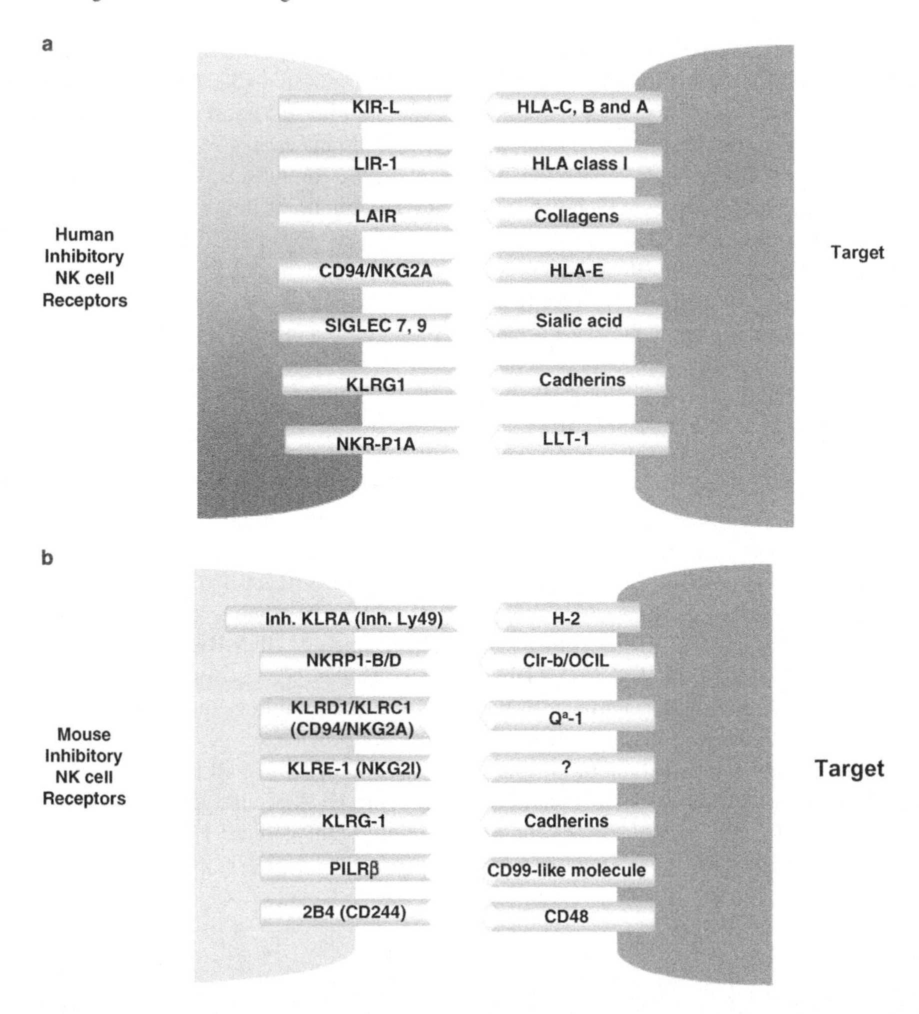

Fig. 3a,b. The "inhibitory NK cell zipper." Inhibitory cell surface receptors and ligands for human (**a**) and mouse (**b**) NK cells. *KIR*, killer Ig-like receptors; *LIR*, immunoglobulin-like transcript; *LAIR*, leukocyte-associated immunoglobulin-like receptor; *SIGLEC*, sialic acid binding Ig-like lectins; *KLRG*, killer cell lectin-like receptor; *NKR-P1A*, NK cell receptor protein 1; *HLA*, human leukocyte antigen; *LLT*, lectin-like transcript

of classical MHC class I molecules (Braud et al. 1997; Kraft et al. 2000; Lopez-Botet et al. 1998). As a result, CD94/NKG2A is a sensor of active MHC class I biosynthesis and presentation. CD94/NKG2 is unique in both its high evolutionary conservation and its means of recognizing classical MHC class I as a "proxy" sensor.

Members of the other NK cell MHC class I receptor families directly recognize classical MHC class I molecules. Natural killer cell recognition of subsets of MHC class I allotypes is mediated by members of the Ly49 family in the mouse and rat.

In contrast, human NK cells use immunoglobulin domain-containing type I transmembrane proteins for the same function. Immunoglobulin-like transcript 2 (ILT2; or LIR1 or CD85j) recognizes a broad range of both classical and non-classical MHC class I molecules (Chapman et al. 1999; Colonna et al. 1997), whilst the KIR family members (or CD158 molecules) are specific for subsets of MHC allotypes. Primate KIR receptors therefore appear to be functional homologues of rodent Ly49 despite their independent evolutionary origins, thus providing an interesting example of convergent evolution. A feature of both KIR and Ly49 recognition of MHC class I is sensitivity to peptides bound in the MHC class I groove. Many KIR and some Ly49 receptors are sensitive to peptide changes, although clearly less sensitive than the TCR (Franksson et al. 1999; Hanke et al. 1999; Hansasuta et al. 2004; Natarajan et al. 2002; Peruzzi et al. 1996; Rajagopalan and Long 1997; Stewart et al. 2005; Zappacosta et al. 1997).

Despite the specificity of KIR and Ly49 receptors for restricted groups of MHC class I allotypes, these receptors endow human and mouse NK cells with a panoramic recognition system of MHC class I. The breadth of this system arises from a number of features. In any individual, multiple *KIR* genes mean that receptors are present for many different types of MHC class I: KIR2DL1 recognizes HLA-C allotypes with a lysine at position 80, KIR2DL2/3 recognize HLA-C allotypes with asparagine at position 80, KIR3DL1 recognizes HLA-Bw4 allotypes, and KIR3DL2 binds HLA-A3 and HLA-A11 molecules (Boyington and Sun 2002; Hansasuta et al. 2004; Parham 2005). Each of these inhibitory receptors is present in most humans (Middleton et al. 2005), meaning that the NK cell pool of most individuals can detect alterations in the expression of almost all HLA-C allotypes, a large proportion of HLA-B allotypes and some HLA-A allotypes. The capacity of the NK cell pool to detect changes in expression of these HLA molecules is also founded on the variegated expression of individual NK cell receptors and resulting repertoire of specificities within the NK cell pool (Anderson 2006; Raulet et al. 2001).

2.3.2 Inhibitory Receptors for Non-MHC Class I Molecules

ITIM-based inhibitory signaling is not restricted to MHC class I recognition. ITIM-bearing receptors recognizing many different molecules are found on cells across the spectrum of hematopoiesis. On NK cells, such inhibitory receptors include carcinoembryonic antigen-related cell adhesion molecule 1 (Ceacam-1), certain sialic acid binding Ig-like lectins (Siglecs), inhibitory NK cell receptor protein 1 (NKR-P1) molecules, and killer cell lectin-like receptor G1 (KLRG1) (Fig. 3a,b). The ligands for some of these receptors are widely expressed, like MHC class I molecules themselves, leading to proposals that these receptors have missing-self function (Kumar and McNerney 2005; Plougastel and Yokoyama 2006).

2.3.2.1 Siglecs

Siglec-7 (p75/AIRM1) is a member of the CD33-related Siglec family that has an ITIM and an ITIM-like sequence in its cytoplasmic tail enabling the binding of SHP-1 and SHP-2 upon tyrosine phosphorylation (Crocker 2005; Falco et al. 1999; Nicoll et al. 1999). Siglec-7 is expressed by all NK cells, all monocytes, and some $CD8^+$ T cells (Falco et al. 1999; Nicoll et al. 1999). Activation of NK cytotoxicity and other forms of cellular activation are inhibited upon ligation of Siglec-7 with specific antibody (Crocker 2005; Falco et al. 1999). Siglec-7 has a ligand preference for α2,8-linked disialic acids that are found on certain gangliosides including GD3 (Ito et al. 2001b; Yamaji et al. 2002). GD3 is expressed at high levels on neurons but also on melanoma and renal cell carcinoma cells (Ito et al. 2001a; Urmacher et al. 1989), potentially making such cells resistant to monocyte and NK cell responses. However, the understanding of Siglec biology is complicated by "masking" interactions in *cis* that can occur between Siglecs and multiple sialylated ligands on the same cell. In this way, NK cells require sialidase treatment, and the consequential unmasking of Siglec-7, for their cytolytic activity to be inhibited by recognition of Siglec-7 ligand, ganglioside GD3, on P815 target cells (Nicoll et al. 2003). Intriguingly, in the presence of masked Siglec-7, the over-expression of ganglioside GD3 on P815 cells rendered them more sensitive to lysis by interleukin-2 (IL-2) activated peripheral blood lymphocytes, suggesting that GD3 may also act as an activating ligand for a currently unknown receptor.

2.3.2.2 KLRG1

Killer cell lectin-like receptor G1 (KLRG1) was originally identified as mast cell function-associated antigen (MAFA). Resting human NK cells (~50%), mouse NK cells (30%) and lower numbers of CD4 and CD8 T cells express KLRG1. A large increase in expression is seen on these cell populations following activation, for example that occurring during viral infection (Blaser et al. 1998; Hanke et al. 1998; Robbins et al. 2002, 2004; Voehringer et al. 2002). KLRG1 expression on human peripheral blood T cells occurs on antigen-experienced T cells with an effector memory phenotype (Voehringer et al. 2001, 2002). Recently the ligands for mouse KLRG1 have been identified as the E-, N- and R-cadherins, of which E-cadherin shows the strongest binding (Grundemann et al. 2006; Ito et al. 2006). Cadherins are ubiquitously expressed on solid tissues, and are localized at tight junctions in epithelia. Many epithelial cancers are known to down-regulate E-cadherin, thus facilitating invasion and metastasis (Cavallaro and Christofori 2004). In addition to their role as cell adhesion molecules involved in homophilic interactions, E-cadherin also serves as a ligand for the integrin α_E(CD103)β7. α_E(CD103)β7 is expressed by very low frequencies of blood cells, instead being found on intraepithelial lymphocytes, mucosal mast cells, mucosal dendritic cell and CD25$^+$ T regulatory cell populations (Andrew et al. 1996; Kilshaw and Higgins 2002; Lehmann et al. 2002). Ligation of KLRG1 inhibits activation of rat basophilic leukemia cells and inhibits cytotoxicity of NK cells in some studies (Guthmann et al. 1995; Ito et al. 2006; Robbins et al. 2002). However, in other NK cell studies the ligation of KLRG1 by antibody or by its E-cadherin ligand did not block cytotoxicity, but did reduce proliferation of T cells in response to antigen stimulation (Grundemann et al. 2006; Hanke et al. 1998).

2.4 Complexity of the "NK Cell Zipper"

The mere lack of MHC class I surface expression is not sufficient to ensure sensitivity to NK cell attack. This is true for red blood cells, but also for MHC class I-deficient lymphocytes that are classically used as in vitro NK targets only during a short time window after mitogenic treatment (2–3 days after ConA treatment for mouse splenocytes or after PHA treatment for human peripheral blood cells). In addition, some MHC class I-positive cells can be sensitive NK cell targets. These observations have prompted the search and the discovery of other NK cell receptors whose engagement with target cell ligands triggers NK cell cytolytic programs, as well as cytokine and/or chemokine secretion. Some of these activating cell surface receptors initiate protein tyrosine kinase (PTK)-dependent pathways through non-covalent association with transmembrane signaling adaptors that harbor ITAMs (immunoreceptor tyrosine-based activation motifs; YxxL/Ix$_{6–8}$YXXL/I). Additional cell surface receptors that are not directly coupled to ITAMs also participate in NK cell activation, including NKG2D which is non-covalently associated with the DAP10 transmembrane signaling adaptor, as well as integrins and cytokine receptors. A variety of activating NK cell receptors (in particular those that are dependent upon ITAMs) can be antagonized by the aforementioned ITIM-bearing inhibitory cell surface receptors. The tyrosine phosphorylation status of several signaling components, that are substrates for both PTKs and PTPs, is therefore key to the propagation of the NK cell effector pathways (Long 1999; MacFarlane and Campbell 2005; Tomasello et al. 2000; Vely and Vivier 2005a,b; Vivier et al. 2004). Consequentially, NK cell activation is controlled by a dynamic balance between activating and inhibitory pathways. A very complex "zipper" thus forms between NK cells and their interacting partner, involving complementary as well as antagonist pathways (Fig. 3a,b, Color Plate 2a,b). Dissecting the integration of these multiple signals represents a current and future challenge in the understanding and manipulation of NK cell effector signaling pathways. Natural killer cells have been instrumental in revealing the function of ITIM-bearing molecules and their dynamic equilibrium with activating receptors. However, it is important to note that these notions are in no way limited to NK cells, but are widely described in a variety of (if not all) hematopoietic cells, as well as some non-hematopoietic cells.

2.4.1 Natural Cytotoxicity Receptors

The natural cytotoxicity receptors (NCR) are NK cell receptors that are critically involved in the human natural cytotoxicity against a broad panel of target cell types in vitro. Members of the NCR, NKp46, NKp44 and NKp30 (also known as NCR1, NCR2, and NCR3 respectively), were identified through the screening of antibodies raised against human NK cell molecules whose cross-linking resulted in redirected NK cytotoxicity against FcγR^+ P815 target cells (Moretta et al. 2001; Pende et al. 1999; Sivori et al. 1997; Vitale et al. 1998). All three NCR are members of the immunoglobulin superfamily, NKp46 containing two C2-type Ig-like domains and both NKp44 and NKp30 containing one V-type Ig-like domain. NKp46 and NKp30 are expressed by all human NK cells and associate with the FcRγ and CD3ζ adaptors. NKp46 or NKp30 cross-linking by antibodies results in a strong NK cell activation, including calcium flux, cytotoxicity, and production of cytokines (Bottino et al. 2005). NKp44 is not expressed by freshly isolated NK cells, but is induced following culture with IL-2. Natural cytotoxicity receptors are mostly NK cell-specific. However, NKp46 has been detected on subsets of activated intraepithelial T lymphocytes (Jabri et al. 2000), and NKp44 has been also found on rare γδ T-cell clones (Vitale et al. 1998). Recently, NKp44 has also been described on a subset of plasmacytoid DC (Fuchs et al. 2005). In contrast to its strong stimulatory function on NK cells, NKp44 ligation inhibits IFN-α in response to cytosine-phosphate-guanosine oligonucleotides (Fuchs et al. 2005), prompting the extensive dissection of ITAM-mediated inhibition of immune responses seen also in other models (Hamerman and Lanier 2006). Indeed, NKp44 signaling is based on its association with the ITAM-containing KARAP/DAP12 adaptor (Tomasello and Vivier 2005).

Natural cytotoxicity receptors have heterogeneous expression levels among individuals. The ability to discriminate NK clones based upon NCR expression level was first described for NKp46 (Sivori et al. 1999) and it was later found for NKp30 and NKp44 expression levels (Pende et al. 1999). The NCR^{bright} and NCR^{dull} phenotypes vary in proportion between the NK populations of different individuals and are unchanged upon culture of NK cells with different cytokines including IL-2, IL-15, IL-12 and IL-18. The NCR^{bright} or NCR^{dull} phenotype correlates with the ability of NK populations and NK clones to perform natural cytotoxicity. In addition, antibodies blocking studies indicate that NCR^{bright} and NCR^{dull} cells have a high and low capacities for NCR mediated natural cytotoxicity respectively (Moretta et al. 2001; Sivori et al. 1999).

The involvement of NCR in NK recognition of cells of both tumoral and non-tumoral origin is demonstrated by the blocking of NK cell activation using antibodies recognizing one or combinations of NCR (Moretta et al. 2001). However, a number of obstacles prevent a complete understanding of NCR biology. Firstly, there is currently only one murine model for NCR function, the NKp46-deficient mice (Gazit et al. 2006). Indeed, there is no known murine homologue of NKp44, and the *Ncr3* gene for mouse NKp30 is a pseudogene in many mouse strains (Hollyoake et al. 2005). Secondly, the tumor ligands recognized by NCR have not yet been conclusively determined. NKp46 and NKp44 have been reported to recognize the hemagglutinin of influenza virus and the hemagglutinin-neuraminidase of Sendai virus (Arnon et al. 2001; Mandelboim et al. 2001). This interaction requires the sialylation of NKp46 and NKp44 (Arnon et al. 2004). These findings are supported by the mortality of Ncr1(NKp46)-deficient mice upon influenza virus infection (Gazit et al. 2006). In addition, NKp30 has been reported to interact with the pp65 major tegument protein of human cytomegalovirus (Arnon et al. 2005). Heparan sulfate proteoglycans have been described as tumoral ligands for NKp30 and NKp46 (Bloushtain et al. 2004; Zilka et al. 2005). However, this finding was not reproducible in other experimental systems (Warren et al. 2005). Therefore the molecular basis of NCR-based NK target recognition remains an open and important question.

2.4.2 NKG2D

NKG2D is a receptor for self ligands that are typically found at low levels on normal cells but at higher levels on infected, stressed or transformed cells (Gonzalez et al. 2006; Raulet 2003). The association of induced NKG2D ligand expression with cellular "stress" has led to the hypothesis that NKG2D responds to "danger" signals associated with tumorigenesis. Indeed, the NKG2D ligands, MICA, MICB and Raet1 are often up-regulated by tumor cells (Cerwenka et al. 2000; Diefenbach et al. 2000; Groh et al. 1999). "Danger signals" that are included in the "infectious nonself" mode of immune recognition have been extensively characterized and include the detection of pathogen associated molecular pattern (PAMP) by Pattern Recognition Receptors, such as the Toll-like receptors (TLR) (Janeway and Medzhitov 2002). However, the molecular definition of "danger signals" that are not microbial molecules has been controversial (Gallucci and Matzinger 2001). Agents inducing DNA damage have recently been shown to result in NKG2D ligand expression (Gasser et al. 2005), identifying a non-microbial "danger signal" that results in "stress-induced self recognition." This mechanism stands not only as a new paradigm in NK cell recognition (Vivier et al. 2002), but more generally as a novel strategy of immune detection.

NKG2D is a member of the NKG2 family of C-type lectins encoded in the human and mouse natural killer gene complexes (NKC) (Trowsdale et al. 2001). Unlike other NKG2 family members, NKG2D forms homodimers and does not associate with CD94. In addition to NK cells, large populations of $CD8^+$ $\alpha\beta$ T cells, $\gamma\delta$ T cells and $NK1.1^+$ T cells also express NKG2D (Bauer et al. 1999; Diefenbach et al. 2000). NKG2D-based stimulation of NK cells leads to strong activation (Bauer et al. 1999; Cerwenka et al. 2000; Diefenbach et al. 2000; Diefenbach et al. 2001). However, NKG2D gives co-stimulatory signals to most $CD8^+$ $\alpha\beta$ T cells and $\gamma\delta$ T cells as effector functions are not induced without a primary activating signal coming from the TCR (Das et al. 2001; Groh et al. 2001; Roberts et al. 2001). Distinct NKG2D-based stimulatory and co-stimulatory activities may be partially due to alternative signaling mechanisms in different cell populations. Human NKG2D and the long splice-variant form of mouse NKG2D (NKG2D-L) associate with the signaling adaptor, DAP10 (Andre et al. 2004; Diefenbach et al. 2002; Gilfillan et al. 2002; Rosen et al. 2004). DAP10 contains a YxxM motif that binds the p85 regulatory subunit of phosphatidylinositol-3 kinase and Grb2 following phosphorylation (Chang et al. 1999; Wu et al. 1999). In mouse, an additional short splice-variant form of NKG2D (NKG2D-S) is able to associate with both DAP10 and KARAP/DAP12, therefore also triggering ITAM-mediated activation of the Syk-family protein tyrosine kinases. However, most murine T cells lack KARAP/DAP12 expression, thus restricting NKG2D activation to the DAP10 pathway in T cells (Diefenbach et al. 2002; Gilfillan et al. 2002). Ectopic expression of KARAP/DAP12 in murine T cells confers direct stimulatory capacity on NKG2D (Diefenbach et al. 2002; Teng et al. 2005), demonstrating that it is the lack of KARAP/DAP12 expression that prevents full T-cell stimulation through NKG2D. In human, pre-activated T cells, such as IL-15-stimulated T cells, are fully activated by NKG2D ligation in absence of KARAP/DAP12 association, maybe because of very high levels of DAP10 in these cells (Meresse et al. 2004).

The numerous NKG2D ligands have structural homology to MHC class I molecules. In humans, these include MICA, MICB and various RAET1/ULBP molecules, whilst murine H60, MULT1 and Raet1 molecules (also known as Rae1) are mouse NKG2D ligands (Cerwenka et al. 2000; Diefenbach et al. 2000; Groh et al. 1999). Human MICA and MICB molecules are encoded within the MHC on chromosome 6. Human RAET1 (or ULBP) molecules are also encoded on chromosome 6, but outside the MHC. The genes for all the known murine NKG2D ligands, the Raet1 family members (Rae1 molecules), H60 and MULT1, are linked on a region of chromosome 10 that is syntenic with the human RAET1 gene cluster (Radosavljevic et al. 2002; Sutherland et al. 2001). Mouse Raet1 molecules and some human RAET1 molecules have no transmembrane domain, but are instead anchored to the cell surface through glycosylphosphatidylinositol (GPI) linkage (Bacon et al. 2004; Cerwenka et al. 2000; Cosman et al. 2001). All NKG2D ligands have $\alpha 1$ and $\alpha 2$ domains that form an MHC class I-like fold. MICA and MICB also have an MHC class I $\alpha 3$-like domain, but they do not interact with $\beta 2$-microglobulin (Li et al. 1999). The crystal structures of single and NKG2D-complexed ligands suggest that NKG2D ligands are unlikely to have peptide-binding grooves (Holmes et al. 2002; Li et al. 1999, 2001, 2002; Radaev et al. 2001). Despite the large molecular differences between NKG2D ligands, in all cases NKG2D binding occurs over their α-helical surfaces in a manner analogous to TCR-MHC class I/peptide interactions (Li et al. 2001, 2002; Radaev et al. 2001).

2.4.3 CD16

CD16 (FcγRIII) is the low-affinity IgG Fc receptor involved in antibody-dependent cellular cytotoxicity (ADCC) of antibody coated (opsonized) target cells. The gene for CD16 expressed by human NK cells (FCGR3A) is also expressed by subsets of mononuclear phagocytes and by $\gamma\delta$ T-cell subsets, whilst polymorphonuclear neutrophils (PMN) express *FCGR3B*. Peripheral blood human NK cells can be divided into CD56^{bright} and CD56^{dim} subsets comprising around 5% and 95% of the circulating NK cell pool. The large majority of blood CD56^{dim} NK cells expresses CD16 and CD56^{bright} cells exhibit low level or negative staining (Cooper et al. 2001). A CD16 positive phenotype is therefore predominant in the peripheral blood NK cell population, but this situation is not necessarily reflected in other tissues. For example, most decidual, lymph node, and skin NK cells lack the expression of CD16 (Ebert et al. 2006; Ferlazzo et al. 2004; King et al. 1997).

FcγRIIIA/CD16, like NKp46 and NKp30, associates with FcRγ and CD3ζ and uses these ITAM-bearing adaptors for signal transduction. As with other NK activating receptors including the NCR, NKG2D and activating MHC class I receptors, triggering of CD16 on IL-2 activated NK cells leads to cytotoxicity against target cells (Bryceson et al. 2005b). Fresh NK cells are also activated by anti-CD16 antibody stimulation, resulting in calcium flux or production of the cytokines tumor necrosis factor (TNF)-α and interferon (IFN)-γ upon CD16 cross-linking, or cytotoxic granule degranulation and cytotoxicity in the context of a cellular target (Bryceson et al. 2005b). Interestingly, the triggering of CD16 in apparent isolation by insect cells (which are not though to express ligands for mammalian NK receptors) coated with rabbit immunoglobulin, leads to degranulation of fresh, otherwise unstimulated, human NK cells (Bryceson et al. 2005a). The high reactivity of CD16 may reflect a physiological requirement of IgG antibodies that have been produced in accordance with rigorous tolerance mechanisms of the adaptive immune system (Bryceson et al. 2005b).

Animals deficient for FcγRIII lack NK cells that can perform ADCC. Other deficiencies in phagocytosis and mast cell degranulation also exist in these mice, and the immunological effects of this deficiency are not purely NK cell mediated (Hazenbos et al. 1996). A number of polymorphisms of FCGR3A exist in the human population (Orange 2002). The L48→H polymorphism has been linked in a few cases to susceptibility to herpes virus infections, though NK cells from individuals homozygous for the 48H allele have normal NK cell ADCC (de Vries et al. 1996; Jawahar et al. 1996). The affinity of 48H CD16 for IgG is frequently higher than the 48L form due to linkage of this polymorphism with another 158V/F FCGR3A polymorphism that directly influences IgG binding (de Haas et al. 1996; Koene et al. 1997). The 158V/F polymorphism also influences the clinical sensitivity to cell-depleting therapeutic antibodies, implicating NK cell ADCC as a potentially important mechanism of their action (Dall'Ozzo et al. 2004).

2.4.4 Activating Homologues of Inhibitory MHC Class I Receptors

The KIR, Ly49, and NKG2 families have activating members that lack the ITIM in the cytoplasmic tail and instead associate with the ITAM-containing signaling adaptor, KARAP/DAP12. These activating receptors are often highly homologous to particular inhibitory receptors in their extracellular domains. The existence of these activating and inhibitory isoforms is a feature of most (if not all) ITIM-bearing receptors (Vély and Vivier 1997). Direct binding of the "activating homologue" to the inhibitory receptor's MHC class I ligand has been demonstrated for certain KIR and CD94/NKG2C (Biassoni et al. 1997; Stewart et al. 2005; Vales-Gomez et al. 1998, 1999, 2000). In addition, the stimulatory murine receptor Ly49D gives activating signals in response to H-2D (George et al. 1999). Phylogenetic analysis of activating *KIR* and *Ly49* suggests that these genes are constantly evolving from the extracellular domains of inhibitory homologues which, through processes of gene duplication and recombination, combine with "activating" transmembrane and cytoplasmic domains (Abi-Rached and Parham 2005). Notably, the affinities of activating interactions are lower than those of inhibitory counterparts. The murine activating Ly49 molecules Ly49H and Ly49P have roles specific to infection with murine cytomegalovirus (MCMV). Ly49H directly binds the product of the MCMV gene m157, and control of infection in some strains of mice is critically dependent upon Ly49H (Arase et al. 2002; Brown et al. 2001; Lee et al. 2001, 2003b; Smith et al. 2002). Ly49P is also involved in recognition of MCMV. However, the recognition of MCMV-infected cells by Ly49P is MHC-restricted, being dependent upon the presence of H-2Dk molecules on the host cell (Desrosiers et al. 2005). The human activating receptor KIR2DS4 has also been reported to bind a non-MHC class I ligand expressed by melanoma cells (Katz et al. 2004).

Except in the context of mouse infection with MCMV infection, the functions of the numerous activating Ly49 and KIR are still incompletely understood. Genetic epidemiologic studies have given some clues. *KIR* and *Ly49* are highly polymorphic. *KIR* haplotypes contain variable numbers of genes with most of this variation relating to the activating *KIR* genes present (Hsu et al. 2002). Individual *KIR* genes are also highly polymorphic with certain allelic polymorphisms are known to affect ligand binding (Carr et al. 2005; Gardiner et al. 2001). A growing number of genetic disease-association studies show roles for *KIR* genes in susceptibility or resistance to disease (Carrington and Martin 2006; Rajagopalan and Long 2005). In particular, activating *KIR* genes have been linked to susceptibility to a number of autoimmune conditions, including psoriatic arthritis, psoriasis, rheumatoid arthritis, scleroderma and type I diabetes (Martin et al. 2002b; Momot et al. 2004; Nelson et al. 2004; Suzuki et al. 2004; van der Slik et al. 2003; Williams et al. 2005; Yen et al. 2001). In addition, activating *KIR* might be beneficial against some microbial infections, as the presence of *KIR3DS1* and *HLA-Bw4* genes is associated to better clinical evolution in HIV-infected individuals (Martin et al. 2002a).

2.4.5 Natural Killer Cell Co-stimulatory and Adhesion Molecules

In addition to cell surface molecules that are coupled to specialized transduction subunits, NK cells express a number of cell surface molecules that are included in the "NK cell zipper" and participate in adhesion and co-stimulatory functions (Fig. 4a, b).

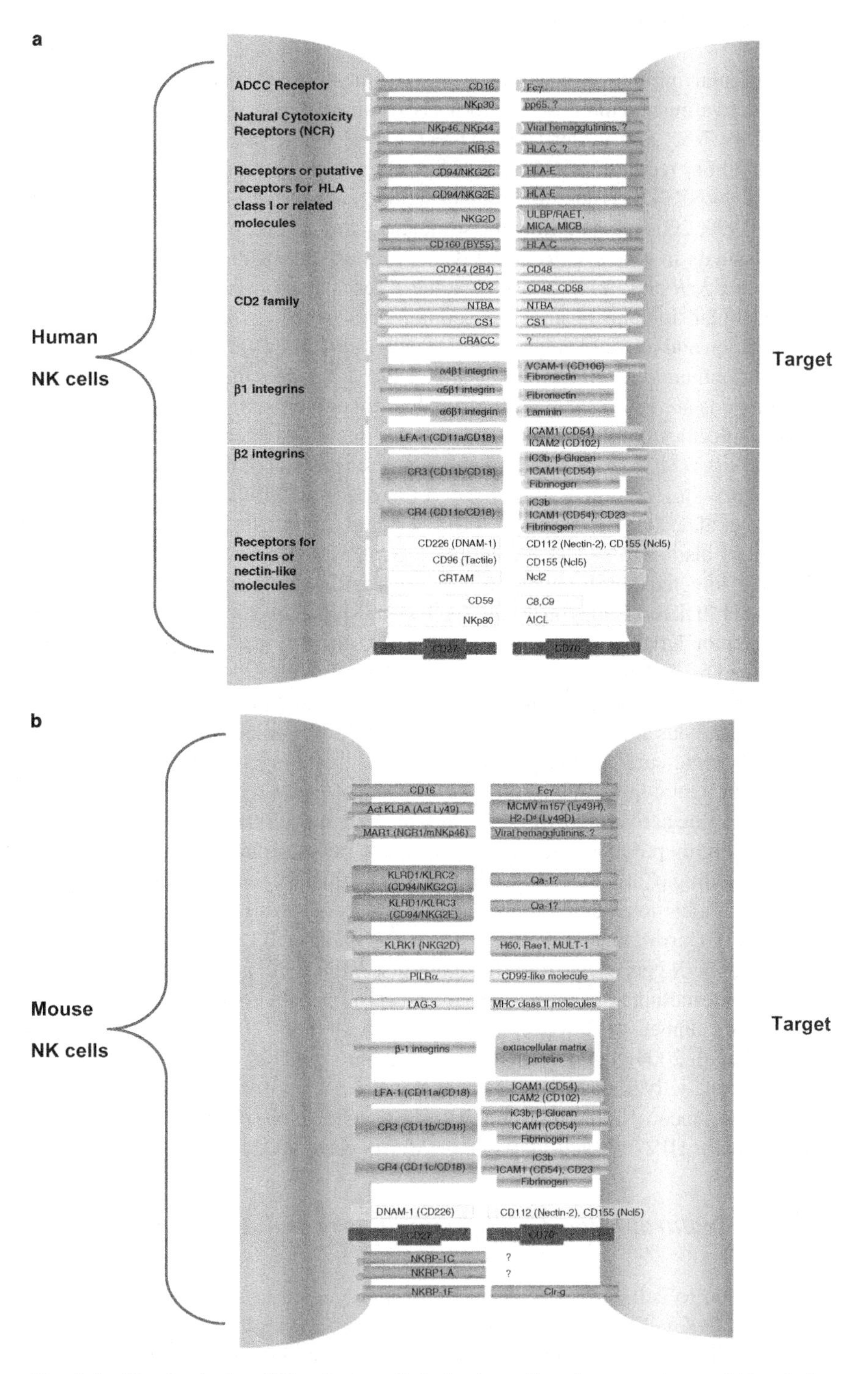

Fig. 4a,b. The "activating NK cell zipper." Activating cell surface receptors and ligands for human (**a**) and mouse (**b**) NK cells. Immunoreceptor tyrosine-based activation motif-bearing transducing signaling subunits are in *red squares*. DAP10 is marked as a *yellow square*. (See Color Plates)

2.4.5.1 The CD2 Family

The CD2 family is a family of immunoglobulin-superfamily molecules with homology to CD2 that are broadly expressed on cells of the immune system. Natural killer cells express CD2, CD48, CD58 (LFA-3), 2B4 (CD244), CD229 (Ly9), CS1 (CD2 subset 1; CRACC, CD2-like receptor activating cytotoxic cells), and NTB-A (NK-T-B antigen) (McNerney and Kumar 2006; Nichols et al. 2005). A small population of NK cells expresses CD150 (or SLAM, signaling lymphocyte activation molecule) following activation (Sayos et al. 2000). All CD2 family members have an N-terminal Ig V-type domain followed by a C-terminal Ig C2-type domain. For most, these are single copy domains and are followed by a transmembrane domain. However, CD48 and one form of CD58 are GPI-linked and CD229 has two pairs of Ig V-type and Ig C2-type domains (McNerney and Kumar 2006; Nichols et al. 2005).

Members of the CD2 family perform signaling functions through a variety of mechanisms. CD2 has a large proline-rich cytoplasmic domain that interacts directly with src family kinases. Members of the CD150 subfamily of CD2 family members (including 2B4, CD229, CS1, NTB-A, and CD150) have a number of copies of the TxYxxV/I immunoreceptor tyrosine-based switch motif (ITSM) in their cytoplasmic tails. ITSM can perform either stimulatory or inhibitory functions. They can associate with signaling molecules classically involved in ITAM and ITIM-based signaling, but they also associate with the Src homology 2 (SH2) containing proteins SH2 domain-containing gene 1A (SH2D1A; also SAP, SLAM-associated protein), and Ewing sarcoma-activated transcript-2 (EAT-2) (Veillette and Latour 2003). Importantly, mutations in SH2D1A cause X-linked lymphoproliferative disorder (XLP). XLP is a fatal progressive variable combined immunodeficiency in which symptoms associated with aberrant proliferation of lymphocytes and macrophages appear upon Epstein–Barr virus (EBV) infection. Disease manifestations include fulminant infectious mononucleosis, lymphomas and hypogammaglobulinemia (Nichols et al. 2005). The other CD2 family members, CD48 and CD58, are believed to have principally ligand function. However, CD48 engagement by 2B4 has been reported to stimulate NK activity and signaling may occur through the GPI-linkage-based association of this molecule with lipid microdomains (Assarsson et al. 2004; Messmer et al. 2006).

CD2 binds CD58 in humans and CD48 in mice (McNerney and Kumar 2006). The CD2-CD58 interaction is involved in activating human NK cells. Transfection of targets with CD58 induces killing by human NK clones and NK redirected lysis can be induced using antibody against CD2. Conversely, activation of NK cells can be blocked using antibodies against CD58 (Lanier et al. 1997). However, CD2-deficient mice have no major defect in NK cell function, possibly due to a redundancy in receptor function (McNerney and Kumar 2006).

2B4 (CD244) binds CD48. The expression of CD48 on target cells and its recognition by 2B4 leads to inhibition of NK cytotoxicity (Lee et al. 2004b). On the contrary, 2B4–CD48 interactions between cells of the same population, i.e. NK cell–NK cell interactions or T cell–T cell interactions, leads to enhanced activation

(Lee et al. 2003a). The bimodal function of 2B4 may be linked to its use of the adaptor SH2D1A, as this is required for NK cell activation and in its absence, 2B4 can mediate inhibitory signaling (McNerney and Kumar 2006). In the mouse, CD48 is the shared ligand of both CD2 and 2B4. 2B4 and CD48 are both important in IL-2 driven proliferation of NK cells along with NK cytotoxicity and cytokine production, whereas CD2 is apparently redundant (Lee et al. 2006), suggesting that 2B4/CD48 interaction is the more physiologically important.

The other CD2 family members expressed by NK cells, CD229, CS1 and NTB-A, all perform homophilic interactions and activate human NK cell cytotoxicity (McNerney and Kumar 2006; Stark and Watzl 2006).

2.4.5.2 Nectin and Nectin-Like Receptors

DNAM-1 (CD226), Tactile (CD96) and class I-restricted T-cell associated molecule (CRTAM) are a group of Ig-family NK cell receptors that bind Nectin and Nectin-like (Necl) adhesion molecules. DNAM-1 is expressed by most NK cells, T cells, monocyte, platelets and some B cells (Kojima et al. 2003; Shibuya et al. 1996), and recognizes Necl-5 (poliovirus receptor, PVR; CD155) and Nectin-2 (CD112) (Bottino et al. 2003; Tahara-Hanaoka et al. 2004). Tactile is expressed by NK cells, T cells and some B cells and binds Necl-5 (Fuchs et al. 2004). CRTAM is expressed by NK cells and T cells following activation and recognizes Necl-2 (also tumor suppressor in lung cancer-1, TSLC1) (Boles et al. 2005).

DNAM-1 triggers NK cell cytotoxicity and its interaction with Necl-5 and Nectin-2 may be critical in regulation of NK cell lysis and adhesion to target cells (Bottino et al. 2003; Shibuya et al. 1996). Tactile and CRTAM also mediate strong adhesion and stimulate cytotoxicity (Boles et al. 2005; Fuchs et al. 2004). Nectins and Necls are expressed by epithelial cells, mediating cell-cell adhesion (Sakisaka and Takai 2004), and also by APCs (Boles et al. 2005; Pende et al. 2006). Additionally, DNAM-1 ligands are over-expressed by certain types of tumor providing a possible recognition structure for NK cell tumor recognition (Castriconi et al. 2004; Pende et al. 2005; Tahara-Hanaoka et al. 2006).

2.4.5.3 Integrins

Mature NK cells express both β1 and β2 integrins. These are critical in NK cell-target adhesion, NK cell-matrix interactions and NK cell effector functions (Helander and Timonen 1998). The engagement of the β2 integrin LFA-1 (CD11a/CD18) on NK cells by antibody cross-linking leads to src tyrosine kinase activation, Vav1 phosphorylation, and MAP kinase activation (Perez et al. 2004). The ligand for LFA-1, ICAM-1 (CD54), either plate-bound or expressed by Drosophila insect cells, permits binding of fresh, unstimulated NK cells in a signaling-dependent manner. This binding through the interaction of LFA-1 with ICAM-1 is enhanced following brief activation of NK cells with IL-2 or IL-15 or in concert with CD2/

CD58 or 2B4/CD48 interaction (Barber and Long 2003). In addition to adhesion function, LFA-1-based signals lead to cytotoxic granule polarization in resting NK cells and cytolytic activity of activated NK cells against ICAM-1 expressing insect cells (Barber et al. 2004; Bryceson et al. 2005a). Therefore β2 integrin signaling plays an important and complex role by mediating multiple aspects of NK cell adhesion and effector functions.

2.5 Coordination of NK Cell Activating and Inhibitory Signals in Cancer

2.5.1 Cancer Immunosurveillance

The cancer immunosurveillance hypothesis, as originally proposed by Ehrlich in 1909 and later extended by Burnet and Thomas, stated that the immune system acts to recognize and remove newly arising transformed cells. Though initially focusing on a role for the adaptive immune system in preventing carcinogenesis, the hypothesis has been extended to encompass the concepts of innate immune recognition and immune tolerance mechanisms (Smyth et al. 2001b). The validity of the cancer immunosurveillance hypothesis has been and still remains contested (Dunn et al. 2002; Willimsky and Blankenstein 2005). However, a wide range of data from genetic models to clinical observations supports a role for the immune system in cancer regulation.

Carcinogenesis is a multi-step process and multiple natural regulatory mechanisms can prevent the generation of cancer (de Visser et al. 2006; Jakobisiak et al. 2003). The six "hallmarks of cancer" that distinguish tumors from normal cell populations are "self-sufficiency in growth signals, insensitivity to growth-inhibitory (antigrowth) signals, evasion of programmed cell death (apoptosis), limitless replicative potential, sustained angiogenesis, and tissue invasion and metastasis" (Hanahan and Weinberg 2000). Each of these features represents a cellular or physiological tumor-defense mechanism that has been breached. It has been estimated that in humans at least four to six mutations or epigenetic changes are required to bypass these checks (Hahn and Weinberg 2002). It is therefore important to consider any role for the immune system in the regulation of cancer within this framework of carcinogenesis, both with respect to other mechanisms that prevent tumorigenesis, and in how the immune system might recognize tumors.

Definitive evidence for cancer immunosurveillance is difficult to obtain because observation of successful cancer immunosurveillance of newly arising tumors is almost invisible. Evidence therefore comes from immunocompromised animals in which immune defects can be correlated with increased incidences of both spontaneous and carcinogen-induced tumors. These studies have the caveat that indirect effects of the immunosuppression, in particular those caused by pathogen infection, e.g., *Helicobacter pylori*, may be influencing cancer incidence. Important additional evidence therefore comes from the higher immunogenicity (and therefore lower success upon tumor transfer) of tumors originating in immunocompromised animals than in wild-type animals, suggesting that such tumors would be subject to immune control in a wild-type situation (Dunn et al. 2002).

Recently the concept of tumor immunosurveillance has been extended to include the capacities of the immune system to shape or modulate, ignore, or even assist in stages of tumor development. The "cancer immunoediting" hypothesis proposed by Dunn and colleagues highlights three potential interactions between a tumor and the immune system: elimination, equilibrium and escape (Dunn et al. 2004a). Elimination represents the original concept of cancer immunosurveillance, whereas equilibrium and escape respectively describe an immune system's ability to control but not destroy certain tumors and a tumor's capacity to finally evade this control. Support for an equilibrium phase comes from a few transplantation cases. In these cases, a tumor was believed to have been eliminated from an organ donor, but the same tumor then reappeared in an organ recipient undergoing pharmacological immunosuppression, suggesting that the tumor had been controlled in a latent state before transplant (Dunn et al. 2004b). Escape is thought to occur through the loss of molecules on which immune control is based. Alternatively, an immune system may be tolerant to a tumor. In a model of spontaneous tumorigenesis in immunocompetent mice, Willimsky and Blankenstein found that highly immunogenic tumors developed, despite the induction of tumor antigen-specific $CD8^+$ T cells and antibody responses, due to immune tolerance associated with non-responsive cytotoxic T cells (CTL), high serum levels of tolerogenic TGF-β and low serum levels of pro-inflammatory IFN-γ (Willimsky and Blankenstein 2005). Furthermore, under some conditions, the actions of the immune system may exacerbate carcinogenesis (de Visser et al. 2006). Agents causing chronic inflammation are known to predispose to cancer. For example, *Helicobacter pylori* is linked to gastric cancer and the hepatitis B and hepatitis C viruses are linked to chronic hepatitis and hepatocellular carcinoma (Balkwill et al. 2005; O'Byrne and Dalgleish 2001). Amongst other factors, the inflammatory environment provides growth factors, reactive oxygen species that promote DNA-damage and cyclooxygenase enzymes that promote angiogenesis, all of which could contribute to carcinogenesis. The interplays between the immune system and carcinogenesis are therefore multiple and complex.

2.5.2 Natural Killer Cell Recognition of Transplanted Tumors

In addition to their ability to kill tumor cells in vitro, NK cells are involved in the in vivo rejection of certain transplanted tumors dependent upon the NK ligands expressed by the tumor. The best-known experiment involving NK cell elimination of transplanted tumor cells was Kärre and colleagues demonstration of missing-self recognition (Kärre et al. 1986). This experiment demonstrated that mutagenized lymphoma cells (RMA), but not derivatives selected for the loss of MHC expression (RMA-S), are able to grow following sub-cutaneous injection in syngeneic

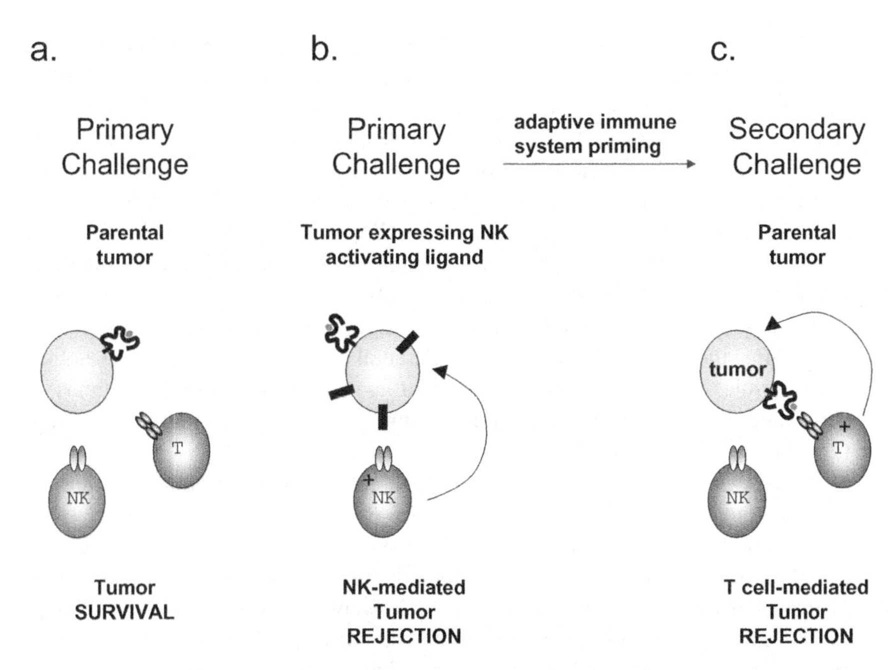

Fig. 5a–c. Natural killer recognition of tumors leads to tumor rejection and adaptive immunity against the tumor cells. (**a**) A parental tumor cell line is resistant to NK and T cells. (**b**) Expression of an activating NK cell ligand on the tumor leads to tumor rejection. (**c**) Following rejection of the tumor in (**b**), hosts are able to reject the parental tumor line through T cell-mediated adaptive immunity

mice. The selection of RMA for loss of MHC expression had rendered these cells sensitive to NK cell mediated cytotoxicity. As detailed previously, the molecular details of this missing-self recognition are now well established.

The roles of a range of NK activating receptors in tumor rejection have been studied in similar tumor-transplant systems. Amongst these, NKG2D is the only "primary" activating receptor studied to date. Ectopic expression of multiple NKG2D ligands, including Rae1β, H60, Rae1δ and Rae1γ, on RMA tumor cells facilitates their rejection from syngeneic mice (Fig. 5) (Cerwenka et al. 2001; Diefenbach et al. 2001). This NKG2D ligand-mediated tumor response occurs for other cell types including the B16-BL6 melanoma cell line, and in addition to preventing subcutaneous or intraperitoneal tumor growth, it can reduce the number and/or growth of lung metastases (Diefenbach et al. 2001). As NKG2D is expressed by γδ T cells and $CD8^+$ αβ T cells in addition to NK cells (Bauer et al. 1999; Diefenbach et al. 2000), any NK-mediated antitumor effect required precision. Diefenbach and colleagues showed that in RAG-deficient animals lacking T cells, NKG2D ligand-expressing tumors were still rejected, but this rejection was prevented by NK cell-depletion using anti-NK1.1 antibodies. However, the growth of

NKG2D ligand-expressing tumors in these NK cell and T cell-deficient animals was still retarded compared with controls, possibly due to incomplete NK cell depletion or alternatively due to the action of other cells such as macrophages (Diefenbach et al. 2001). Cerwenka and colleagues' approach to confirm the role of NK cells was similar in its use of anti-NK1.1 antibodies, additionally showing that certain NK1.1 expressing T-cell populations (invariant NKT cells) were not relevant by the use of animals deficient in CD1 (against which these T cells are restricted) (Cerwenka et al. 2001). In addition, under some experimental conditions (Diefenbach et al. 2001; Westwood et al. 2004), but not others (Cerwenka et al. 2001), the rejection of NKG2D ligand-expressing tumors leads to antitumor immunity against the NKG2D ligand-negative parental cell lines (see Fig. 5). In summary, these studies effectively demonstrate that expression of NKG2D ligands can contribute to NK cell-mediated innate immune responses against transplanted tumors.

NKG2D may not be sufficient to mediate antitumor innate immune responses. The EL4 and RMA tumor lines used in these studies are precisely those tumors used in the demonstration of missing-self recognition (Kärre et al. 1986), indicating that these tumors express stimulatory ligands for NK cells. As the lack of staining of these cells with recombinant NKG2D (Cerwenka et al. 2001; Diefenbach et al. 2001) indicates they do not express high levels of NKG2D ligands, their rejection in the context of missing-self is probably due to other NK activating receptors. Additionally, there may be other receptors for NKG2D ligands (Kriegeskorte et al. 2005).

A number of co-receptor and ligand pairs have been studied in similar tumor rejection systems, using NK or T-cell depletion to confirm roles for NK cells. The expression of the DNAM-1 ligands, CD155 and CD112, by RMA tumor cells leads to increased rejection and better survival of transplanted animals. Depletion of $CD8^+$ cells or $NK1.1^+$ cells with antibodies implicated both cell types in this DNAM-1-mediated rejection (Tahara-Hanaoka et al. 2005). In other models, expression of either CD70 or CD80 on tumor cells results in more efficient rejection of the tumor by syngeneic mice (Kelly et al. 2002a,b). CD70 is recognized by CD27 that is constitutively expressed by most murine NK cells, whilst CD80 (also known as B7.1) is recognized by CD28 and CD152 (CTLA-4). However, on human NK cells, a receptor other than CD28 may be involved in recognition of CD80 (Wilson et al. 1999). Tumor rejection in these models is dependent upon NK cells as suggested by anti-asialo-GM1 antibody-depletion of NK cells (Kelly et al. 2002a,b). An additional T cell-dependent component of rejection is shown by the higher tumor incidence in RAG-deficient animals. Support for direct interaction between tumor-expressed CD70 or CD80 and NK cells and its influence on tumor rejection is given by the NK cell-dependent reduced growth of CD70-transduced or CD80-transduced tumors in RAG-deficient animals. Interestingly, the immune responses against RMA-S expressing either CD70 or CD80 led to adaptive immunity against the parental RMA tumor line in a similar manner to that seen upon NKG2D ligand transduction (see Fig. 5) (Diefenbach et al. 2001; Kelly et al. 2002a,b). The priming of adaptive immunity was blocked if NK cell or IFN-γ depletion was performed by antibody administration, demonstrating a crucial role for NK cells in T-cell priming (Kelly et al. 2002a,b).

2.5.3 Natural Killer Cell Recognition of Spontaneous and Induced Tumors

Roles of NK cells as mediators of cellular innate immunity and regulators of adaptive immune responses make them pertinent candidates as effectors of cancer immunosurveillance. In human, an 11-year follow-up epidemiologic survey conducted on 3625 individuals has shown that the level of NK cell activity in peripheral blood is associated with reduced cancer risk, whereas low activity is associated with increased cancer risk (Imai et al. 2000). This important study suggests a role for NK cells against cancer in human. In the mouse, initial evidence for a role of NK cells in tumor immunosurveillance came from beige mice that are deficient in natural cytotoxicity (Talmadge et al. 1980). In these mice, higher levels of spontaneous and carcinogen-induced tumors are observed (Haliotis et al. 1985). However, the beige defect results from a generalized defect in the lysosomal secretory pathway and is not restricted to the NK cell compartment. Though alternative models of NK cell deficiency have been reported, a clean model that results in selective deficiency of the NK compartment has been lacking (Kim et al. 2000, 2005; Yokota et al. 1999). The recent innovative use of the NKp46 promoter to selectively target NK cells for depletion will provide an important tool for future studies (Walzer et al. 2007).

Owing to the lack of a selective NK-deficiency model, many studies elucidating the role of NK cells in spontaneous or induced carcinogenesis rely on antibody-mediated depletion of NK cells. Antibodies directed against either NK1.1 or the glycolipid asialo-GM1 are typically used for selective depletion of NK cells, but a number of side effects may also occur. Depletion of NK cells with anti-NK1.1 antibody (PK136) may also affect populations of invariant-TCR ($V\alpha14J\alpha281$) NKT cells and other NK1.1$^+$ T-cell populations, making interpretation of depletion experiments difficult. For example, in a model of methylcholanthrene (MCA)-induced fibrosarcoma the incidence of tumors in $J\alpha281$-deficient mice (which lack NKT cells) was not altered by depletion of NK1.1$^+$ cells. This contrasts with wild-type mice in which depletion of NK1.1$^+$ cells leads to an increase in fibrosarcoma incidence, suggesting that NK1.1$^+$ cell-based tumor immunosurveillance requires NKT cells and may be NK cell independent (Smyth et al. 2000a). In some models, depletion using antibody against asialo-GM1 may be a more selective means of NK cell depletion. Expression of asialo-GM1 is higher on NK cells than NKT or T cells and NK cell depletion with antibody against asialo-GM1 does not always drastically affect the size or function of the NKT-cell population (Smyth et al. 2001a). Using this calibrated approach of NK depletion Smyth and colleagues showed that both NK cells and NKT cells are important in control of MCA-induced fibrosarcoma. Antibody-mediated depletion of asialo-GM1$^+$ (selective for NK) cells, NK1.1$^+$ (NK and NKT) cells, or genetic deficiency in NKT cells ($J\alpha281$-deficient mice) or T cells (RAG1-deficient mice), all lead to increased incidence of induced tumors (Smyth et al. 2001a). Although asialo-GM1 expression may be selective for NK cells over invariant NKT cells, other cell populations including activated macrophages and both naïve and virus specific $CD8^+$ T cells also express asialo-GM1 (Lee et al. 1996; Slifka et al. 2000). Caution is therefore required when interpreting studies based upon antibody depletion both because of the lack of specificity of antibody treatment against NK populations and because the depletion process itself, through cell surface molecule targeting and destruction of cell populations, may have other effects on the immune system.

In addition to a generic role for NK cells in tumor recognition, the roles of individual NK cell activation mechanisms and effector functions in protection against spontaneous and induced tumors have been addressed using a number of genetic or depletion models. A role for NK cell-mediated cytokine production is supported by analysis of IFN-γ signaling-deficient animals, one for direct NK cell cytotoxicity is supported by perforin-deficient mice and a role for inducers of apoptosis is suggested by models of TRAIL deficiency. In addition, means of NK cell activation through the NKG2D receptor and through type I IFN are also implicated as pathways of NK cell immunosurveillance.

2.5.3.1 Interferon-γ

NK cells and T cells are major producers of IFN-γ, and IFN-γ produced by NK cells and NKT cells plays an important role in antitumor immunity (Hayakawa et al. 2001). Selective deficiency in either IFN-γ or the α chain of the IFN-γ receptor (IFNGR1) results in increased incidences of MCA carcinogen-induced tumors or spontaneous tumors in a number of different models (Kaplan et al. 1998; Shankaran et al. 2001; Street et al. 2001, 2002). Street and colleagues demonstrated that C57BL/6 (B6) mice deficient in IFN-γ ($IFN\text{-}\gamma^{-/-}$) or perforin ($pfp^{-/-}$), but not mice deficient in the cytokines IL-12, IL-18, or TNF, have a higher incidence of spontaneous tumor development than wild-type mice (Street et al. 2002). 50% (16/32) of $IFN\text{-}\gamma^{-/-}$ B6 mice developed tumors over a 750-day observation period. Whilst most of these tumors from B6 mice were disseminated lymphomas, a similar experiment in BALB/c $IFN\text{-}\gamma^{-/-}$ mice resulted in the development of spontaneous lung adenocarcinomas and other tumors, but not lymphomas. A previous study by the same group (Street et al. 2001) showed an increased incidence of MCA-induced fibrosarcoma in B6 mice deficient for perforin, IFN-γ, or Jα281 compared with wild-type. Interestingly, the speed of tumor growth was faster when the mice were deficient for IFN-γ, but not in animals deficient for perforin. However, this increased tumor growth was also observed in NKT cell-deficient animals ($J\alpha 281^{-/-}$) suggesting that NKT cells are mediators of this IFN-γ -dependent growth inhibition. Schreiber's group has investigated the mechanism of IFN-γ-mediated tumor suppression. 129/Sv/Ev p53-deficient animals that are also deficient for the IFN-γ receptor α-chain ($p53^{-/-}IFNGR1^{-/-}$) develop tumors more rapidly than $p53^{-/-}$ mice. Additionally, a different spectrum of tumor types occurs in these animals. Transplantation of a selection of MCA-induced tumor lines into wild-type or $IFN\text{-}\gamma^{-/-}$ animals led to equivalent growth of these tumors, suggesting that the IFN-γ signaling deficiency was at the level of the tumor cell and not components of the immune system. Furthermore, reconstituted expression of the IFN-γRα chain in one tumor line led to de novo rejection of this line from wild-type mice, suggesting a direct effect of IFN-γ in controlling tumor cell growth (Kaplan et al. 1998). Rejection of the same tumor line could also be induced by over-expression of TAP1 in the tumor cell (Shankaran et al. 2001), suggesting that deficiency in IFN-γ-induction of the MHC class I processing and presentation pathway allowed this tumor to avoid immune surveillance. Tumor rejection was T-cell dependent and led to immunity against the parental tumor cell line. These results suggest that one role of IFN-γ in tumor immunosurveillance is to induce MHC class I expression on the tumor cell itself, supporting the following model: MHC class I-negative tumor cells→IFN-γ production by NK cells→MHC class I-positive tumors→tumor cell recognition by T cells. However, this role is one of many in tumor immunosurveillance, as IFN-γ also induces other tumor suppressor effects, including sensitization to apoptosis, and additional stimulation of immune responses (Farrar and Schreiber 1993; Schroder et al. 2004; Tanaka and Taniguchi 2000).

2.5.3.2 Perforin

Expression of perforin is more restricted than that of IFN-γ, being limited to NK cell and T-cell subsets. An increased incidence of MCA-induced or spontaneous tumor has been well documented for perforin (pfp)-deficient animals (Smyth et al. 2000a,b; Street et al. 2001; van den Broek et al. 1996). Van den Broek and colleagues first described an increased frequency and faster occurrence of MCA-induced fibrosarcomas in pfp-deficient mice (van den Broek et al. 1996). An increased fibrosarcoma incidence was not observed in $CD8^{-/-}$ animals, suggesting that CD8-negative perforin-expressing cells were responsible for the protection. Importantly, perforin deficiency had no effect on incidence of skin papillomas induced by the application of 12-*O*-tetradecanoylphorbol-13-acetate (TPA) and 7,12-dimethylbenzanthracene (DMBA) to the skin. A tumor-specific susceptibility of perforin-deficient mice was also found with spontaneous tumors occurring in $pfp^{-/-}$ animals (Smyth et al. 2000b). 10 of 20 $pfp^{-/-}$ animals developed lymphomas during late life compared to 1 in 36 of perforin-sufficient controls. An additional experiment using mice deficient in p53 or doubly deficient in both p53 and perforin showed that the effect of perforin deficiency is largely restricted to the control of disseminated lymphomas. p53-deficient animals developed a wide range of tumors, including disseminated lymphomas, sarcomas and thymic lymphomas. Only in the case of disseminated lymphomas was the age of tumor morbidity lower for mice additionally deficient for perforin. The same group later studied the immune responses mediated against such lymphomas upon transfer to tumor-free mice (Street et al. 2004). Primary B-cell lymphomas were derived from C57BL/6 mice deficient in both perforin and β2-microglobulin. These tumors were rejected from wild-type but not pfp-deficient mice. Additionally, mice deficient for TRAIL, TNF, FasL, IFN-γ, IL-12, or IL-18 rejected these B-cell lymphomas. Through a combination of knockout animals and antibody depletion, the rejection of lymphoma was shown to be mediated by either γδ T cells or $NK1.1^{+}$ cells, suggesting that γδ T cells and NK cells are each capable of this rejection. Importantly, this experiment was performed using lymphoma cells lacking β2m, meaning that CD1d-restricted NKT cells were not involved in the rejection, but also introducing a potential for missing-self recognition by NK cells due to the lack of MHC class I by the lymphomas. Whether the rejection afforded by NK cells was due to missing-self recognition was not addressed.

2.5.3.3 Other Cytolytic Molecules

The granzyme serine proteases found within cytotoxic granules are released into target cells during cytolysis. The prototypical granzymes A and B cooperate with perforin to induce apoptosis associated with DNA-fragmentation and granzyme B leads to caspase activation (Russell and Ley 2002). The functions of multiple other known granzymes have yet to be fully determined. Though these effectors are variously important in the control of virus infections, there is currently no evidence that they are required in tumor immunosurveillance processes (Smyth et al. 2003a).

Tumor necrosis factor-related apoptosis-inducing ligand (TRAIL) induces apoptotic cell death in various target cells. TRAIL is not generally detectable on freshly isolated NK cells, T cells, NKT cells, B cells, DCs or monocytes. However, a population of mouse liver NK cells, but not human liver NK cells (Ishiyama et al. 2006), expresses TRAIL under normal conditions. Stimulation of NK cells, T cells, DCs or monocytes with various cytokines leads to up-regulation of TRAIL expression (Smyth et al. 2003b). Importantly, NK cells activated through IL-2 or IL-15 treatment (LAK cells) are capable of killing through perforin, FasL and TRAIL-dependent pathways (Kayagaki et al. 1999). TRAIL-deficient mice are more susceptible to MCA-induced fibrosarcoma (Cretney et al. 2002). An increase in MCA-induced fibrosarcomas is similarly observed upon treatment of mice with anti-TRAIL antibodies (Takeda et al. 2002). In $p53^{-/+}$ animals, treatment with either anti-TRAIL antibodies or anti-asialo-GM1 antibodies during MCA-induced tumorigenesis leads to the outgrowth of tumor cells that are more susceptible to TRAIL-mediated lysis than those from mock-treated animals, supporting a role for NK cells in TRAIL-mediated tumor immunosurveillance. Animals treated with both anti-asialo-GM1 and anti-TRAIL antibodies were even more susceptible to MCA-induced tumorigenesis, suggesting that NK cells can also use other effector pathways. However, anti-TRAIL treatment did not affect tumorigenesis in IFN-γ-deficient mice, indicating a requirement for IFN-γ in the protection afforded by TRAIL (Takeda et al. 2002).

2.5.3.4 NKG2D

A role for NK cells in the direct detection of neoplasia is best exemplified by the role of NKG2D in carcinogen-induced neoplasia. C57BL/6 and BALB/c mice treated with nondepleting blocking antibody against NKG2D have a higher incidence of MCA-induced fibrosarcoma than wild-type controls (Smyth et al. 2005). Perforin-deficient animals were unaffected by anti-NKG2D antibody treatment, indicating that NKG2D function in prevention of neoplasia is dependent upon perforin. However, IFN-γ- and TRAIL-deficient animals were more susceptible to MCA if treated with anti-NKG2D antibody. A similar dependence of NKG2D-mediated antitumor effects on perforin but not IFN-γ or TRAIL was previously demonstrated in models of metastasis affected by cytokine treatment (Smyth et al. 2004). Strong support that NKG2D mediates its effects in direct recognition of tumor cells was provided through examination of NKG2D ligand expression on tumors derived from MCA-treated mice. Most tumors derived from wild-type mice were Rae1 negative (10/16), whereas all tumors (7/7) from perforin-deficient animals all had medium to high expression of Rae1 (Smyth et al. 2005). Furthermore, transfer of a Rae1 positive tumor to recipients resulted in rejection that was dependent upon NKG2D and perforin, suggesting that direct NKG2D-based recognition of tumor cells acts as a protective mechanism against cancer. A simple role for NK cells in this NKG2D-based recognition was not confirmed as blocking NKG2D did not affect the incidence of MCA-induced fibrosarcoma in T-cell-deficient animals (RAG-1-deficient). However, fibrosarcoma incidence in NK cell-deficient animals (anti-asialo-GM1-treated) was also refractory to anti-NKG2D antibody treatment (Smyth et al. 2005). It is therefore possible that both NK cells and T cells play non-redundant roles in NKG2D-mediated tumor-suppression.

Roles for other NK recognition pathways are less well understood. NKG2D is not the only murine receptor involved in triggering NK cytolysis of targets as some NK-susceptible cell lines, including RMA-S, do not express NKG2D ligand. In addition, NK cytolysis of NKG2D ligand-positive cells is not completely abrogated by the addition of antibody against NKG2D (Jamieson et al. 2002). The expression of NKG2D ligand by multiple different tumors indicates that this receptor is not sufficient to mediate antitumor immunity. Type I IFNs may play an important role as these cytokines can be produced by nonhematopoietic cells and they induce activation of NK cytotoxic activity. Deficiency in type I IFN signaling results in increased susceptibility to MCA-induced fibrosarcoma, with hematopoietic cells rather than cancer cells being the relevant targets of type I IFN in protection against tumors (Dunn et al. 2005).

2.5.4 *Other Antitumor Innate Effectors*

Together, the higher incidence of spontaneous or carcinogen-induced neoplasia in multiple models of NK cell-deficiencies, and the higher immunogenicity of tumors derived within these models, strongly support a role for NK cells in protection against neoplastic disease. Redundancy may also be in place both within pathways of NK recognition through multiple different receptors, and in the NK effector mechanisms used. For example, RMA-S cells are not killed by perforin-deficient NK cells in short 4 hour assays, but up-regulation of death receptors including Fas on RMA-S occurs during longer incubations and results in killing (Screpanti et al. 2001). Alternatively, different newly arising tumors may be sensitive to different effector mechanisms (Smyth et al. 2004). Models of selective deficiency in the NK cell population are required in order to understand the full extent of NK cell-based control of spontaneous or carcinogen-induced neoplasia. Yet, non-NK cell populations also play a major role in tumor elimination, and in particular in NKG2D-mediated tumor recognition. For example, induction of skin tumors through initiation with DMBA and promotion with TPA is enhanced in mice deficient for γδ T cells. This tumorigenic treatment with TPA induces expression of the NKG2D ligands H60 and Rae1 both 24 hours after treatment and in resulting papillomas and carcinomas. Additionally, γδ T-cell killing is sensitive to blocking of NKG2D (Girardi et al. 2001). Unlike MCA-induced tumorigenesis, perforin-deficient animals are not more sensitive to DMBA/TPA-treatment than wild-type mice (van den Broek et al. 1996). Therefore, depending upon the cancer and its tissue location, different immune cell types and different effector mechanisms may be important in immunosurveillance.

Very recently, a population of hematopoietic cells with NK and DC features was described in the mouse (Chan et al. 2006; Taieb et al. 2006). These IFN-producing killer dendritic cells (IKDCs; MHC class II^+ $Ly6C^-$ $B220^+$ $NK1.1^+$) produce type I IFN, IL-12 and IFN-γ. They kill typical NK target cells using NK-activating receptors such as NKG2D and can present antigen to T cells. Although these cells are still an enigma, their antitumoral function prompts further characterization in mouse and man.

2.6 Programming a "Natural" Killer: Steady-State and Situation-Specific Regulation of NK Effector Functions

Despite their fame as mediators of spontaneous "natural" cytotoxicity (Herberman et al. 1975; Kiessling et al. 1975), fresh unstimulated NK cells from specific pathogen-free mice or healthy human donors are limited in their capacity to kill classical NK targets. For this reason, mouse studies often use polyinosinic polycytidylic acid (poly(I:C)) effector NK cells or lymphokine activated killer (LAK) cells produced by NK culture with IL-2. Similarly, human studies frequently employ IL-2 to observe responses of primed, activated NK cells. These types of stimulation can produce large and divergent changes in NK cell responsiveness and phenotype (Bryceson et al. 2005b; Chiesa et al. 2006). In addition to soluble mediators, direct interactions of NK cells with other cell populations such as DCs, results in major phenotypic changes (Degli-Esposti and Smyth 2005; Moretta et al. 2005; Walzer et al. 2005). The power of cytokine stimulation to activate NK cells suggests that priming may be an important regulator of NK cells in vivo and illustrates that their "natural" effector functions remain highly sensitive to environmental control.

2.6.1 Natural Killer Responsiveness to MHC Class I and Missing-Self

Inhibitory MHC class I receptors have been recently shown to contribute to NK cell education via the recognition of self-MHC class I molecules. As described above, missing-self recognition of MHC class I is mediated by multiple inhibitory receptors that are expressed in a variegated fashion by the NK cell population leading a repertoire of specificity for MHC class I (Raulet et al. 2001). This NK repertoire must remain tolerant to the MHC class I molecules expressed by self cells. One proposed mechanism for self-tolerance was that all NK cells express "at least one" inhibitory receptor for self-MHC class I. This hypothesis was supported by Valiante et al. who found that virtually all NK clones grown from two human donors were inhibited by individual self-MHC class I molecules (Valiante et al. 1997). However, using combinations of antibodies, populations of fresh murine or human NK cells can be identified that do not express any known inhibitory NK receptor (Anfossi et al. 2006; Fernandez et al. 2005). Additionally, in mice and humans who lack normal cell surface expression of MHC class I, NK cells remain tolerant to self, whilst the fraction of the NK pool expressing individual Ly49 or KIR receptors is largely unchanged (Furukawa et al. 1999; Salcedo et al. 1998; Zimmer et al. 1998). The existence of NK cells lacking inhibitory MHC class I receptor expression argues against the "at least one" hypothesis and suggests that Valiante et al.'s results may be peculiar to cloned NK cells (Valiante et al. 1997).

Coincident with their self-tolerance, the NK cells from mice and humans lacking normal MHC class I expression (due to deficiencies in β2-microglobulin or in TAP) are hyporesponsive to the classical NK cell targets such as YAC-1 and K562 respectively (Furukawa et al. 1999; Liao et al. 1991; Zimmer et al. 1998). However, there is no evidence that this hyporesponsiveness is due to the decreased expression of activating receptors or the increased expression of other inhibitory receptors (Vitale et al. 2002). Additionally, the NK cell hyporesponsiveness can be (at least partially) overcome in vitro by long term stimulation of NK cells with cytokines. Thus, activation of human TAP-deficient NK cells with IL-2, IL-12, or IL-15 renders them responsive against classical NK targets (Furukawa et al. 1999; Zimmer et al. 1998). In addition, IL-2 treatment of TAP2-deficient NK cells allows them to kill autologous B-lymphoblastoid cell lines (B-LCL) or IFN-treated fibroblasts, both cell types being protected from lysis under normal circumstances due to expression of MHC class I (Zimmer et al. 1998, 1999). This break of tolerance might not, however, be universal; as autologous PHA-blasts were not susceptible to lysis by IL-2 activated TAP2-deficient NK cells (Vitale et al. 2002). A break in hyporesponsiveness and/or tolerance can also be achieved through treatment of murine NK cells with IL-2. Natural killer cells from mosaic mice expressing an MHC class I transgene (Dd/Ld) on a subpopulation of hematopoietic cells are tolerant to cells lacking the MHC class I determinant. However, separation of NK cells into Dd/Ld–positive and Dd/Ld–negative populations and culture of these in IL-2, endows the Dd/Ld–positive population of NK cells with reactivity against Dd/Ld–negative targets (Johansson et al. 1997). Therefore tolerance to an MHC class I-determinant can be broken in its absence.

It has recently been shown that the mechanism resulting in NK cell tolerance to self-MHC class I affects the responsiveness of NK cells per se. Specifically, the stimulation of NK cells in a target-cell free system shows a generalized hyporesponsiveness of NK cells that do not express a self-MHC class I-specific inhibitory receptor (Anfossi et al. 2006; Fernandez et al. 2005; Kim et al. 2005a,b). Stimulation of fresh mouse NK cells with plate-bound antibody against NK1.1 (the activating receptor Nkrp1c) led to IFN-γ production by B6 NK cells that express Ly49C, which recognizes H-2K^b present in B6 mice, but not B6 NK cells that express Ly49A, which has no known B6 ligand. The higher responsiveness of NK cells expressing a self-MHC class I specific inhibitory receptor was termed "licensing" and was found to be dependent upon the presence of the correct MHC class I determinant and the ITIM of the Ly49 receptor, but not the PTP SHP-1 (Kim et al. 2005a,b). Similarly, murine NK cells lacking the expression of all known MHC class I-specific inhibitory receptors are hyporesponsive to stimulation with antibody against NKG2D or Ly49D (Fernandez et al. 2005). This phenomenon also extends to human NK cells. Stimulation with antibody against CD16 results in a higher response from NK cells expressing a self-specific inhibitory KIR than NK cells expressing a KIR with no self-ligand (Anfossi et al. 2006). In all these systems, the phenomenon of licensing or education applies to NK cell responses against a variety of cellular targets and other stimuli (Anfossi et al. 2006; Fernandez et al. 2005; Kim et al. 2005). However, the presence of this phenomenon in the absence of target cells rules out the possibility of other inhibitory receptors compensating for the lack of MHC class I-inhibitory receptor interaction. It further suggests that this mechanism of tolerance acts directly on the "programming" of individual NK cells.

Education of NK cells on their MHC class I background may serve to optimize missing-self responsiveness (see Fig. 6). Maturation of NK cells in the context of inhibitory KIR or Ly49 ligation would increase the potential of activating signaling circuits. Natural killer cells would therefore be fully responsive upon subsequent encounter of MHC class I$^{dim/-}$ cells. HLA class I ligands were recently reported to increase the frequencies of NK cells expressing cognate KIR (Yawata et al. 2006), demonstrating an amplification of MHC class I-inhibitory receptor bearing NK cells during a still unknown stage of NK cell maturation. Therefore selective proliferation of educated NK cells could be an additional mechanism providing optimal missing-self responsiveness. Furthermore, the hyporesponsiveness of "uneducated" or "unlicensed" NK cells does not prevent NK function in all circumstances as NK cells from MHC class I-deficient mice are fully protective against mouse cytomegalovirus infection (Tay et al. 1995).

The heightened activation state of MHC class I educated NK cells has implications for our understanding of NK biology in the context of autoimmunity and immunotherapies. Genetic epidemiologic studies have revealed a role for KIR and HLA in the susceptibility or resistance to a variety of pathologies, including autoimmune syndromes, cancer, and infectious diseases (Carrington and Martin 2006).

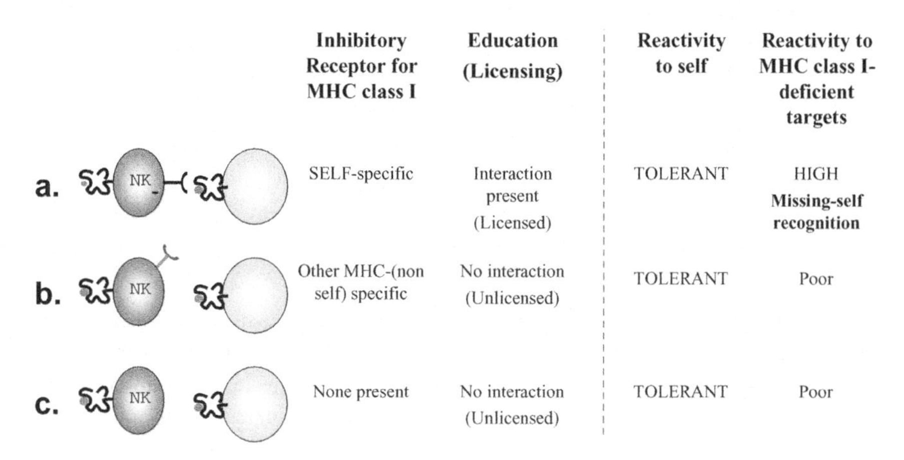

Fig. 6a–c. Natural killer cell (*NK*) Education (or Licensing) enables effective missing-self recognition. Natural killer cells that express a self-specific inhibitory major histocompatibility (*MHC*) class I receptor (**a**) are educated to be reactive to cells lacking this particular MHC class I molecule. Natural killer cells lacking self-specific inhibitory MHC class I receptors (**b,c**) do not gain this reactivity during their education and are hyporesponsive. Education processes therefore maintain self-tolerance. It is not yet clear whether NK cell education is a continuous process in vivo, or whether it only occurs at specific stages of NK cell development

The current interpretation of these associations relies on the role of KIR-HLA interaction in the modulation of NK cell effector function during interaction with target cells. One can now suggest that the association between KIR, HLA and human disease might also be the consequence of the role of KIR-HLA in NK cell education. Similarly, pioneering work has shown that donor-versus-recipient NK cell alloreactivity can eliminate leukemia relapse and graft rejection, while protecting patients against graft-versus-host disease (Ruggeri et al. 2002). In this regime and model, it appears that during the first few months post-hematopoietic transplant, donor NK cells develop in the recipient in the same way as they would develop in the donor (Parham 2005). Under certain KIR-HLA donor/recipient combinations, some NK cells become alloreactive to the recipient. If these alloreactive NK cells were instead educated on the MHC class I profile of the recipient, they would be hyporesponsive. The instruction of NK cell education may be dependent upon the recipient's conditioning regimen as well as the dose of donor hematopoietic progenitors (i.e. the likelihood of interaction with donor or recipient MHC class I molecules), providing a possible basis for the variable outcome of these MHC-mismatched hematopoietic transplantation protocols (Parham 2005).

It is likely that other inhibitory receptors expressed by NK cells contribute to their education through MHC class I-dependent or -independent mechanisms. As discussed above, a variety of other inhibitory receptors have been described on hematopoietic cells, several of them being expressed at early stages of cell differentiation (Daeron and Vivier 1999; Long 1999; Ravetch and Lanier 2000). Thus, inhibitory receptors, other than MHC class I-specific receptors, might also educate a wider array of hematopoietic cells to discriminate between interacting cells that do, or do not, express self-ligands.

2.6.2 Natural Killer Responsiveness to Activating NK Ligands

Pathways of NK cell activation are also sensitive to tolerogenic mechanisms that may affect their responsiveness in tumor settings. The sensitivity of the NKG2D recognition system is modulated through a number of mechanisms. In addition to membrane bound forms of NKG2D ligands that can stimulate immune system cells, tumor cells may also produce soluble or membrane forms of these ligands that promote the down-regulation of NKG2D on NK and T cells (Coudert et al. 2006; Groh et al. 2002; Salih et al. 2003). Alternatively, transforming growth factor β (TGF-β) can induce down-regulation of the NKG2D receptor (Castriconi et al. 2003; Lee et al. 2004a). TGF-β is produced by some cancers, but is also an important product and mediator of $CD4^{+}CD25^{+}$ regulatory T-cell populations.

In vivo, the effect of altered NKG2D responsiveness due to tolerogenic mechanisms is best illustrated by transgenic mice ectopically expressing the Rae1ε NKG2D ligand (Oppenheim et al. 2005). In these mice, ectopic expression of Rae1ε on all cells (conferred by a β-actin promoter) or selectively on epithelial cells (using an involucrin promoter) resulted in lower expression levels of the NKG2D receptor by NK cells and T cells, a reduced capacity of NK cells to kill Rae1 bearing targets, and a higher susceptibility to DMBA and TPA-induced tumors. The in vivo killing of MHC class I-deficient RMA-S targets was also reduced in transgenic mice, suggesting that NKG2D ligand-induced tolerance produces a wider suppression of NK responsiveness. Furthermore, the suppression of NK functions could be partially overcome by poly(I:C) treatment of transgenic mice (Oppenheim et al. 2005).

The suppression of NK cell effector functions by $CD4^{+}CD25^{+}$ regulatory T cells (T regs) and TGF-β has been recently described by a number of groups (Barao et al. 2006; Ghiringhelli et al. 2005; Laouar et al. 2005; Smyth et al. 2006). Scurfy mice, which lack the Foxp3 transcription factor necessary for development of T regs, have NK cells that have higher basal activation and are more cytolytic against YAC-1 targets (Ghiringhelli et al. 2005). In addition, the killing of NKG2D ligand-bearing targets is regulated by T reg cells in a TGF-β dependent fashion (Ghiringhelli et al. 2005; Smyth et al. 2006). Interestingly, the presence of T regs in tumors may prevent in vivo rejection of tumors by NK cells, as depletion of T regs can reveal potent NK cell-dependent cytotoxicity against autologous tumor cells (Ghiringhelli et al. 2005).

2.7 Conclusions and Perspectives: NK Cells, "Self Versus Nonself," and Cancer

How do NK cells discriminate "self" from "nonself"? NK cells can recognize infectious nonself ligands, stress-induced self ligands, and missing-self (Fig. 1). We have discussed above how the recognition of stress-induced ligands and missing-self might be involved in innate immunosuppression mechanisms to control cancer formation. We have also discussed how these NK cell recognition pathways can be modulated by other homeostatic processes such as the effect of T reg cells. Definitive proof of NK cell involvement in cancer immunosurveillance has not yet been obtained due to inherent difficulties. However, the mass of correlative evidence discussed here suggests that NK cells are important mediators of cancer resistance, reinforcing a need for fine dissection of their biology and highlighting potential therapeutic application.

Acknowledgments Thanks to T. Walzer for advice. E.V. is supported by specific grants from European Union ("ALLOSTEM"), Agence Nationale de la Recherche (ANR MIME, ANR RIB, INCa), Ligue Nationale contre le Cancer ("Equipe labellisée La Ligue") and institutional grants from INSERM, CNRS and Ministère de l'Enseignement Supérieur et de la Recherche. C.A.S. is supported by ALLOSTEM.

References

Abi-Rached L, Parham P (2005) Natural selection drives recurrent formation of activating killer cell immunoglobulin-like receptor and Ly49 from inhibitory homologues. J Exp Med 201:1319–1332

Anderson SK (2006) Transcriptional regulation of NK cell receptors. Curr Top Microbiol Immunol 298:59–75

Andre P, Castriconi R, Espeli M, Anfossi N, Juarez T, Hue S, Conway H, Romagne F, Dondero A, Nanni M, et al (2004) Comparative analysis of human NK cell activation induced by NKG2D and natural cytotoxicity receptors. Eur J Immunol 34:961–971

Andrew DP, Rott LS, Kilshaw PJ, Butcher EC (1996) Distribution of alpha 4 beta 7 and alpha E beta 7 integrins on thymocytes, intestinal epithelial lymphocytes and peripheral lymphocytes. Eur J Immunol 26:897–905

Anfossi N, André P, Guia S, Falk C, Stewart CA, Breso V, Roetynck S, Frassati C, Reviron D, Middleton D, et al (2006) Human NK cell education by inhibitory receptors for MHC class I. Immunity 25:331–342

Arase H, Mocarski ES, Campbell AE, Hill AB, Lanier LL (2002) Direct recognition of cytomegalovirus by activating and inhibitory NK cell receptors. Science 296:1323–1326

Arnon TI, Lev M, Katz G, Chernobrov Y, Porgador A, Mandelboim O (2001) Recognition of viral hemagglutinins by NKp44 but not by NKp30. Eur J Immunol 31:2680–2689

Arnon TI, Achdout H, Lieberman N, Gazit R, Gonen-Gross T, Katz G, Bar-Ilan A, Bloushtain N, Lev M, Joseph A, et al (2004) The mechanisms controlling the recognition of tumor- and virus-infected cells by NKp46. Blood 103:664–672

Arnon TI, Achdout H, Levi O, Markel G, Saleh N, Katz G, Gazit R, Gonen-Gross T, Hanna J, Nahari E, et al (2005) Inhibition of the NKp30 activating receptor by pp65 of human cytomegalovirus. Nat Immunol 6:515–523

Assarsson E, Kambayashi T, Schatzle JD, Cramer SO, von Bonin A, Jensen PE, Ljunggren HG, Chambers BJ (2004) NK cells stimulate proliferation of T and NK cells through 2B4/CD48 interactions. J Immunol 173:174–180

Bacon L, Eagle RA, Meyer M, Easom N, Young NT, Trowsdale J (2004) Two human ULBP/RAET1 molecules with transmembrane regions are ligands for NKG2D. J Immunol 173:1078–1084

Balkwill F, Charles KA, Mantovani A (2005) Smoldering and polarized inflammation in the initiation and promotion of malignant disease. Cancer Cell 7:211–217

Barao I, Hanash AM, Hallett W, Welniak LA, Sun K, Redelman D, Blazar BR, Levy RB, Murphy WJ (2006) Suppression of natural killer cell-mediated bone marrow cell rejection by CD4+CD25+ regulatory T cells. Proc Natl Acad Sci USA 103:5460–5465

Barber DF, Long EO (2003) Coexpression of CD58 or CD48 with intercellular adhesion molecule 1 on target cells enhances adhesion of resting NK cells. J Immunol 170:294–299

Barber DF, Faure M, Long EO (2004) LFA-1 contributes an early signal for NK cell cytotoxicity. J Immunol 173:3653–3659

Bauer S, Groh V, Wu J, Steinle A, Phillips JH, Lanier LL, Spies T (1999) Activation of NK cells and T cells by NKG2D, a receptor for stress-inducible MICA. Science 285:727–729

Beutler B (2004) Inferences, questions and possibilities in Toll-like receptor signaling. Nature 430:257–263

Biassoni R, Pessino A, Malaspina A, Cantoni C, Bottino C, Sivori S, Moretta L, Moretta A (1997) Role of amino acid position 70 in the binding affinity of p50.1 and p58.1 receptors for HLA-Cw4 molecules. Eur J Immunol 27:3095–3099

Blaser C, Kaufmann M, Pircher H (1998) Virus-activated CD8 T cells and lymphokine-activated NK cells express the mast cell function-associated antigen, an inhibitory C-type lectin. J Immunol 161:6451–6454

Bloushtain N, Qimron U, Bar-Ilan A, Hershkovitz O, Gazit R, Fima E, Korc M, Vlodavsky I, Bovin NV, Porgador A (2004) Membrane-associated heparan sulfate proteoglycans are involved in the recognition of cellular targets by NKp30 and NKp46. J Immunol 173:2392–2401

Boles KS, Barchet W, Diacovo T, Cella M, Colonna M (2005) The tumor suppressor TSLC1/NECL-2 triggers NK-cell and CD8+ T-cell responses through the cell-surface receptor CRTAM. Blood 106:779–786

Borrego F, Ulbrecht M, Weiss EH, Coligan JE, Brooks AG (1998) Recognition of human histocompatibility leukocyte antigen (HLA)-E complexed with HLA class I signal sequence-derived peptides by CD94/NKG2 confers protection from natural killer cell-mediated lysis. J Exp Med 187:813–818

Bottino C, Castriconi R, Pende D, Rivera P, Nanni M, Carnemolla B, Cantoni C, Grassi J, Marcenaro S, Reymond N, et al (2003) Identification of PVR (CD155) and Nectin-2 (CD112) as cell surface ligands for the human DNAM-1 (CD226) activating molecule. J Exp Med 198:557–567

Bottino C, Castriconi R, Moretta L, Moretta A (2005) Cellular ligands of activating NK receptors. Trends Immunol 26:221–226

Boyington JC, Sun PD (2002) A structural perspective on MHC class I recognition by killer cell immunoglobulin-like receptors. Mol Immunol 38:1007–1021

Braud V, Jones EY, McMichael M (1997) The human major histocompatibility complex class Ib molecule HLA-E binds signal sequence-derived peptides with primary anchor residues at positions 2 and 9. Eur J Immunol 27:1164–1169

Braud VM, Allan DS, O'Callaghan CA, Soderstrom K, D'Andrea A, Ogg GS, Lazetic S, Young NT, Bell JI, Phillips JH, Lanier LL, McMichael AJ (1998) HLA-E binds to natural killer cell receptors CD94/NKG2A B and C. Nature 391:795–799

Brown MG, Dokun AO, Heusel JW, Smith HR, Beckman DL, Blattenberger EA, Dubbelde CE, Stone LR, Scalzo AA, Yokoyama WM (2001) Vital involvement of a natural killer cell activation receptor in resistance to viral infection. Science 292:934–937

Bryceson YT, March ME, Barber DF, Ljunggren HG, Long EO (2005a) Cytolytic granule polarization and degranulation controlled by different receptors in resting NK cells. J Exp Med 202:1001–1012

Bryceson YT, March ME, Ljunggren HG, Long EO (2005b) Synergy among receptors on resting NK cells for the activation of natural cytotoxicity and cytokine secretion. Blood 107: 159–66

Burshtyn DN, Scharenberg AM, Wagtmann N, Rajagopalan S, Berrada K, Yi T, Kinet J-P, Long EO (1996) Recruitment of tyrosine phosphatase HCP by the killer cell inhibitory receptor. Immunity 4:77–85

Carayannopoulos LN, Yokoyama WM (2004) Recognition of infected cells by natural killer cells. Curr Opin Immunol 16:26–33

Carr WH, Pando MJ, Parham P (2005) KIR3DL1 polymorphisms that affect NK cell inhibition by HLA-Bw4 ligand. J Immunol 175:5222–5229

Carrington M, Martin MP (2006) The impact of variation at the KIR gene cluster on human disease. Curr Top Microbiol Immunol 298:225–257

Castriconi R, Cantoni C, Della Chiesa M, Vitale M, Marcenaro E, Conte R, Biassoni R, Bottino C, Moretta L, Moretta A (2003) Transforming growth factor beta 1 inhibits expression of NKp30 and NKG2D receptors: consequences for the NK-mediated killing of dendritic cells. Proc Natl Acad Sci USA 100:4120–4125

Castriconi R, Dondero A, Corrias MV, Lanino E, Pende D, Moretta L, Bottino C, Moretta A (2004) Natural killer cell-mediated killing of freshly isolated neuroblastoma cells: critical role of DNAX accessory molecule-1-poliovirus receptor interaction. Cancer Res 64:9180–9184

Cavallaro U, Christofori G (2004) Cell adhesion and signalling by cadherins and Ig-CAMs in cancer. Nat Rev Cancer 4:118–132

Cerwenka A, Bakker ABH, McClanahan T, Wagner J, Wu J, Phillips JH, Lanier LL (2000) Retinoic acid early inducible genes define a ligand family for the activating NKG2D receptor in mice. Immunity 12:721–727

Cerwenka A, Baron JL, Lanier LL (2001) Ectopic expression of retinoic acid early inducible-1 gene (RAE-1) permits natural killer cell-mediated rejection of a MHC class I-bearing tumor in vivo. Proc Natl Acad Sci USA 98:11521–11526

Chan CW, Crafton E, Fan HN, Flook J, Yoshimura K, Skarica M, Brockstedt D, Dubensky TW, Stins MF, Lanier LL, et al (2006) Interferon-producing killer dendritic cells provide a link between innate and adaptive immunity. Nat Med 12:207–213

Chang C, Dietrich J, Harpur AG, Lindquist JA, Haude A, Loke YW, King A, Colonna M, Trowsdale J, Wilson MJ (1999) Cutting edge: KAP10, a novel transmembrane adapter protein genetically linked to DAP12 but with unique signaling properties. J Immunol 163:4651–4654

Chapman TL, Heikeman AP, Bjorkman PJ (1999) The inhibitory receptor LIR-1 uses a common binding interaction to recognize class I MHC molecules and the viral homolog UL18. Immunity 11:603–613

Chiesa S, Mingueneau M, Fuseri N, Malissen B, Raulet DH, Malissen M, Vivier E, Tomasello E (2006) Multiplicity and plasticity of natural killer cell signaling pathways. Blood 107: 2364–2372

Colonna M, Navarro F, Bellon T, Liano M, Garcia P, Samaridis J, Angman L, Cella M, Lopez-Botet M (1997) A common inhibitory receptor for major histocompatibility complex class I molecules on human lymphoid and myelomonocytic cells. J Exp Med 186:1809–1818

Cooper MA, Fehniger TA, Caligiuri MA (2001) The biology of human natural killer-cell subsets. Trends Immunol 22:633–640

Cooper MA, Fehniger TA, Fuchs A, Colonna M, Caligiuri MA (2004) NK cell and DC interactions. Trends Immunol 25:47–52

Cosman D, Mullberg J, Sutherland CL, Chin W, Armitage R, Fanslow W, Kubin M, Chalupny NJ (2001) ULBPs, novel MHC class I-related molecules, bind to CMV glycoprotein UL16 and stimulate NK cytotoxicity through the NKG2D receptor. Immunity 14:123–133

Coudert JD, Zimmer J, Tomasello E, Cebecauer M, Colonna M, Vivier E, Held W (2005) Altered NKG2D function in NK cells induced by chronic exposure to NKG2D-ligand expressing tumor cells. Blood 106:1711–1717

Cretney E, Takeda K, Yagita H, Glaccum M, Peschon JJ, Smyth MJ (2002) Increased susceptibility to tumor initiation and metastasis in TNF-related apoptosis-inducing ligand-deficient mice. J Immunol 168:1356–1361

Crocker PR (2005) Siglecs in innate immunity. Curr Opin Pharmacol 5:431–437

Daeron M, Vivier E (1999) Biology of immunoreceptor tyrosine-based inhibition motif-bearing molecules. Curr Top Microbiol Immunol 244:1–12

Dall'Ozzo S, Tartas S, Paintaud G, Cartron G, Colombat P, Bardos P, Watier H, Thibault G (2004) Rituximab-dependent cytotoxicity by natural killer cells: influence of FCGR3A polymorphism on the concentration-effect relationship. Cancer Res 64:4664–4669

Das H, Groh V, Kuijl C, Sugita M, Morita CT, Spies T, Bukowski JF (2001) MICA engagement by human Vgamma2Vdelta2 T cells enhances their antigen-dependent effector function. Immunity 15:83–93

de Haas M, Koene HR, Kleijer M, de Vries E, Simsek S, van Tol MJ, Roos D, von dem Borne AE (1996) A triallelic Fc gamma receptor type IIIA polymorphism influences the binding of human IgG by NK cell Fc gamma RIIIa. J Immunol 156:3948–3955

de Visser KE, Eichten A, Coussens LM (2006) Paradoxical roles of the immune system during cancer development. Nat Rev Cancer 6:24–37

de Vries E, Koene HR, Vossen JM, Gratama JW, von dem Borne AE, Waaijer JL, Haraldsson A, de Haas M, van Tol MJ (1996) Identification of an unusual Fc gamma receptor IIIa (CD16) on natural killer cells in a patient with recurrent infections. Blood 88:3022–3027

Degli-Esposti MA, Smyth MJ (2005) Close encounters of different kinds: dendritic cells and NK cells take centre stage. Nat Rev Immunol 5:112–124

Desrosiers MP, Kielczewska A, Loredo-Osti JC, Adam SG, Makrigiannis AP, Lemieux S, Pham T, Lodoen MB, Morgan K, Lanier LL, Vidal SM (2005) Epistasis between mouse Klra and major histocompatibility complex class I loci is associated with a new mechanism of natural killer cell-mediated innate resistance to cytomegalovirus infection. Nat Genet 37:593–599

Di Santo JP (2006) Natural killer cell developmental pathways: a question of balance. Annu Rev Immunol 24:257–286

Diefenbach A, Jamieson AM, Liu SD, Shastri N, Raulet DH (2000) Ligands for the murine NKG2D receptor: expression by tumor cells and activation of NK cells and macrophages. Nat Immunol 1:119–126

Diefenbach A, Jensen ER, Jamieson AM, Raulet DH (2001) Rae1 and H60 ligands of the NKG2D receptor stimulate tumour immunity. Nature 413:165–171

Diefenbach A, Tomasello E, Lucas M, Jamieson AM, Hsia JK, Vivier E, Raulet DH (2002) Selective associations with signaling proteins determine stimulatory versus costimulatory activity of NKG2D. Nat Immunol 3:1142–1149

Dunn GP, Bruce AT, Ikeda H, Old LJ, Schreiber RD (2002) Cancer immunoediting: from immunosurveillance to tumor escape. Nat Immunol 3:991–998

Dunn GP, Old LJ, Schreiber RD (2004a) The three Es of cancer immunoediting. Annu Rev Immunol 22:329–360

Dunn GP, Old LJ, Schreiber RD (2004b) The immunobiology of cancer immunosurveillance and immunoediting. Immunity 21:137–148

Dunn GP, Bruce AT, Sheehan KC, Shankaran V, Uppaluri R, Bui JD, Diamond MS, Koebel CM, Arthur C, White JM, Schreiber RD (2005) A critical function for type I interferons in cancer immunoediting. Nat Immunol 6:722–729

Ebert LM, Meuter S, Moser B (2006) Homing and function of human skin $\gamma\delta$ T cells and NK cells: relevance for tumor surveillance. J Immunol 176:4331–4336

Falco M, Biassoni R, Bottino C, Vitale M, Sivori S, Augugliaro R, Moretta L, Moretta A (1999) Identification and molecular cloning of p75/AIRM1 a novel member of the sialoadhesin family

that functions as an inhibitory receptor in human natural killer cells. J Exp Med 190:793–802
Farrar MA, Schreiber RD (1993) The molecular cell biology of interferon-gamma and its receptor. Annu Rev Immunol 11:571–611
Ferlazzo G, Munz C (2004) NK cell compartments and their activation by dendritic cells. J Immunol 172:1333–1339
Ferlazzo G, Thomas D, Lin SL, Goodman K, Morandi B, Muller WA, Moretta A, Munz C (2004) The abundant NK cells in human secondary lymphoid tissues require activation to express killer cell Ig-like receptors and become cytolytic. J Immunol 172:1455–1462
Fernandez NC, Treiner E, Vance RE, Jamieson AM, Lemieux S, Raulet DH (2005) A subset of natural killer cells achieves self-tolerance without expressing inhibitory receptors specific for self-MHC molecules. Blood 105:4416–4423
Franksson L, Sundback J, Achour A, Bernlind J, Glas R, Kärre K (1999) Peptide dependency and selectivity of the NK cell inhibitory receptor Ly-49C. Eur J Immunol 29:2748–2758
Fuchs A, Cella M, Giurisato E, Shaw AS, Colonna M (2004) Cutting edge: CD96 (tactile) promotes NK cell-target cell adhesion by interacting with the poliovirus receptor (CD155). J Immunol 172:3994–3998
Fuchs A, Cella M, Kondo T, Colonna M (2005) Paradoxic inhibition of human natural interferon-producing cells by the activating receptor NKp44. Blood 106:2076–2082
Furukawa H, Yabe T, Watanabe K, Miyamoto R, Miki A, Akaza T, Tadokoro K, Tohma S, Inoue T, Yamamoto K, Juji T (1999) Tolerance of NK and LAK activity for HLA class I-deficient targets in a TAP1-deficient patient (bare lymphocyte syndrome type I). Hum Immunol 60:32–40
Gallucci S, Matzinger P (2001) Danger signals: SOS to the immune system. Curr Opin Immunol 13:114–119
Gardiner CM, Guethlein LA, Shilling HG, Pando M, Carr WH, Rajalingam R, Vilches C, Parham P (2001) Different NK cell surface phenotypes defined by the DX9 antibody are due to KIR3DL1 gene polymorphism. J Immunol 166:2992–3001
Gasser S, Orsulic S, Brown EJ, Raulet DH (2005) The DNA damage pathway regulates innate immune system ligands of the NKG2D receptor. Nature 436:1186–1190
Gazit R, Gruda R, Elboim M, Arnon TI, Katz G, Achdout H, Hanna J, Qimron U, Landau G, Greenbaum E, et al (2006) Lethal influenza infection in the absence of the natural killer cell receptor gene Ncr1. Nat Immunol 7:517–523
George TC, Mason LH, Ortaldo JR, Kumar V, Bennett M (1999) Positive recognition of MHC class I molecules by the Ly49D receptor of murine NK cells. J Immunol 162:2035–2043
Ghiringhelli F, Menard C, Terme M, Flament C, Taieb J, Chaput N, Puig PE, Novault S, Escudier B, Vivier E, et al (2005) CD4+CD25+ regulatory T cells inhibit natural killer cell functions in a transforming growth factor-beta-dependent manner. J Exp Med 202:1075–1085
Gilfillan S, Ho EL, Cella M, Yokoyama WM, Colonna M (2002) NKG2D recruits two distinct adapters to trigger NK cell activation and costimulation. Nat Immunol 3:1150–1155
Girardi M, Oppenheim DE, Steele CR, Lewis JM, Glusac E, Filler R, Hobby P, Sutton B, Tigelaar RE, Hayday AC (2001) Regulation of cutaneous malignancy by gammadelta T cells. Science 294:605–609
Gonzalez S, Groh V, Spies T (2006) Immunobiology of human NKG2D and its ligands. Curr Top Microbiol Immunol 298:121–138
Groh V, Rhinehart R, Secrist H, Bauer S, Grabstein KH, Spies T (1999) Broad tumor-associated expression and recognition by tumor-derived gammadelta T cells of MICA and MICB. Proc Natl Acad Sci USA 96:6879–6884
Groh V, Rhinehart R, Randolph-Habecker J, Topp MS, Riddell SR, Spies T (2001) Costimulation of CD8alphabeta T cells by NKG2D via engagement by MIC induced on virus-infected cells. Nat Immunol 2:255–260
Groh V, Wu J, Yee C, Spies T (2002) Tumour-derived soluble MIC ligands impair expression of NKG2D and T-cell activation. Nature 419:734–738

Grundemann C, Bauer M, Schweier O, von Oppen N, Lassing U, Saudan P, Becker KF, Karp K, Hanke T, Bachmann MF, Pircher H (2006) Cutting edge: identification of E-cadherin as a ligand for the murine killer cell lectin-like receptor G1. J Immunol 176:1311–1315

Guthmann MD, Tal M, Pecht I (1995) A secretion inhibitory signal transduction molecule on mast cells is another C-type lectin. Proc Natl Acad Sci USA 92:9397–9401

Hahn WC, Weinberg RA (2002) Modelling the molecular circuitry of cancer. Nat Rev Cancer 2:331–341

Haliotis T, Ball JK, Dexter D, Roder JC (1985) Spontaneous and induced primary oncogenesis in natural killer (NK)-cell-deficient beige mutant mice. Int J Cancer 35:505–513

Hamerman JA, Lanier LL (2006) Inhibition of immune responses by ITAM-bearing receptors. Sci STKE 2006:re1

Hanahan D, Weinberg RA (2000) The hallmarks of cancer. Cell 100:57–70

Hanke T, Corral L, Vance RE, Raulet DH (1998) 2F1 antigen, the mouse homolog of the rat "mast cell function-associated antigen", is a lectin-like type II transmembrane receptor expressed by natural killer cells. Eur J Immunol 28:4409–4417

Hanke T, Takizawa H, McMahon CW, Busch DH, Pamer EG, Miller JD, Altman JD, Liu Y, Cado D, Lemonnier FA, et al (1999) Direct assessment of MHC class I binding by seven Ly49 inhibitory NK cell receptors. Immunity 11:67–77

Hansasuta P, Dong T, Thananchai H, Weekes M, Willberg C, Aldemir H, Rowland-Jones S, Braud VM (2004) Recognition of HLA-A3 and HLA-A11 by KIR3DL2 is peptide-specific. Eur J Immunol 34:1673–1679

Hayakawa Y, Takeda K, Yagita H, Kakuta S, Iwakura Y, Van Kaer L, Saiki I, Okumura K (2001) Critical contribution of IFN-gamma and NK cells but not perforin-mediated cytotoxicity, to anti-metastatic effect of alpha-galactosylceramide. Eur J Immunol 31:1720–1727

Hazenbos WL, Gessner JE, Hofhuis FM, Kuipers H, Meyer D, Heijnen IA, Schmidt RE, Sandor M, Capel PJ, Daeron M, et al (1996) Impaired IgG-dependent anaphylaxis and Arthus reaction in Fc gamma RIII (CD16) deficient mice. Immunity 5:181–188

Helander TS, Timonen T (1998) Adhesion in NK cell function. Curr Top Microbiol Immunol 230:89–99

Herberman RB, Nunn ME, Lavrin DH (1975) Natural cytotoxic reactivity of mouse lymphoid cells against syngeneic acid allogeneic tumors I. Distribution of reactivity and specificity. Int J Cancer 16:216–229

Hollyoake M, Campbell RD, Aguado B (2005) NKp30 (NCR3) is a pseudogene in 12 inbred and wild mouse strains, but an expressed gene in *Mus caroli*. Mol Biol Evol 22:1661–1672

Holmes MA, Li P, Petersdorf EW, Strong RK (2002) Structural studies of allelic diversity of the MHC class I homolog MIC-B, a stress-inducible ligand for the activating immunoreceptor NKG2D. J Immunol 16:1395–1400

Hsu KC, Chida S, Geraghty DE, Dupont B (2002) The killer cell immunoglobulin-like receptor (KIR) genomic region: gene-order, haplotypes and allelic polymorphism. Immunol Rev 190:40–52

Imai K, Matsuyama S, Miyake S, Suga K, Nakachi K (2000) Natural cytotoxic activity of peripheral-blood lymphocytes and cancer incidence: an 11-year follow-up study of a general population. Lancet 356:1795–1799

Ishiyama K, Ohdan H, Ohira M, Mitsuta H, Arihiro K, Asahara T (2006) Difference in cytotoxicity against hepatocellular carcinoma between liver and periphery natural killer cells in humans. Hepatology 43:362–372

Ito A, Handa K, Withers DA, Satoh M, Hakomori S (2001a) Binding specificity of siglec7 to disialogangliosides of renal cell carcinoma: possible role of disialogangliosides in tumor progression. FEBS Lett 504:82–86

Ito A, Handa K, Withers DA, Satoh M, Hakomori S (2001b) Binding specificity of siglec7 to disialogangliosides of renal cell carcinoma: possible role of disialogangliosides in tumor progression. FEBS Lett 498:116–120

Ito M, Maruyama T, Saito N, Koganei S, Yamamoto K, Matsumoto N (2006) Killer cell lectin-like receptor G1 binds three members of the classical cadherin family to inhibit NK cell cytotoxicity. J Exp Med 203:289–295

Jabri B, De Serre NP, Cellier C, Evans K, Gache C, Carvalho C, Mougenot JF, Allez M, Jian R, Desreumaux P, et al (2000) Selective expansion of intraepithelial lymphocytes expressing the HLA-E-specific natural killer receptor CD94 in celiac disease. Gastroenterology 118:867–879

Jakobisiak M, Lasek W, Golab J (2003) Natural mechanisms protecting against cancer. Immunol Lett 90:103–122

Jamieson AM, Diefenbach A, McMahon CW, Xiong N, Carlyle JR, Raulet DH (2002) The role of the NKG2D immunoreceptor in immune cell activation and natural killing. Immunity 17:19–29

Janeway CA, Jr, Medzhitov R (2002) Innate immune recognition. Annu Rev Immunol 20:197–216

Jawahar S, Moody C, Chan M, Finberg R, Geha R, Chatila T (1996) Natural Killer (NK) cell deficiency associated with an epitope-deficient Fc receptor type IIIA (CD16-II). Clin Exp Immunol 103:408–413

Johansson MH, Bieberich C, Jay G, Kärre K, Hoglund P (1997) Natural killer cell tolerance in mice with mosaic expression of major histocompatibility complex class I transgene. J Exp Med 186:353–364

Johansson S, Berg L, Hall H, Hoglund P (2005) NK cells: elusive players in autoimmunity. Trends Immunol 26:613–618

Kaplan DH, Shankaran V, Dighe AS, Stockert E, Aguet M, Old LJ, Schreiber RD (1998) Demonstration of an interferon gamma-dependent tumor surveillance system in immunocompetent mice. Proc Natl Acad Sci USA 95:7556–7561

Kärre K (1997) How to recognize a foreign submarine. Immunol Rev 155:5–9

Kärre K, Ljunggren HG, Piontek G, Kiessling R (1986) Selective rejection of H-2-deficient lymphoma variants suggests alternative immune defence strategy. Nature 319:675–678

Katz G, Gazit R, Arnon TI, Gonen-Gross T, Tarcic G, Markel G, Gruda R, Achdout H, Drize O, Merims S, Mandelboim O (2004) MHC class I-independent recognition of NK-activating receptor KIR2DS4. J Immunol 173:1819–1825

Kayagaki N, Yamaguchi N, Nakayama M, Takeda K, Akiba H, Tsutsui H, Okamura H, Nakanishi K, Okumura K, Yagita H (1999) Expression and function of TNF-related apoptosis-inducing ligand on murine activated NK cells. J Immunol 163:1906–1913

Kelly JM, Darcy PK, Markby JL, Godfrey DI, Takeda K, Yagita H, Smyth MJ (2002a) Induction of tumor-specific T-cell memory by NK cell-mediated tumor rejection. Nat Immunol 3:83–90

Kelly JM, Takeda K, Darcy PK, Yagita H, Smyth MJ (2002b) A role for IFN-gamma in primary and secondary immunity generated by NK cell-sensitive tumor-expressing CD80 in vivo. J Immunol 168:4472–4479

Kiessling R, Klein E, Wigzell H (1975a) "Natural" killer cells in the mouse. I. Cytotoxic cells with specificity for mouse Moloney leukemia cells. Specificity and distribution according to genotype. Eur J Immunol 5:112–117

Kilshaw PJ, Higgins JM (2002) Alpha E: no more rejection? J Exp Med 196:873–875

Kim DD, Song WC (2006) Membrane complement regulatory proteins. Clin Immunol 118:127–136

Kim S, Iizuka K, Aguila HL, Weissman IL, Yokoyama WM (2000) In vivo natural killer cell activities revealed by natural killer cell-deficient mice. Proc Natl Acad Sci USA 97: 2731–2736

Kim S, Poursine-Laurent J, Truscott SM, Lybarger L, Song YJ, Yang L, French AR, Sunwoo JB, Lemieux S, Hansen TH, Yokoyama WM (2005b) Licensing of natural killer cells by host major histocompatibility complex class I molecules. Nature 436:709–713

Kim S, Song YJ, Higuchi DA, Kang HP, Pratt JR, Yang L, Hong CM, Poursine-Laurent J, Iizuka K, French AR, et al (2005a) Arrested natural killer cell development associated with transgene insertion into the Atf2 locus. Blood 107:1024–1030

King A, Loke YW, Chaouat G (1997) NK cells and reproduction. Immunol Today 18:64–66

Koene HR, Kleijer M, Algra J, Roos D, von dem Borne AE, de Haas M (1997) Fc gammaRIIIa-158V/F polymorphism influences the binding of IgG by natural killer cell Fc gammaRIIIa, independently of the Fc gammaRIIIa-48L/R/H phenotype. Blood 90:1109–1114

Kojima H, Kanada H, Shimizu S, Kasama E, Shibuya K, Nakauchi H, Nagasawa T, Shibuya A (2003) CD226 mediates platelet and megakaryocytic cell adhesion to vascular endothelial cells. J Biol Chem 278:36748–36753

Korbel DS, Finney OC, Riley EM (2004) Natural killer cells and innate immunity to protozoan pathogens. Int J Parasitol 34:1517–1528

Kraft JR, Vance RE, Pohl J, Martin AM, Raulet DH, Jensen PE (2000) Analysis of Qa-1(b) peptide binding specificity and the capacity of CD94/NKG2A to discriminate between Qa-1-peptide complexes. J Exp Med 192:613–624

Kriegeskorte AK, Gebhardt FE, Porcellini S, Schiemann M, Stemberger C, Franz TJ, Huster KM, Carayannopoulos LN, Yokoyama WM, Colonna M, et al (2005) NKG2D-independent suppression of T-cell proliferation by H60 and MICA. Proc Natl Acad Sci USA 102:11805–11810

Kumar V, McNerney ME (2005) A new self: MHC-class-I-independent natural-killer-cell self-tolerance. Nat Rev Immunol 5:363–374

Lanier LL (2005) NK cell recognition. Annu Rev Immunol 23:225–274

Lanier LL, Phillips JH, Hackett J Jr, Tutt M, Kumar V (1986) Natural killer cells: Definition of a cell type rather than a function. J Immunol 137:2735–2739

Lanier LL, Corliss B, Phillips JH (1997) Arousal and inhibition of human NK cells. Immunol Rev 155:145–154

Laouar Y, Sutterwala FS, Gorelik L, Flavell RA (2005) Transforming growth factor-beta controls T helper type 1 cell development through regulation of natural killer cell interferon-gamma. Nat Immunol 6:600–607

Lee U, Santa K, Habu S, Nishimura T (1996) Murine asialo GM1+CD8+ T cells as novel interleukin-12-responsive killer T-cell precursors. Jpn J Cancer Res 87:429–432

Lee N, Llano M, Carretero M, Ishitani A, Navarro F, Lopez-Botet M, Geraghty DE (1998) HLA-E is a major ligand for the natural killer inhibitory receptor CD94/NKG2A. Proc Natl Acad Sci USA 95:5199–5204

Lee SH, Girard S, Macina D, Busa M, Zafer A, Belouchi A, Gros P, Vidal SM (2001) Susceptibility to mouse cytomegalovirus is associated with deletion of an activating natural killer cell receptor of the C-type lectin superfamily. Nat Genet 28:42–45

Lee KM, Bhawan S, Majima T, Wei H, Nishimura MI, Yagita H, Kumar V (2003a) Cutting edge: the NK cell receptor 2B4 augments antigen-specific T-cell cytotoxicity through CD48 ligation on neighboring T cells. J Immunol 170:4881–4885

Lee SH, Zafer A, de Repentigny Y, Kothary R, Tremblay ML, Gros P, Duplay P, Webb JR, Vidal SM (2003b) Transgenic expression of the activating natural killer receptor Ly49H confers resistance to cytomegalovirus in genetically susceptible mice. J Exp Med 197:515–526

Lee JC, Lee KM, Kim DW, Heo DS (2004a) Elevated TGF-beta1 secretion and down-modulation of NKG2D underlies impaired NK cytotoxicity in cancer patients. J Immunol 172:7335–7340

Lee KM, McNerney ME, Stepp SE, Mathew PA, Schatzle JD, Bennett M, Kumar V (2004b) 2B4 acts as a non-major histocompatibility complex binding inhibitory receptor on mouse natural killer cells. J Exp Med 199:1245–1254

Lee KM, Forman JP, McNerney ME, Stepp S, Kuppireddi S, Guzior D, Latchman YE, Sayegh MH, Yagita H, Park CK, et al (2006) Requirement of homotypic NK-cell interactions through 2B4(CD244)/CD48 in the generation of NK effector functions. Blood 107:3181–3188

Lehmann J, Huehn J, de la Rosa M, Maszyna F, Kretschmer U, Krenn V, Brunner M, Scheffold A, Hamann A (2002) Expression of the integrin alpha Ebeta 7 identifies unique subsets of CD25+ as well as CD25– regulatory T cells. Proc Natl Acad Sci USA 99:13031–13036
Li P, Willie ST, Bauer S, Morris DL, Spies T, Strong RK (1999) Crystal structure of the MHC class I homolog MIC-A, a gammadelta T-cell ligand. Immunity 10:577–584
Li P, Morris DL, Willcox BE, Steinle A, Spies T, Strong RK (2001) Complex structure of the activating immunoreceptor NKG2D and its MHC class I-like ligand MICA. Nat Immunol 2:443–451
Li P, McDermott G, Strong RK (2002) Crystal structures of RAE-1beta and its complex with the activating immunoreceptor NKG2D. Immunity 16:77–86
Liao NS, Bix M, Zijlstra M, Jaenisch R, Raulet D (1991) MHC class I deficiency: susceptibility to natural killer (NK) cells and impaired NK activity. Science 253:199–202
Lodoen MB, Lanier LL (2005) Viral modulation of NK cell immunity. Nat Rev Microbiol 3:59–69
Long EO (1999) Regulation of immune responses through inhibitory receptors. Annu Rev Immunol 17:875–904
Lopez-Botet M, Carretero M, Bellon T, Perez-Villar JJ, Llano M, Navarro F (1998) The CD94/NKG2 C-type lectin receptor complex. Curr Top Microbiol Immunol 230:41–52
MacFarlane AW IV, Campbell KS (2005) Signal transduction in Natural Killer cells. Curr Top Microbiol Immunol 298:23–57
Mandelboim O, Lieberman N, Lev M, Paul L, Arnon TI, Bushkin Y, Davis DM, Strominger JL, Yewdell JW, Porgador A (2001) Recognition of haemagglutinins on virus-infected cells by NKp46 activates lysis by human NK cells. Nature 409:1055–1060
Martin MP, Gao X, Lee JH, Nelson GW, Detels R, Goedert JJ, Buchbinder S, Hoots K, Vlahov D, Trowsdale J, et al (2002a) Epistatic interaction between KIR3DS1 and HLA-B delays the progression to AIDS. Nat Genet 31:429–434
Martin MP, Nelson G, Lee JH, Pellett F, Gao X, Wade J, Wilson MJ, Trowsdale J, Gladman D, Carrington M (2002b) Cutting edge: susceptibility to psoriatic arthritis: influence of activating killer Ig-like receptor genes in the absence of specific HLA-C alleles. J Immunol 169: 2818–2822
McNerney ME, Kumar V (2006) The CD2 family of natural killer cell receptors. Curr Top Microbiol Immunol 298:91–120
Meresse B, Chen Z, Ciszewski C, Tretiakova M, Bhagat G, Krausz TN, Raulet DH, Lanier LL, Groh V, Spies T, et al (2004) Coordinated induction by IL-15 of a TCR-independent NKG2D signaling pathway converts CTL into lymphokine-activated killer cells in celiac disease. Immunity 21:357–366
Messmer B, Eissmann P, Stark S, Watzl C (2006) CD48 Stimulation by 2B4 (CD244)-Expressing Targets Activates Human NK Cells. J Immunol 176:4646–4650
Middleton D, Williams F, Halfpenny IA (2005) KIR genes. Transpl Immunol 14:135–142
Momot T, Koch S, Hunzelmann N, Krieg T, Ulbricht K, Schmidt RE, Witte T (2004) Association of killer cell immunoglobulin-like receptors with scleroderma. Arthritis Rheum 50:1561–1565
Moretta A, Bottino C, Vitale M, Pende D, Cantoni C, Mingari MC, Biassoni R, Moretta L (2001) Activating receptors and coreceptors involved in human natural killer cell-mediated cytolysis. Annu Rev Immunol 19:197–223
Moretta A, Bottino C, Mingari MC, Biassoni R, Moretta L (2002) What is a natural killer cell? Nat Immunol 3:6–8
Moretta A, Marcenaro E, Sivori S, Della Chiesa M, Vitale M, Moretta L (2005) Early liaisons between cells of the innate immune system in inflamed peripheral tissues. Trends Immunol 26:668–675
Natarajan K, Dimasi N, Wang J, Mariuzza RA, Margulies DH (2002) Structure and function of natural killer cell receptors: multiple molecular solutions to self, nonself discrimination. Annu Rev Immunol 20:853–885

Nelson GW, Martin MP, Gladman D, Wade J, Trowsdale J, Carrington M (2004) Cutting edge: heterozygote advantage in autoimmune disease: hierarchy of protection/susceptibility conferred by HLA and killer Ig-like receptor combinations in psoriatic arthritis. J Immunol 173:4273–4276

Nichols KE, Ma CS, Cannons JL, Schwartzberg PL, Tangye SG (2005) Molecular and cellular pathogenesis of X-linked lymphoproliferative disease. Immunol Rev 203:180–199

Nicoll G, Ni J, Liu D, Klenerman P, Munday J, Dubock S, Mattei MG, Crocker PR (1999) Identification and characterization of a novel siglec, siglec-7, expressed by human natural killer cells and monocytes. J Biol Chem 274:34089–34095

Nicoll G, Avril T, Lock K, Furukawa K, Bovin N, Crocker PR (2003) Ganglioside GD3 expression on target cells can modulate NK cell cytotoxicity via siglec-7-dependent and -independent mechanisms. Eur J Immunol 33:1642–1648

O'Byrne KJ, Dalgleish AG (2001) Chronic immune activation and inflammation as the cause of malignancy. Br J Cancer 85:473–483

Olcese L, Lang P, Vély F, Cambiaggi A, Marguet D, Blery M, Hippen KL, Biassoni R, Moretta A, Moretta L, et al (1996) Human and mouse killer-cell inhibitory receptors recruit PTP1C, PTP1D protein tyrosine phosphatases. J Immunol 156:4531–4534

Oppenheim DE, Roberts SJ, Clarke SL, Filler R, Lewis JM, Tigelaar RE, Girardi M, Hayday AC (2005) Sustained localized expression of ligand for the activating NKG2D receptor impairs natural cytotoxicity in vivo and reduces tumor immunosurveillance. Nat Immunol 6:928–937

Orange JS (2002) Human natural killer cell deficiencies and susceptibility to infection. Microbes Infect 4:1545–1558

Orange JS, Ballas ZK (2006) Natural killer cells in human health and disease. Clin Immunol 118:1–10

Pancer Z, Cooper MD (2006) The evolution of adaptive immunity. Annu Rev Immunol 24:497–518

Parham P (2005) MHC class I molecules and KIRs in human history, health and survival. Nat Rev Immunol 5:201–214

Pende D, Parolini S, Pessino A, Sivori S, Augugliaro R, Morelli L, Marcenaro E, Accame L, Malaspina A, Biassoni R, et al (1999) Identification and molecular characterization of NKp30, a novel triggering receptor involved in natural cytotoxicity mediated by human natural killer cells. J Exp Med 190:1505–1516

Pende D, Spaggiari GM, Marcenaro S, Martini S, Rivera P, Capobianco A, Falco M, Lanino E, Pierri I, Zambello R, et al (2005) Analysis of the receptor-ligand interactions in the natural killer-mediated lysis of freshly isolated myeloid or lymphoblastic leukemias: evidence for the involvement of the Poliovirus receptor (CD155) and Nectin-2 (CD112). Blood 105:2066–2073

Pende D, Castriconi R, Romagnani P, Spaggiari GM, Marcenaro S, Dondero A, Lazzeri E, Lasagni L, Martini S, Rivera P, et al (2006) Expression of the DNAM-1 ligands, Nectin-2 (CD112) and poliovirus receptor (CD155), on dendritic cells: relevance for natural killer-dendritic cell interaction. Blood 107:2030–2036

Perez OD, Mitchell D, Jager GC, Nolan GP (2004) LFA-1 signaling through p44/42 is coupled to perforin degranulation in CD56+CD8+ natural killer cells. Blood 104:1083–1093

Peruzzi M, Parker KC, Long EO, Malnati MS (1996) Peptide sequence requirements for the recognition of HLA-B.2705 specific natural killer cells. J Immunol 157:3350–3356

Plougastel BF, Yokoyama WM (2006) Extending missing-self? Functional interactions between lectin-like NKrp1 receptors on NK cells with lectin-like ligands. Curr Top Microbiol Immunol 298:77–89

Radaev S, Rostro B, Brooks AG, Colonna M, Sun PD (2001) Conformational plasticity revealed by the cocrystal structure of NKG2D and its class I MHC-like ligand ULBP3. Immunity 15:1039–1049

Radosavljevic M, Cuillerier B, Wilson MJ, Clement O, Wicker S, Gilfillan S, Beck S, Trowsdale J, Bahram S (2002) A cluster of ten novel MHC class I related genes on human chromosome 6q24.2-q25.3. Genomics 79:114–123
Rajagopalan S, Long EO (1997) The direct binding of a p58 killer cell inhibitory receptor to human histocompatibility leukocyte antigen (HLA)-Cw4 exhibits peptide selectivity. J ExpMed 185:1523–1528
Rajagopalan S, Long EO (2005) Understanding how combinations of HLA and KIR genes influence disease. J Exp Med 201:1025–1029
Raulet DH (2003) Roles of the NKG2D immunoreceptor and its ligands. Nat Rev Immunol 3:781–790
Raulet DH, Vance RE, McMahon CW (2001) Regulation of the natural killer cell receptor repertoire. Annu Rev Immunol 19:291–330
Ravetch JV, Lanier LL (2000) Immune inhibitory receptors. Science 290:84–89
Robbins SH, Nguyen KB, Takahashi N, Mikayama T, Biron CA, Brossay L (2002) Cutting edge: inhibitory functions of the killer cell lectin-like receptor G1 molecule during the activation of mouse NK cells. J Immunol 168:2585–2589
Robbins SH, Tessmer MS, Mikayama T, Brossay L (2004) Expansion and contraction of the NK cell compartment in response to murine cytomegalovirus infection. J Immunol 173: 259–266
Roberts AI, Lee L, Schwarz E, Groh V, Spies T, Ebert EC, Jabri B (2001) Cutting edge: NKG2D receptors induced by IL-15 costimulate CD28-negative effector CTL in the tissue microenvironment. J Immunol 167:5527–5530
Rosen DB, Araki M, Hamerman JA, Chen T, Yamamura T, Lanier LL (2004) A Structural basis for the association of DAP12 with mouse but not human NKG2D. J Immunol 173:2470–2478
Ruggeri L, Capanni M, Urbani E, Perruccio K, Shlomchik WD, Tosti A, Posati S, Rogaia D, Frassoni F, Aversa F, et al (2002) Effectiveness of donor natural killer cell alloreactivity in mismatched hematopoietic transplants. Science 295:2097–2100
Ruggeri L, Capanni M, Mancusi A, Perruccio K, Burchielli E, Martelli MF, Velardi A (2005) Natural killer cell alloreactivity in haploidentical hematopoietic stem cell transplantation. Int J Hematol 81:13–17
Russell JH, Ley TJ (2002) Lymphocyte-mediated cytotoxicity. Annu Rev Immunol 20:323–370
Sakisaka T, Takai Y (2004) Biology and pathology of nectins and nectin-like molecules. Curr Opin Cell Biol 16:513–521
Salcedo M, Andersson M, Lemieux S, Van Kaer L, Chambers BJ, Ljunggren HG (1998) Fine tuning of natural killer cell specificity and maintenance of self tolerance in MHC class I-deficient mice. Eur J Immunol 28:1315–1321
Salih HR, Antropius H, Gieseke F, Lutz SZ, Kanz L, Rammensee HG, Steinle A (2003) Functional expression and release of ligands for the activating immunoreceptor NKG2D in leukaemia. Blood 102:1389–1396
Sayos J, Nguyen KB, Wu C, Stepp SE, Howie D, Schatzle JD, Kumar V, Biron CA, Terhorst C (2000) Potential pathways for regulation of NK and T-cell responses: differential X-linked lymphoproliferative syndrome gene product SAP interactions with SLAM and 2B4. Int Immunol 12:1749–1757
Schroder K, Hertzog PJ, Ravasi T, Hume DA (2004) Interferon-gamma: an overview of signals mechanisms and functions. J Leukoc Biol 75:163–189
Screpanti V, Wallin RP, Ljunggren HG, Grandien A (2001) A central role for death receptor-mediated apoptosis in the rejection of tumors by NK cells. J Immunol 167:2068–2073
Shankaran V, Ikeda H, Bruce AT, White JM, Swanson PE, Old LJ, Schreiber RD (2001) IFN-gamma and lymphocytes prevent primary tumour development and shape tumour immunogenicity. Nature 410:1107–1111

Shibuya A, Campbell D, Hannum C, Yssel H, Franz-Bacon K, McClanahan T, Kitamura T, Nicholl J, Sutherland GR, Lanier LL, Phillips JH (1996) DNAM-1, a novel adhesion molecule involved in the cytolytic function of T lymphocytes. Immunity 4:573–581

Sivori S, Vitale M, Morelli L, Sanseverino L, Augugliaro R, Bottino C, Moretta L, Moretta A (1997) p46, a novel natural killer cell-specific surface molecule that mediates cell activation. J Exp Med 186:1129–1136

Sivori S, Pende D, Bottino C, Marcenaro E, Pessino A, Biassoni R, Moretta L, Moretta A (1999) NKp46 is the major triggering receptor involved in the natural cytotoxicity of fresh or cultured human NK cells. Correlation between surface density of NKp46 and natural cytotoxicity against autologous, allogeneic or xenogeneic target cells. Eur J Immunol 29:1656–1666

Slifka MK, Pagarigan RR, Whitton JL (2000) NK markers are expressed on a high percentage of virus-specific CD8+ and CD4+ T cells. J Immunol 164:2009–2015

Smith HR, Heusel JW, Mehta IK, Kim S, Dorner BG, Naidenko OV, Iizuka K, Furukawa H, Beckman DL, Pingel JT, et al (2002) Recognition of a virus-encoded ligand by a natural killer cell activation receptor. Proc Natl Acad Sci USA 99:8826–8831

Smyth MJ, Thia KY, Street SE, Cretney E, Trapani JA, Taniguchi M, Kawano T, Pelikan SB, Crowe NY, Godfrey DI (2000a) Differential tumor surveillance by natural killer (NK) and NKT cells. J Exp Med 191:661–668

Smyth MJ, Thia KY, Street SE, MacGregor D, Godfrey DI, Trapani JA (2000b) Perforin-mediated cytotoxicity is critical for surveillance of spontaneous lymphoma. J Exp Med 192:755–760

Smyth MJ, Crowe NY, Godfrey DI (2001a) NK cells and NKT cells collaborate in host protection from methylcholanthrene-induced fibrosarcoma. Int Immunol 13:459–463

Smyth MJ, Godfrey DI, Trapani JA (2001b) A fresh look at tumor immunosurveillance and immunotherapy. Nat Immunol 2:293–299

Smyth MJ, Street SE, Trapani JA (2003a) Cutting edge: granzymes A and B are not essential for perforin-mediated tumor rejection. J Immunol 171:515–518

Smyth MJ, Takeda K, Hayakawa Y, Peschon JJ, van den Brink MR, Yagita H (2003b) Nature's TRAIL—on a path to cancer immunotherapy. Immunity 18:1–6

Smyth MJ, Swann J, Kelly JM, Cretney E, Yokoyama WM, Diefenbach A, Sayers TJ, Hayakawa Y (2004) NKG2D recognition and perforin effector function mediate effective cytokine immunotherapy of cancer. J Exp Med 200:1325–1335

Smyth MJ, Swann J, Cretney E, Zerafa N, Yokoyama WM, Hayakawa Y (2005) NKG2D function protects the host from tumor initiation. J Exp Med 202:583–588

Smyth MJ, Teng MW, Swann J, Kyparissoudis K, Godfrey DI, Hayakawa Y (2006) CD4+CD25+ T regulatory cells suppress NK cell-mediated immunotherapy of cancer. J Immunol 176:1582–1587

Stark S, Watzl C (2006) 2B4 (CD244) NTB-A and CRACC (CS1) stimulate cytotoxicity but no proliferation in human NK cells. Int Immunol 18:241–247

Stewart CA, Laugier-Anfossi F, Vely F, Saulquin X, Riedmuller J, Tisserant A, Gauthier L, Romagne F, Ferracci G, Arosa FA, et al (2005) Recognition of peptide-MHC class I complexes by activating killer immunoglobulin-like receptors. Proc Natl Acad Sci USA 102:13224–13229

Stewart CA, Vivier E, Colonna M (2006) Strategies of natural killer cell recognition and signaling. Curr Top Microbiol Immunol 298:1–21

Street SE, Cretney E, Smyth MJ (2001) Perforin and interferon-gamma activities independently control tumor initiation, growth, and metastasis. Blood 97:192–197

Street SE, Trapani JA, MacGregor D, Smyth MJ (2002) Suppression of lymphoma and epithelial malignancies effected by interferon gamma. J Exp Med 196:129–134

Street SE, Hayakawa Y, Zhan Y, Lew AM, MacGregor D, Jamieson AM, Diefenbach A, Yagita H, Godfrey DI, Smyth MJ (2004) Innate immune surveillance of spontaneous B-cell lymphomas by natural killer cells and {gamma}{delta} T cells. J Exp Med 199:879–884

Sutherland CL, Chalupny NJ, Cosman D (2001) The UL16-binding proteins, a novel family of MHC class I-related ligands for NKG2D, activate natural killer cell functions. Immunol Rev 181:185–192

Suzuki Y, Hamamoto Y, Ogasawara Y, Ishikawa K, Yoshikawa Y, Sasazuki T, Muto M (2004) Genetic polymorphisms of killer cell immunoglobulin-like receptors are associated with susceptibility to psoriasis vulgaris. J Invest Dermatol 122:1133–1136

Tahara-Hanaoka S, Shibuya K, Onoda Y, Zhang H, Yamazaki S, Miyamoto A, Honda S, Lanier LL, Shibuya A (2004) Functional characterization of DNAM-1 (CD226) interaction with its ligands PVR (CD155) and nectin-2 (PRR-2/CD112). Int Immunol 16:533–538

Tahara-Hanaoka S, Shibuya K, Kai H, Miyamoto A, Morikawa Y, Ohkochi N, Honda S, Shibuya A (2006) Tumor rejection by the poliovirus receptor family ligands of the DNAM-1 (CD226) receptor. Blood 107:1491–1496

Taieb J, Chaput N, Menard C, Apetoh L, Ullrich E, Bonmort M, Pequignot M, Casares N, Terme M, Flament C, et al (2006) A novel dendritic cell subset involved in tumor immunosurveillance. Nat Med 12:214–219

Takeda K, Smyth MJ, Cretney E, Hayakawa Y, Kayagaki N, Yagita H, Okumura K (2002) Critical role for tumor necrosis factor-related apoptosis-inducing ligand in immune surveillance against tumor development. J Exp Med 195:161–169

Talmadge JE, Meyers KM, Prieur DJ, Starkey JR (1980) Role of NK cells in tumour growth and metastasis in beige mice. Nature 284:622–624

Tanaka N, Taniguchi T (2000) The interferon regulatory factors and oncogenesis. Semin Cancer Biol 10:73–81

Tay CH, Welsh RM, Brutkiewicz RR (1995) NK cell response to viral infections in beta 2-microglobulin-deficient mice. J Immunol 154:780–789

Teng MW, Kershaw MH, Hayakawa Y, Cerutti L, Jane SM, Darcy PK, Smyth MJ (2005) T cells gene-engineered with DAP12 mediate effector function in an NKG2D-dependent and major histocompatibility complex-independent manner. J Biol Chem 280:38235–38241

Tomasello E, Vivier E (2005) KARAP/DAP12/TYROBP: three names and a multiplicity of biological functions. Eur J Immunol 35:1670–1677

Tomasello E, Blery M, Vély E, Vivier E (2000) Signaling pathways engaged by NK cell receptors: double concerto for activating receptors, inhibitory receptors and NK cells. Semin Immunol 12:139–147

Trinchieri G (1989) Biology of natural killer cells. Adv Immunol 47:187–376

Trowsdale J, Barten R, Haude A, Stewart CA, Beck S, Wilson MJ (2001) The genomic context of natural killer receptor extended gene families. Immunol Rev 181:20–38

Urmacher C, Cordon-Cardo C, Houghton AN (1989) Tissue distribution of GD3 ganglioside detected by mouse monoclonal antibody R24. Am J Dermatopathol 11:577–581

Vales-Gomez M, Reyburn H, Strominger J (2000) Interaction between the human NK receptors and their ligands. Crit Rev Immunol 20:223–244

Vales-Gomez M, Reyburn HT, Erskine RA, Strominger J (1998) Differential binding to HLA-C of p50-activating and p58-inhibitory natural killer cell receptors. Proc Natl Acad Sci USA 95:14326–14331

Vales-Gomez M, Reyburn HT, Erskine RA, Lopez-Botet M, Strominger JL (1999) Kinetics and peptide dependency of the binding of the inhibitory NK receptor CD94/NKG2-A and the activating receptor CD94/NKG2-C to HLA-E. EMBO J 18:4250–4260

Valiante NM, Uhrberg M, Shilling HG, Lienert-Weidenbach K, Arnett KL, D'Andrea A, Phillips JH, Lanier LL, Parham P (1997) Functionally and structurally distinct NK cell receptor repertoires in the peripheral blood of two human donors. Immunity 7:739–751

van den Broek ME, Kagi D, Ossendorp F, Toes R, Vamvakas S, Lutz WK, Melief CJ, Zinkernagel RM, Hengartner H (1996) Decreased tumor surveillance in perforin-deficient mice. J Exp Med 184:1781–1790

van der Slik AR, Koeleman BP, Verduijn W, Bruining GJ, Roep BO, Giphart MJ (2003) KIR in type 1 diabetes: disparate distribution of activating and inhibitory natural killer cell receptors in patients versus HLA-matched control subjects. Diabetes 52:2639–2642

Vance RE, Kraft JR, Altman JD, Jensen PE, Raulet DH (1998) Mouse CD94/NKG2A is a natural killer cell receptor for the nonclassical major histocompatibility complex (MHC) class I molecule Qa-1(b). J Exp Med 188:1841–1848

Veillette A, Latour S (2003) The SLAM family of immune-cell receptors. Curr Opin Immunol 15:277–285

Vely F, Vivier E (2005a) Natural killer cell receptor signaling pathway. Sci STKE 2005:cm6

Vely F, Vivier E (2005b) Natural killer cell receptor signaling pathway in mammals. Sci STKE 2005:cm7

Vély F, Vivier E (1997) Commentary: conservation of structural features reveals the existence of a large family of inhibitory cell surface receptors and non-inhibitory/activatory counterparts. J Immunol 159:2075–2077

Vitale M, Bottino C, Sivori S, Sanseverino L, Castriconi R, Marcenaro E, Augugliaro R, Moretta L, Moretta A (1998) NKp44 a novel triggering surface molecule specifically expressed by activated natural killer cells, is involved in non-major histocompatibility complex-restricted tumor cell lysis. J Exp Med 187:2065–2072

Vitale M, Zimmer J, Castriconi R, Hanau D, Donato L, Bottino C, Moretta L, de la Salle H, Moretta A (2002) Analysis of natural killer cells in TAP2-deficient patients: expression of functional triggering receptors and evidence for the existence of inhibitory receptor(s) that prevent lysis of normal autologous cells. Blood 99:1723–1729

Vivier E, Biron CA (2002) A pathogen receptor on natural killer cells. Science 296:1248–1249

Vivier E, Daëron M (1997) Immunoreceptor tyrosine-based inhibition motifs (ITIMs). Immunol Today 18:286–291

Vivier E, Malissen B (2005) Innate and adaptive immunity: specificities and signaling hierarchies revisited. Nat Immunol 6:17–21

Vivier E, Tomasello E, Paul P (2002) Lymphocyte activation via NKG2D: towards a new paradigm in immune recognition? Curr Opin Immunol 14:306–311

Vivier E, Nunes JA, Vely F (2004) Natural killer cell signaling pathways. Science 306: 1517–1519

Voehringer D, Blaser C, Brawand P, Raulet DH, Hanke T, Pircher H (2001) Viral infections induce abundant numbers of senescent CD8 T cells. J Immunol 167:4838–4843

Voehringer D, Koschella M, Pircher H (2002) Lack of proliferative capacity of human effector and memory T cells expressing killer cell lectinlike receptor G1 (KLRG1). Blood 100:3698–3702

Walzer T, Dalod M, Robbins SH, Zitvogel L, Vivier E (2005) Natural killer cells and dendritic cells: "l'union fait la force". Blood 106:2252–2258

Walzer T, Bléry M, Chaix J, Fuseri N, Chasson C, Robbins SH, Jaeger S, André P, Gauthier L, Daniel L, Chemin K, Morel Y. Dalod M, Imbert J, Pierres M, Moretta A, Romagné F, Vivier E (2007) Identification, activation and selective in vivo ablation of mouse NK cells via NKp46. Proc. Natl Acad Sci 104:3384–3389

Warren HS, Jones AL, Freeman C, Bettadapura J, Parish CR (2005) Evidence that the cellular ligand for the human NK cell activation receptor NKp30 is not a heparan sulfate glycosaminoglycan. J Immunol 175:207–212

Westwood JA, Kelly JM, Tanner JE, Kershaw MH, Smyth MJ, Hayakawa Y (2004) Cutting edge: novel priming of tumor-specific immunity by NKG2D-triggered NK cell-mediated tumor rejection and Th1-independent CD4(+) T-cell pathway. J Immunol 172:757–761

Williams F, Meenagh A, Sleator C, Cook D, Fernandez-Vina M, Bowcock AM, Middleton D (2005) Activating killer cell immunoglobulin-like receptor gene KIR2DS1 is associated with psoriatic arthritis. Hum Immunol 66:836–841

Willimsky G, Blankenstein T (2005) Sporadic immunogenic tumours avoid destruction by inducing T-cell tolerance. Nature 437:141–146

Wilson JL, Charo J, Martin-Fontecha A, Dellabona P, Casorati G, Chambers BJ, Kiessling R, Bejarano MT, Ljunggren HG (1999) NK cell triggering by the human costimulatory molecules CD80 and CD86. J Immunol 163:4207–4212

Wu J, Song Y, Bakker AB, Bauer S, Spies T, Lanier LL, Phillips JH (1999) An activating immunoreceptor complex formed by NKG2D and DAP10. Science 285:730–732

Yamaji T, Teranishi T, Alphey MS, Crocker PR, Hashimoto Y (2002) A small region of the natural killer cell receptor, Siglec-7, is responsible for its preferred binding to alpha 2,8-disialyl and branched alpha 2,6-sialyl residues. A comparison with Siglec-9. J Biol Chem 277:6324–6332

Yawata M, Yawata N, Draghi M, Little AM, Partheniou F, Parham P (2006) Roles for HLA and KIR polymorphisms in natural killer cell repertoire selection and modulation of effector function. J Exp Med 203:633–645

Yen JH, Moore BE, Nakajima T, Scholl D, Schaid DJ, Weyand CM, Goronzy JJ (2001) Major histocompatibility complex class I-recognizing receptors are disease risk genes in rheumatoid arthritis. J Exp Med 193:1159–1167

Yokota Y, Mansouri A, Mori S, Sugawara S, Adachi S, Nishikawa S, Gruss P (1999) Development of peripheral lymphoid organs and natural killer cells depends on the helix-loop-helix inhibitor Id2. Nature 397:702–706

Zappacosta F, Borrego F, Brooks AG, Parker KC, Coligan JE (1997) Peptides isolated from HLA-CW 0304 confer different degrees of protection from natural killer cell-mediated lysis. Proc Natl Acad Sci USA 94:6313–6318

Zhang J, Croy BA, Tian Z (2005) Uterine natural killer cells: their choices, their missions. Cell Mol Immunol 2:123–129

Zilka A, Landau G, Hershkovitz O, Bloushtain N, Bar-Ilan A, Benchetrit F, Fima E, van Kuppevelt TH, Gallagher JT, Elgavish S, Porgador A (2005) Characterization of the heparin/heparan sulfate binding site of the natural cytotoxicity receptor NKp46. Biochemistry 44:14477–14485

Zimmer J, Donato L, Hanau D, Cazenave JP, Tongio MM, Moretta A, de la Salle H (1998) Activity and phenotype of natural killer cells in peptide transporter (TAP)-deficient patients (type I bare lymphocyte syndrome). J Exp Med 187:117–122

Zimmer J, Donato L, Hanau D, Cazenave JP, Moretta A, Tongio MM, de la Salle H (1999) Inefficient protection of human TAP-deficient fibroblasts from autologous NK cell-mediated lysis by cytokines inducing HLA class I expression. Eur J Immunol 29:1286–1291

3
Recent Progress on Paired Immunoglobulin-Like Receptors

Hiromi Kubagawa[1], Ching-Cheng Chen[1], Ikuko Torii[1], Max D. Cooper[2], Kyoko Masuda[3], Yoshimoto Katsura[4], and Hiroshi Kawamoto[3]

3.1 Introduction

Almost 10 years ago the Takai and Kubagawa laboratories independently identified the paired immunoglobulin-like receptors (PIRs) in mice based on limited homology with the human Fcα receptor/CD89 (Hayami et al. 1997; Kubagawa et al. 1997). Two PIR isoforms were identified on the basis of their signaling properties as activating (PIR-A) and inhibitory (PIR-B) types. *Pir* is a multigene family located on the proximal end of mouse chromosome 7 (Kubagawa et al. 1997; Tun et al. 2003; Yamashita et al. 1998a), in a region syntenic with the human chromosome 19q13 where a cluster of structurally related gene families called the leukocyte receptor complex resides. Among these human genes are the closest PIR homologs, the immunoglobulin (Ig)-like transcripts (ILTs) [also called leukocyte Ig-like receptors (LIRs), monocyte/macrophage Ig-like receptors or CD85; see the new LILR nomenclature at www.gene.ucl.ac.uk/nomenclature/genefamily/lilr.html] (Arm et al. 1997; Barten et al. 2001; Colonna et al. 1999; Cosman et al. 1999; Long 1999; Martin et al. 2002; Wagtmann et al. 1997). Paired immunoglobulin-like receptor-A and PIR-B are cell surface glycoproteins with very similar extracellular regions (>92% homology) containing six Ig-like domains, but with structurally and functionally distinct transmembrane and cytoplasmic regions (see Fig. 1). There are multiple PIR-A isoforms (>6), each encoded by a different *Pira* gene. Paired

[1]Division of Developmental and Clinical Immunology, University of Alabama at Birmingham, SHEL 506, 1825 University Blvd. Birmingham, AL 35294-2182, USA

[2]Howard Hughes Medical Institute, Birmingham, AL, USA

[3]Laboratory for Lymphocyte Development, RIKEN Research Center for Allergy and Immunology, Yokohama, Japan

[4]Division of Cell Regeneration and Transplantation, Advanced Medical Research Center, Nihon University School of Medicine, Tokyo, Japan

Correspondence to: H. Kubagawa

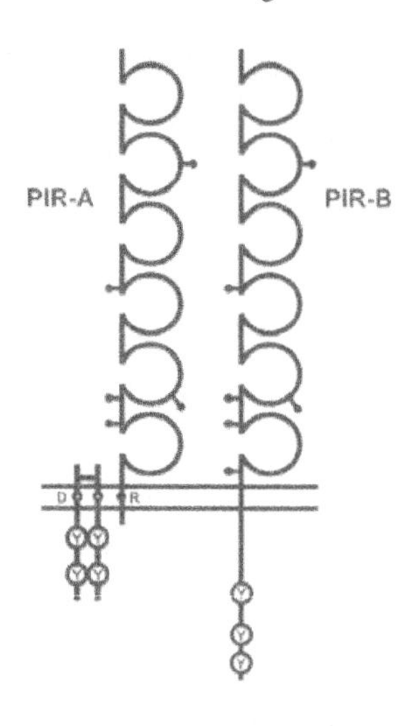

Fig. 1. Schematic presentation of paired immunoglobulin-like receptor (*PIR*)-A and PIR-B. Both PIR-A and PIR-B cDNAs encode type I transmembrane proteins consisting of similar extracellular regions with six Ig-like domains, but having distinctive trans-membrane and cytoplasmic regions. The ectodomain has five or six potential sites for N-linked glycosylation (*bars with closed circles*). The predicted PIR-A has a short cytoplasmic tail and a positively charged arginine (*R*) residue in the transmembrane segment, which is noncovalently associated with a negatively charged aspartic acid (*D*) in the transmembrane domain of the disulfide-linked homodimer of the Fc receptor common γ chain (*FcRγc*) carrying immunoreceptor tyrosine-based activation motifs (*ITAMs*). In contrast, the PIR-B protein has a typical uncharged transmembrane region and a long cytoplasmic tail with immunoreceptor tyrosine-based inhibitory motifs (ITIMs).

immunoglobulin-like receptor-As associate non-covalently with the Fc receptor common γ chain (FcRγc), a transmembrane signal transducer that contains immunoreceptor tyrosine-based *activation* motif (ITAM) "D/ExxYxxL/Ix_{6-8}YxxL/I" (single amino acid code, where x represents any amino acid) in the cytoplasmic tail, to form a cell activation complex (Kubagawa et al. 1999a; Maeda et al. 1998b; Ono et al. 1999; Taylor and McVicar 1999). In contrast, PIR-B is encoded by a single gene and contains three functional immunoreceptor tyrosine-based *inhibitory* motifs (ITIM) "I/VxYxxL/V" in its cytoplasmic tail, thereby negatively regulating cellular activity via the SHP-1 and SHP-2 tyrosine phosphatases (Bléry et al. 1998; Maeda et al. 1998a; Uehara et al. 2001; Yamashita et al. 1998b). It has been suggested recently that PIR-B may also have an additional SH2-binding motif called the immunoreceptor tyrosine-based *switch* motif (ITSM) "TxYxxV/I", as characterized in the signaling lymphocytic activation molecule or CD150 protein family (Siderenko and Clark 2003). Paired immunoglobulin-like receptor-A and PIR-B are expressed by many hematopoietic cell types. These include B lymphocytes, dendritic cells (DCs), monocyte/macrophages, granulocytes, mast cells, and megakaryocyte/platelets. Paired immunoglobulin-like receptors are not expressed by T lymphocytes, natural killer (NK) cells or erythrocytes (Kubagawa et al. 1999a), a feature that distinguishes PIRs from human ILTs/LIRs some of which are expressed by T lymphocytes and NK cells as well (Colonna et al. 1999; Cosman et al. 1999; Long 1999; Arm et al. 1997). The cell surface levels of PIR often increase as a function of cellular differentiation, suggesting that PIR is involved primarily in mature cell function. Our recent findings, however, suggest that PIR is also expressed by early hematopoietic progenitors. In addition, several interesting results regarding PIR-B ligands and functions have recently been reported. Since several review articles describing PIR are now available (Kubagawa et al. 1999b; Takai 2005a,b; Takai and Ono 2001), we will focus this discourse on recent findings.

3.2 PIR Expression by Early Hematopoietic Cells

To determine whether PIR is expressed by early hematopoietic progenitors, we examined progenitor populations that were enriched from adult bone marrow based on the lack of expression of lineage markers (Lin^-) as defined by monoclonal antibodies (mAbs) against Ter119, B220, Mac-1, Gr1, CD3, CD4, CD8, and DX5 antigens and the expression of c-kit and Sca-1 antigens. The c-kit^+/Sca-1^+/Lin^- cells, which represent ~0.05% of the mononuclear bone marrow cells, were found to express PIR proteins at variable levels on their cell surface (Fig. 2). Since mAbs discriminating PIR-A and PIR-B are unavailable, the levels of cell surface PIR-A and PIR-B proteins were then determined by comparing the PIR staining intensity on cells among mice deficient for PIR-B (PIR-$B^{-/-}$) or FcRγc (FcR$\gamma c^{-/-}$) and wild-type control mice. Since FcRγc has been shown to be prerequisite for cell surface expression of most PIR-A family members, FcR$\gamma c^{-/-}$ mice are deficient in cell surface PIR-A. The surface PIR intensity on the c-kit^+/Sca-1^+/Lin^- cells was found to be greatly reduced in PIR-$B^{-/-}$ mice, but not in FcR$\gamma c^{-/-}$ mice when compared with wild-type control mice, suggesting the predominance of PIR-B isoform on those early hematopoietic cells. This data is supported by the results of reverse transcription–polymerase chain reaction (RT-PCR) analysis which demonstrated higher levels of PIR-B transcripts in the c-kit^+/Sca-1^+/Lin^- cells. The predominance of inhibitory activity seems a general rule for other pairs of activating and

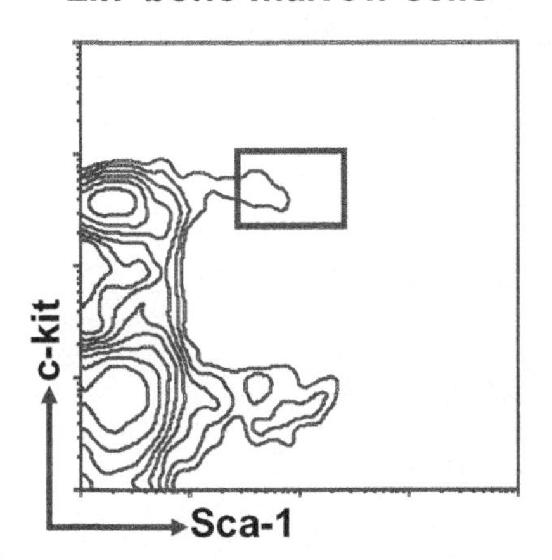

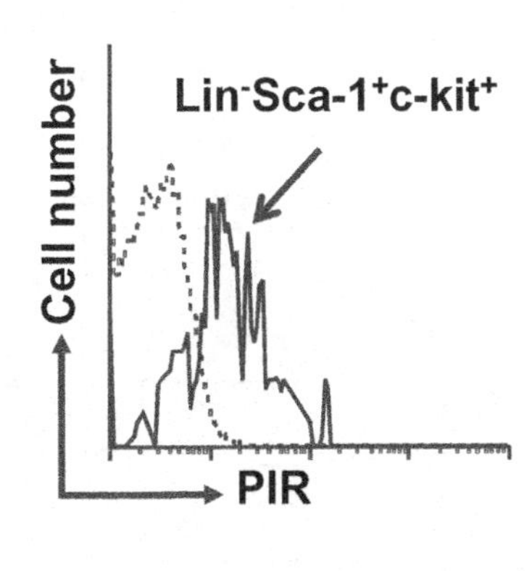

Fig. 2. Identification of PIR^+ progenitors in adult bone marrow. Adult (6 weeks) bone marrow cells were stained with PE-labeled mAbs specific for lineage markers (Lin: Ter119, B220, Mac-1, Gr1, CD4, CD8 and DX5) and anti-PE antibody-coupled magnetic beads before depletion of Lin^+ cells using a magnetic sorter. Enriched Lin^- cells were stained with a combination of FITC-anti-Sca-1, APC-anti-c-kit and PE-anti-PIR mAbs and analyzed by a FACSCalibur. Both c-kit and Sca-1 positive cells (*box*) were examined for PIR expression (*solid lines*). *Dotted lines* indicate the background staining with an isotype-matched control mAb. Note that hematopoietic stem cells express variable levels of PIR on their cell surface

inhibitory receptors (Lanier 2001; Ravetch and Lanier 2000). Like adult bone marrow, a significant fraction of the c-kit$^+$/CD45$^+$/Lin$^-$ hematopoietic progenitors in fetal liver at 11–15 days post coitus (dpc) were also found to express PIR proteins at variable levels (Masuda et al. 2005). A more detailed analysis of the progenitor subpopulations in fetal liver revealed that PIR is expressed by multipotent (c-kit$^+$/Sca-1^{hi}/Lin$^-$) and lymphoid (c-kit$^+$/Sca-1$^+$/IL-7Rα^+/Lin$^-$) progenitors, and not by myeloid (c-kit$^+$/FcγRII/III^{hi}/Lin$^-$) and erythroid (c-kit$^+$/CD45$^-$/Lin$^-$) progenitors (Masuda et al. 2005). Thus, PIR is selectively (or preferentially) expressed by early hematopoietic cells in adult bone marrow and fetal liver tissues.

3.3 PIR Expression by Thymocyte Progenitors

When adult thymocytes subdivided on the basis of CD4 and CD8 expression were examined for PIR expression, a small subpopulation of the CD4 and CD8 double-negative (DN) cells was found to express cell surface PIR at varying levels.

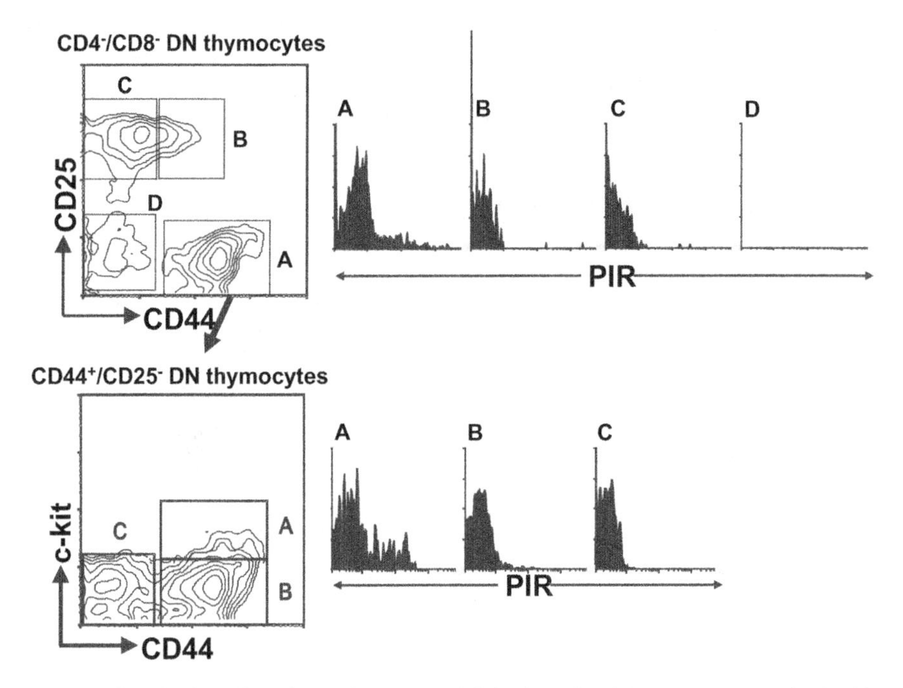

Fig. 3. Identification of PIR$^+$ cells in thymus. Adult (6 weeks) thymocytes were stained with PE-labeled mAbs specific for CD4 and CD8 and anti-PE mAb-coupled magnetic beads before depletion of CD4$^+$ and/or CD8$^+$ cells by a magnetic sorter. Enriched CD4$^-$/CD8$^-$ DN thymocytes were stained with a combination of fluorescein isothiocyanate (FITC)-anti-CD44, APC-anti-c-kit, Cy5-anti-CD25, and PE-anti-PIR mAbs, and analyzed by a FACSCalibur. Cells in each box (in the *left panel*) are examined for PIR expression (*solid profile*) in the *right panel*. Note that small subpopulations of the CD44$^+$/CD25$^-$ cells (Fx A in top panel) and the CD44$^+$/c-kit$^+$ cells (Fx A in bottom panel) express PIR

The PIR$^+$ cells were positive for c-kit and CD44 and negative for CD25 (c-kit$^+$/CD44$^+$/CD25$^-$). Thus, they resided in the most immature thymocyte compartment, designated DN1, which includes newly arrived progenitor cells (Fig. 3). A similar predominance of cell surface PIR-B expression was observed for the neonatal DN1 thymocytes (not shown). Remarkably, the great majority of DN1 cells in fetal thymus at 12 dpc expressed PIR proteins at relatively high levels. The cell surface PIR levels on this subset of thymocytes decreased dramatically beginning around 13 dpc, prior to the expression of CD25, which coincided with the early progression phase of newly arrived T-cell progenitors (Masuda et al. 2005). Paired immunoglobulin-like receptor expression is thus down-regulated within the DN1 stage, a finding that has been confirmed using an in vitro culture system with stromal cells expressing the Notch ligands, wherein the PIR$^+$/c-kit$^+$/CD44$^+$/CD25$^-$ cells give rise to conventional DN1 cells. Immunohistochemical analysis of fetal and neonatal thymi revealed that PIR$^+$ cells are selectively localized in the corticomedullary junction, the site of thymic entry for T-cell precursors. Paired immunoglobulin-like receptor is thus expressed by subpopulations of cells in the most immature compartment of thymocytes. These findings also raise the possibility that down-regulation of the surface PIR-B expression on newly arriving thymocyte progenitors is a prerequisite for their subsequent T-lineage differentiation.

3.4 Differentiation Potential of PIR$^+$ and PIR$^-$ Lymphoid Progenitors

To define the functional potential of PIR$^+$ and PIR$^-$ progenitor populations in hematopoietic tissues, we initially focused on the differentiation capability of PIR$^+$ and PIR$^-$ lymphoid (c-kit$^+$/IL-7R$^+$/Lin$^-$) progenitors in fetal liver by using a high oxygen submersion fetal thymic organ culture for assaying multi-linage progenitors (MLPs) (Masuda et al. 2005). This MLP assay has proven to be an excellent method for allowing a single progenitor cell to generate different progeny over a 10 day culture period with appropriate cytokine stimulation (Katsura 2002; Kawamoto et al. 1997, 2000). To our surprise, the PIR-positive population was greatly enriched for T-lineage progenitors as evidenced by the generation of progeny including CD4 and CD8 double-negative, double-positive, and single-positive $\alpha\beta$ T cells as well as $\gamma\delta$ T cells (Masuda et al. 2005; see Fig. 4). In a modified MLP assay system the PIR$^+$ T-cell progenitors in fetal liver were also capable of giving rise to NK cells and DCs, thereby defining them as common progenitors for T cells, NK cells and DC (T/NK/DC progenitors). The PIR$^+$ T/NK/DC progenitors appear to be selectively released from fetal liver into the circulation during 11–14 dpc, as PIR$^-$/IL-7R$^+$ progenitors are rarely observed in fetal blood. On their entry into the thymus, PIR expression is rapidly down-regulated prior to the onset of CD25 expression.

In contrast to the PIR$^+$ progenitor population, the PIR-negative lymphoid progenitors (PIR$^-$/IL-7R$^+$/Lin$^-$) contained B-lineage progenitors as evidenced by their ability to give rise to B220$^+$/IgM$^-$ pro-B/pre-B-cell as well as IgM$^+$ B-cell progeny (Masuda et al. 2005). Notably, these PIR$^-$ lymphoid progenitors can give rise to B cells within fetal thymic lobes even without adding exogenous cytokines. The

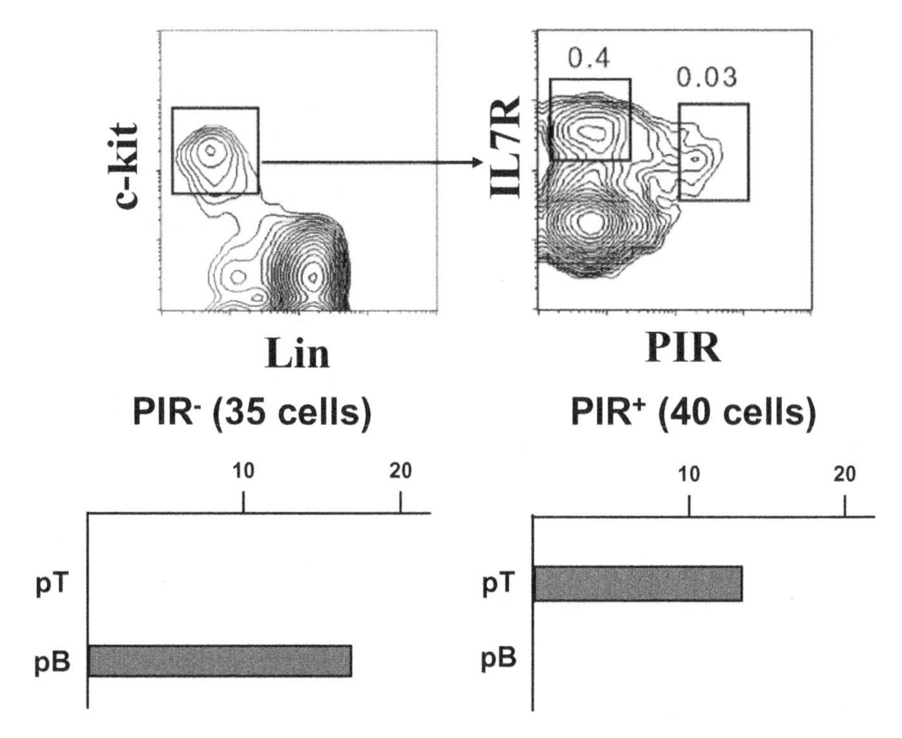

Fig. 4. Differentiation potential of PIR$^+$ and PIR$^-$ lymphoid progenitors. Fetal (15 dpc) liver cells were stained with FITC-mAbs specific for lineage markers (Lin), APC-anti-c-kit, Cy5-anti-IL-7R and PE-anti-PIR mAbs, and the c-kit$^+$/Lin$^-$ cells (box in the left panel) were examined for their expression of IL-7R and PIR (*right panel*). The IL-7R$^+$/PIR$^-$ cells and IL-7R$^+$/PIR$^+$ cells were sorted and a single cell was placed in fetal thymic organ cultures for assaying T-lineage or B-lineage committed progenitors (pT or pB). Note that among 35 isolated PIR$^-$ lymphoid progenitors, 17 are pB and the remainder are untyped because no progeny were generated. In contrast, among 40 PIR$^+$ lymphoid progenitors, 13 are pT and the remainder are untyped

commitment of these cells to the B-lineage differentiation therefore appears to be sustained even under conditions biased toward T-cell generation. The expression profiles of lineage-associated transcription factors correlate well with the differentiation capability of PIR$^+$ and PIR$^-$ lymphoid progenitors. While the expression levels of Ikaros, PU.1, Gata2 and Gata3 are comparable between the PIR$^+$ and PIR$^-$ population, transcription factors specific for T-lineage (Tcf-1) and B-lineage (EBF, Pax5, mb-1/Igα, λ5) are exclusively detected in PIR$^+$ and PIR$^-$ lymphoid progenitors, respectively (Masuda et al. 2005). Taken together, these findings indicate that PIR provides an excellent marker for pre-thymic progenitors with potential to give rise to T cells, NK cells and DCs in fetal liver and blood. We are currently conducting similar analyses for adult bone marrow in order to determine whether or not the PIR$^+$ progenitors in the fetus and adults have the same hematopoietic potential. More importantly, these findings raise the question of PIR function during the development of hematopoietic cells.

3.5 PIR Expression by Dendritic Cell Precursors

Another unique population of cells expressing PIR was found among early hematopoietic cells when adult bone marrow cells were examined. Approximately two-thirds of the $B220^{+}/CD19^{-}/DX5^{-}$ cells, which comprise ~0.5% of the nucleated bone marrow cells, were found to express high levels of PIR. The $PIR^{hi}/B220^{+}/CD19^{-}/DX5^{-}$ cells also expressed CD11c, Ly6C, CD43, and variable levels of CD4 and major histocompatibility complex (MHC) class II (Fig. 5). RT-PCR analysis of the $PIR^{+}/B220^{+}/CD19^{-}$ cells revealed the coordinate expression of PIR-A and PIR-B transcripts as well as FcRγc and early B-lineage gene transcripts (e.g., B29/

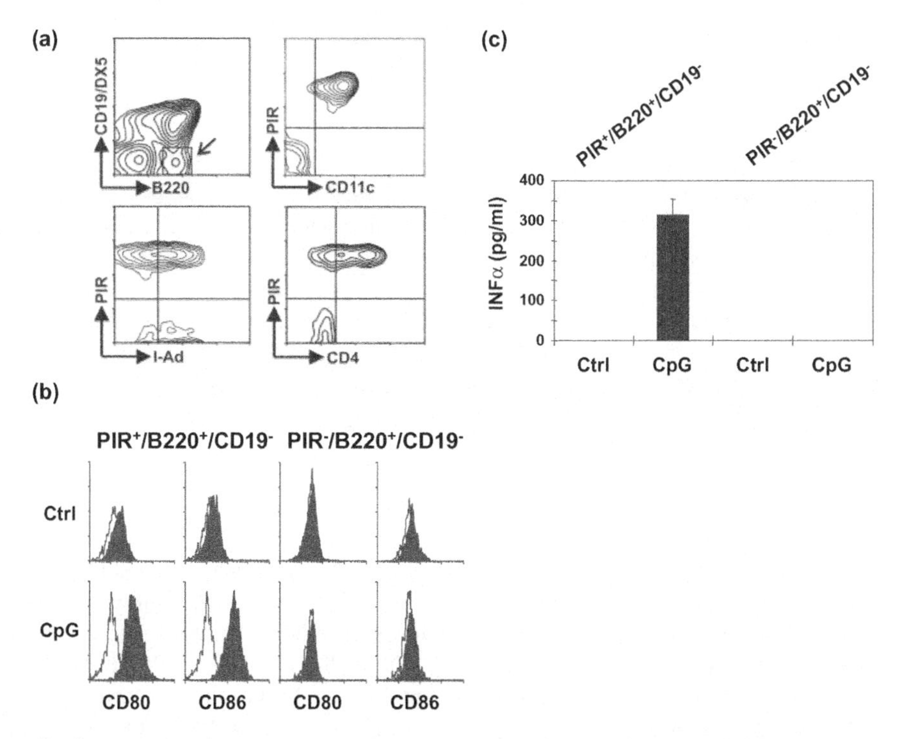

Fig. 5a–c. Paired immunoglobulin-like receptor expression by plasmacytoid DC precursors, pre-DC2. (**a**) Bone marrow nucleated cells were stained with a combination of FITC-anti-CD19, FITC-anti-DX5, Cy-anti-B220, PE-anti-PIR mAbs and APC-labeled mAb specific for CD11c, class II or CD4 before analyzing the $B220^{+}/CD19^{-}/DX5^{-}$ cell population. Note that $B220^{+}/CD19^{-}/PIR^{+}$ cells are positive for CD11c and express variable levels of class II and CD4. (**b**) The $PIR^{+}/B220^{+}/CD19^{-}$ and $PIR^{-}/B220^{+}/CD19^{-}/PIR^{-}$ subpopulations were sorted and cultured for 24 h in the presence or absence of CpG oligodeoxynucleotides. To compensate for DNase-mediated degradation, 10 μl of 100 μM CpG oligo were added at multiple time points during culture. The cultured cells were stained with biotin-labeled mAbs to CD80 or CD86, then with PE-streptavidin before flow cytometric analysis. (**c**) The interferon alpha (*INFα*) levels (pg/ml) in the culture supernatants were determined by an INFα enzyme linked immunosorbent assay

Igβ, germ-line μ, IL-7Rα, bcl-2). Generation of the $PIR^+/B220^+/CD19^-$ cells was unaffected in bone marrow samples from mice deficient for Ig μ heavy chain, recombination activating gene 1, IL-7Rα or common cytokine receptor γ. When the $B220^+/CD19^-$ cells were divided into PIR^+ and PIR^- sub-populations and cultured in the presence of CpG oligodeoxynucleotides to stimulate DC development, the PIR^+ cells were induced to express the co-stimulatory molecules CD80 and CD86, while the PIR^- cells were unresponsive. The PIR^+ population also produced interferon-α (IFNα), whereas the PIR^- population did not. Unlike the $PIR^-/B220^+/CD19^-$ cells, the $PIR^+/B220^+/CD19^-$ cells failed to proliferate in vitro under the B-lineage supportive conditions (Chen CC et al., unpublished). Collectively, these findings suggest that the $PIR^+/B220^+/CD19^-$ cells population is enriched for DC precursors, most likely for pre-DC2 or plasmacytoid DC.

3.6 PIR Ligands

We have postulated that PIR may recognize multiple ligands based upon the sequence variability observed in the extracellular Ig-like domains of the PIR isoforms. While PIR-A and PIR-B have very similar extracellular regions, amino acid sequence diversity in the second amino terminal Ig-like domain of PIR-A is concentrated in regions predicted to be on the loops between β sheets by comparative modeling based on the structure of killer cell Ig-like receptor (KIR/CD158). At least four different patterns of diversity are recognized in such regions of PIR, thereby implying different recognition specificities.

Major histocompatibility complex class I or class I-like molecule specificity of PIRs has been suggested by several findings. Paired immunoglobulin-like receptor-B isolated from splenic B cells and macrophages were found to be constitutively tyrosine-phosphorylated, irrespective of cell activation status (e.g., small resting versus large activated B cells) and to be associated with SHP-1 tyrosine phosphatase (Ho et al. 1999). Recent studies of Pereira and Lowell indicated that PIR-B on resting, non-adherent bone marrow-derived monocytes was negligibly tyrosine-phosphorylated and that its phosphorylation was induced by adherence to plates (Pereira and Lowell 2003; Pereira et al. 2004). In $Lyn^{-/-}$ mice, PIR-B tyrosine phosphorylation was greatly reduced, suggesting that Lyn is a major participant in the constitutive tyrosine phosphorylation of PIR-B (Ho et al. 1999). In contrast to the in vivo findings, tyrosine phosphorylation of PIR-B was not observed in most myeloid and B-cell lines, but could be induced by ligation of the PIR molecules, implying that the constitutive tyrosine phosphorylation of PIR-B is a consequence of interaction with self-ligands. Since some members of the human ILTs/LIRs (the closest relatives of mouse PIR) and KIR/CD158 have been shown to have binding specificity for different MHC class I alleles (Arm et al. 1999; Barten et al. 2001; Colonna et al. 1999; Cosman et al. 1999; Long 1999; Martin et al. 2002; Wagtmann et al. 1997), the PIR-B tyrosine phosphorylation status was examined in mice lacking β2 microglobulin (β2m), the transporter associated with antigen processing

(TAP-1) or MHC class II. The level of PIR-B tyrosine phosphorylation was reduced by ~50% in β2m$^{-/-}$ mice, but was not significantly altered in TAP-1$^{-/-}$ or MHC class II$^{-/-}$ mice, suggesting that PIR-B may recognize an endogenous β2m-associated protein which does not require TAP-1 to traffic to the cell surface, for example MHC class I-like molecules (Ho et al. 1999). In support of this idea, the PIR-B tyrosine phosphorylation status was reduced by ~80% in CD1$^{-/-}$ mice. However, we have not observed an in vitro interaction of recombinant soluble form of PIR-B/Fcγ chimeric protein, which is composed of the extracellular region of PIR-B and the Fc region of human IgG, with CD1 transfected cells (unpublished).

More compelling evidence of an interaction between PIR and MHC class I has come from surface plasmon resonance assays (Nakamura et al. 2004). Recombinant soluble PIR-B protein corresponding with its extracellular region (solPIR-B) was found to bind recombinant MHC class I monomers and tetramers (H-2L^{d}, H-2D^{d}, H-2K^{k}, H-2K^{b}, H-2K^{d}) with an affinity (K_d) of ~300nM and ~10nM, respectively. These affinities are higher than those seen for the ILT/LIR interaction with the classical and non-classical MHC class I molecules (Chapman et al. 1999; Shiroishi et al. 2003). Interestingly, the solPIR-B also bound isolated β2m at a similar affinity range (~50nM), suggesting that the PIR-B specificity is directed toward the β2m on MHC class I molecules (Nakamura et al. 2004). This could account for the apparently broad binding specificity of solPIR-B to MHC class I molecules. Consistent with this finding, incubation of MHC class I tetramers with splenic B cells and macrophages up-regulated the tyrosine phosphorylation status of PIR-B and FcRγc molecules by 10–80%. The binding of MHC class I tetramer to PIR-B expressed on normal splenic B cells and to PIR-A on PIR-B-deficient peritoneal macrophages was verified by confocal microscopic analysis using fluorochrome-labeled MHC class I tetramers and anti-PIR mAb, implying that both native PIR-B and PIR-A cell-surface molecules bind recombinant MHC class I tetramers (Chapman et al. 1999).

In HLA transgenic mouse models, mouse PIR-B has been shown to react with HLA-G, an MHC class I-like molecule expressed on fetal trophoblasts and thymic epithelial cells (Liang et al. 2002). This raises the interesting possibility that PIR-B may recognize the mouse HLA-G homolog, the Qa-2 antigen. Interestingly, the human PIR-B homologues, ILT2/LIR1 and ILT4/LIR2, were also found by surface plasmon resonance assays to bind HLA-G with a 3- to 4-fold higher affinity than classical MHC class I (Shiroishi et al. 2003). In collaborative studies conducted with Dr. Paul Bowness and his colleagues (Medical Research Council, Oxford, UK), the β2m-free heavy chain homodimers of HLA-B27, an allele strongly associated with ankylosing spondylitis (Bird et al. 2002), were found to react with the cell surface of transfectants stably expressing mouse PIR-B and PIR-A4 (Kollnberger et al. 2004). The latter finding was confirmed by an enzyme-linked immunosorbent assay in which the recombinant β2m-free HLA-B27 heavy chain homodimers (B27H_2) bound to the solPIR-B or solPIR-A3-coated wells in a dose dependent fashion, whereas the recombinant HLA-B27 heterodimers (B27/β2m) did not (unpublished). This suggests that unlike the interaction of PIR with mouse MHC class I, the β2m molecule is not required for the binding of PIR-B and

PIR-A3 to HLA-B27. Notably, it has been shown that β2m-free HLA-B27 heavy chains bind ILT4/LIR2-transfected but not ILT2/LIR1-transfected cells (Allen et al. 2001). The fact that ILT2/LIR1 binding is more dependent on β2m than ILT4/LIR2 could be explained by the finding that ILT2/LIR1 recognizes the side of MHC class I molecule with two contact surfaces, the non-polymorphic α3 domain of heavy chain and the β2m, in a 1:1 ILT2/LIR1-MHC class I stoichiometry (Chapman et al. 1999; Shiroishi et al. 2003; Willcox et al. 2003). Collectively, these findings suggest that, like ILT/LIR, PIR can recognize both classical and non-classical MHC class I molecules. It is also quite possible that PIR may recognize additional ligands including pathogen-derived ligands. In this regard, ILT2/LIR1 was originally identified as a receptor for UL18, an MHC class I homolog encoded by human cytomegalovirus (Cosman et al. 1997).

3.7 Paired Immunoglobulin-Like Receptor Function

The initial in vitro analysis of PIR function was conducted by using anti-PIR mAb as a surrogate ligand. While the anti-PIR mAbs do not discriminate PIR-B and PIR-A molecules, B cells and mast cells, unlike myeloid cells, express predominantly PIR-B on their cell surface. Paired immunoglobulin-like receptor co-ligation was found to inhibit B-cell receptor (BCR)-mediated Ca^{2+} mobilization and IgE-mediated mast cell activation (e.g., serotonin release), and this inhibition was observed only when the PIR-B and BCR or FcεRI were brought into physical proximity by ligation with a common secondary reagent (Bléry et al. 1998; Maeda et al. 1998a; Uehara et al. 2001; Yamashita et al. 1998b). It remains to be determined whether both types of receptors, PIR-B and BCR for B cells and PIR-B and FcεRI for mast cells, are brought together following ligation with their respective natural ligands. A different scenario was suggested for c-kit, a prototypic growth factor receptor tyrosine kinase controlling the development and differentiation of hematopoietic progenitor cells and mast cells. Simultaneous co-ligation of PIR-B and c-kit inhibited the c-kit ligand (stem cell growth factor)-induced Ca^{2+} mobilization and proliferative responses of mast cells, but this inhibition did not require bringing PIR-B and c-kit receptors into physical proximity with a secondary reagent (Chen et al. 2002). In the IL-3 dependent BaF/3 pro-B-cell line, IL-3 stimulation led to increased recruitment of SHP-1 tyrosine phosphatase to PIR-B, implying a functional link between PIR-B and cytokine receptor signaling (Wheadon et al. 2002). These findings suggest that the negative regulation of ITIM-bearing receptors like PIR-B may extend to non-ITAM-mediated cell activation, as observed for FcγRIIB and gp49 (Feldweg et al. 2003; Malbec et al. 1999). These findings also suggest that PIR-B may regulate a broad spectrum of cell activation responses (see other examples of leukocyte adhesion and chemotaxis below).

The role of PIR-B in granulocyte and macrophage functions has been studied by Lowell and colleagues (Pereira et al. 2004; Zhang et al. 2005). In their studies, PIR-B deficient granulocytes and macrophages were found to be hyper-responsive to integrin cross-linkage. The PIR-B deficient phagocytes displayed enhanced adhesion, superoxide production and secondary granule release when plated on surfaces coated with integrin ligands. These findings are consistent with the biochemical evidence of enhanced phosphorylation of proteins initiated by integrin signaling in PIR-B deficient cells, thereby suggesting that PIR-B plays an essential role in attenuating the signaling cascade initiated by integrin ligation (Pereira et al. 2004).

PIR-B has also been suggested to be an important regulator of signaling through chemokine receptors, heterotrimeric G-protein-coupled receptors (GPCRs), in granulocytes and DCs (Zhang et al. 2005). As in B cells and macrophages, PIR-B on resting granulocytes was noted to be constitutively tyrosine phosphorylated by myeloid cell-specific Src family tyrosine kinases Fgr and Hck and was associated with SHP-1/SHP-2 phosphatases. This constitutive phosphatase association appeared to restrain chemokine-mediated granulocyte activation, since PIR-B deficient granulocytes and DCs were hyper-responsive to chemokine stimulation as determined by Ca^{2+} mobilization, Erk1/2 activation, actin polymerization and chemotactic responses (Zhang et al. 2005). Moreover, enhanced responses were observed with $Fgr^{-/-}/Hck^{-/-}$ granulocytes and $Fgr^{-/-}$ DCs. Notably, chemokine binding to GPCRs on normal granulocytes and DCs led to transient dephosphorylation of PIR-B (possibly by SHP-2), release of the SHP-1/SHP-2 phosphatases and reversal of tonic inhibition of signaling. This chemokine-induced dephosphorylation appeared to be selective for PIR-B, because another ITIM-bearing protein, signal regulatory phosphatase binding protein-1α, was unaffected. Antibody-mediated cross-linkage of PIR-B (and PIR-A) on normal granulocytes was also shown to lead to transient dephosphorylation of PIR-B (Zhang et al. 2005). Thus, like Lyn in B cells (Chang et al. 1997; Katsuta et al. 1998; Wang et al. 1996), Src family tyrosine kinases Fgr and Hck function as negative regulators in chemokine signaling of granulocytes and DCs by maintaining the tonic phosphorylation of PIR-B. This analysis also raises the question of how chemokine signaling leads to selective PIR-B dephosphorylation and dissociation from SHP-1/SHP-2.

PIR-B-deficient mice have been found to have normal T and B-cell development, except for slightly higher levels of peritoneal B-1 cells (Ujike et al. 2002). As expected, however, PIR-B deficient B cells were hyper-responsive to BCR ligation and to T-cell independent antigens. Surprisingly, $PIR\text{-}B^{-/-}$ mice had significantly augmented IgG1 and IgE responses to T-cell dependent antigens and produced more IL-4 and less IFNγ than wild-type control mice, suggesting an enhanced Th_2 response. The skewed Th_2 response in the $PIR\text{-}B^{-/-}$ mice may be due to the immature phenotype of GM-CSF-induced, bone marrow-derived DCs. Interestingly, PIR-B deficient DCs had several constitutively tyrosine phosphorylated proteins and, upon GM-CSF stimulation, exhibited very transient tyrosine phosphorylation of the GM-CSF receptor common β chain when compared to wild-type control DCs, suggesting the PIR-B involvement in GM-CSF mediated signaling (Ujike et al. 2002). Thus, PIR-B may play an important regulatory role in B-cell responses and DC maturation, and indirectly may balance Th_1/Th_2 immune responses.

Another interesting feature of PIR-$B^{-/-}$ mice is that they exhibit an exaggerated graft-versus-host (GVH) reaction (Nakamura et al. 2004). When sublethally irradiated PIR-$B^{-/-}$ and control mice received allogeneic splenocytes, almost all PIR-$B^{-/-}$ recipients died within 2 weeks, whereas approximately a half of the wild-type recipients survived over a 3-week period. The numbers of IFNγ-producing, donor-derived $CD4^+$ T cells and $CD8^+$ T cells were significantly increased in the PIR-$B^{-/-}$ recipients. Recipient $CD11c^+$ DCs in both types of hosts were activated and expressed high levels of MHC class I, co-stimulatory molecules as well as PIR. However, the DC population in PIR-$B^{-/-}$ recipients was found to contain more IFNγ-producing cells than the DC population in wild-type recipients. This suggests that DCs are hyper-activated in PIR-B-deficient recipients, possibly due to the interaction between PIR-A on PIR-$B^{-/-}$ DCs and allogeneic MHC class I on donor T cells. This abnormal interaction leads to increased production of IFNγ, a critical cytokine in lethal GVH disease, as well as to increased proliferation of donor cytotoxic T cells (Nakamura et al. 2004). Thus, PIR-A and PIR-B serve as counteracting receptors for allogeneic immune regulation.

It is interesting to compare results obtained in humans and rats with those mouse PIR-A/PIR-B studies. $CD8^+$/$CD28^-$ alloantigen-specific suppressor or regulatory T cells were found in humans to induce up-regulation of the ILT4/LIR2 and ILT3/LIR5 inhibitory receptors on antigen-presenting cells (APCs), thereby rendering them tolerogenic (Chang et al. 2002). The tolerogenic APCs displayed reduced expression of co-stimulatory molecules and induced antigen-specific unresponsiveness in $CD4^+$ T helper cells. Notably, the cell surface levels of PIR in mice are also altered during cell activation. For example, IL-4, a multifunctional cytokine produced by Th_2 cells, was shown to reduce the expression of inhibitory receptors, including PIR-B on activated B cells, thereby releasing B cells from inhibitory receptor suppression (Rudge et al. 2001). In a rat heart transplant model, similar regulatory T cells ($CD8^+$/$FOXP3^+$) mediated tolerance to allogeneic heart transplants by inducing up-regulation of PIR-B in allogeneic DCs and heart endothelial cells and by rendering the grafts invulnerable to rejection (Liu et al. 2004). Collectively, these findings suggest that PIR has a regulatory function in the interaction between T cells and APCs, as a crucial priming step for T-cell mediated immune responses.

3.8 Conclusion

While PIR-A and PIR-B are among the earliest identified paired receptors with opposing signaling capabilities, more than 20 such related receptors have now been identified. The pairing of activation and inhibition is thought to be essential modulators for the initiation, amplification and termination of immune responses. Given the wide cellular distribution of PIR (B, monocyte/macrophages, DCs, granulocytes, mast cells, and megakaryocyte/platelets), it has been postulated that PIR-A and PIR-B play specific regulatory roles in host defense, including inflammatory, coagulative, antigen-presenting, allergic, and humoral immune responses. In addition to their expression on these mature cell types, PIR is also expressed by: (i) early hematopoietic progenitors in both fetal liver and adult bone marrow, (ii) subpopulations of the most immature compartment of thymocytes, and (iii) subpopulations of $B220^{+}/CD19^{-}$ cells enriched for DC precursors, especially plasmacytoid DCs. Paired immunoglobulin-like receptor provides an excellent marker for prethymic progenitors in fetal liver and blood with potential to give rise to T cells, NK cells, and DCs. There is now compelling evidence that PIR recognizes classical and non-classical MHC class I molecules, analogous to the human PIR homologs ILT/LIR. Several remarkable functional alterations are observed in PIR-$B^{-/-}$ mice, including (i) hyper-responsiveness of PIR-B-deficient B cells, (ii) enhanced Th_2 response to T-dependent antigens, (iii) exaggerated GVH disease, and (iv) hyper-responsiveness of granulocytes and macrophages to integrin ligation. Paired immunoglobulin-like receptor-B also proves to be an important regulator of signaling through chemokine receptors in granulocytes and DCs. The functional significance of the PIR proteins has been gradually unveiled since identification of the Pir genes in 1997, but we are still far away from seeing the complete picture of PIR biology.

Acknowledgments We thank colleagues and collaborators for sharing their data, Dr. Peter D. Burrows for advice and helpful suggestions, and Ms. E. Ann Brookshire for help in preparing the manuscript. This work was supported in part by NIH grants (AI042127, AI052243).

References

Allen RL, Raine T, Haude A, Trowsdale J, Wilson MJ (2001) Leukocyte receptor complex-encoded immunomodulatory receptors show differing specificity for alternative HLA-B27 structures. J Immunol 167:5543–5547

Arm JP, Nwankwo C, Austen KF (1997) Molecular identification of a novel family of human Ig superfamily members that possess immunoreceptor tyrosine-based inhibition motifs and homology to the mouse gp49B1 inhibitory receptor. J Immunol 159:2342–2349

Barten R, Torkar M, Haude A, Trowsdale J, Wilson MJ (2001) Divergent and convergent evolution of NK-cell receptors. Trends Immunol 22:52–57

Bird LA, Peh CA, Kollnberger S, Sun MY, Elliott T, McMichael A, Bowness P (2003) The spondyloarthrropathy-associated class I allele HLA-B27 forms homodimers both within the ER and at the cell surface. Eur J Immunol 33:748–759

Bléry M, Kubagawa H, Chen CC, Vély F, Cooper MD, Vivier E (1998) The paired Ig-like receptor PIR-B is an inhibitory receptor that recruits the protein-tyrosine phosphatase SHP-1. Proc Natl Acad Sci USA 95:2446–2451

Chan VW, Meng F, Soriano P, DeFranco AL, Lowell CA (1997) Characterization of the B lymphocyte populations in Lyn-deficient mice and the role of Lyn in signal initiation and down-regulation. Immunity 7:69–81

Chang CC, Ciubotariu R, Manavalan JS, Yuan J, Colovai AI, Piazza F, Lederman S, Colonna M, Cortesini R, Dalla-Favera R, Suciu-Foca N (2002) Tolerization of dendritic cells by Ts cells: the critical role of inhibitory receptors ILT3 and ILT4. Nat Immunol 3:237–243

Chapman TL, Heikema AP, Bjorkman PJ (1999) The inhibitory receptor LIR-1 uses a common binding interaction to recognize class I MHC molecules and the viral homolog UL18. Immunity 11:603–613

Chen CC, Kang DW, Cooper MD, Kubagawa H (2002) Mast cell regulation via the paired immunoglobulin-like receptor PIR-B. Immunol Res 26:191–197

Colonna M, Nakajima H, Navarro F, Lopez-Botet M (1999) A novel family of Ig-like receptors for HLA class I molecules that modulate function of lymphoid and myeloid cells. J Leukoc Biol 66:375–381

Cosman D, Fanger N, Borges L, Kubin M, Chin W, Peterson L, Hsu ML (1997) A novel immunoglobulin superfamily receptor for cellular and viral MHC class I molecules. Immunity 7:273–282

Cosman D, Fanger N, Borges L (1999) Human cytomegalovirus, MHC class I and inhibitory signaling receptors: more questions than answers. Immunol Rev 168:177–185

Feldweg AM, Friend DS, Zhou JS, Kanaoka Y, Daheshia M, Li L, Austen F, Katz HR (2003) gp49B1 suppresses stem cell factor-induced mast cell activation-secretion and attendant inflammation in vivo. Eur J Immunol 33:2262–2268

Hayami K, Fukuta D, Nishikawa Y, Yamashita Y, Inui M, Ohyama Y, Hikida M, Ohmori H, Takai T (1997) Molecular cloning of a novel murine cell-surface glycoprotein homologous to killer cell inhibitory receptors. J Biol Chem 272:7320–7327

Ho LH, Uehara T, Chen CC, Kubagawa H, Cooper MD (1999) Constitutive tyrosine phosphorylation of the inhibitory paired immunoglobulin-like receptor PIR-B. Proc Natl Acad Sci USA. 96:15086–15090

Katsura Y (2002) Redefinition of lymphoid progenitors. Nat Rev Immunol 12:127–132

Katsuta H, Tsuji S, Niho Y, Kurosaki T, Kitamura D (1998) Lyn-mediated down-regulation of B-cell antigen receptor signaling: inhibition of protein kinase C activation by Lyn in a kinase-independent fashion. J Immunol 160:1547–1551

Kawamoto H, Ohmura K, Katsura Y (1997) Direct evidence for the commitment of hematopoietic stem cells to T, B, and myeloid lineages in murine fetal liver. Int Immunol 9:1011–1019

Kawamoto H, Ikawa T, Ohmura K, Fujimoto S, Katsura Y (2000) T-cell progenitors emerge earlier than B-cell progenitor in the murine fetal liver. Immunity 12:441–450

Kollnberger S, Bird LA, Roddis M, Hacquard-Bouder C, Kubagawa H, Bodmer HC, Breban M, McMichael AJ, Bowness P (2004) HLA-B27 heavy chain homodimers are expressed in HLA-B27 transgenic rodent models of spondyloarthritis and are ligands for paired Ig-like receptors. J Immunol 173:1699–1710

Kubagawa H, Burrows PD, Cooper MD (1997) A novel pair of immunoglobulin-like receptors expressed by B cells and myeloid cells. Proc Natl Acad Sci USA 94:5261–5266

Kubagawa H, Chen CC, Ho LH, Shimada T, Gartland L, Mashburn C, Uehara T, Ravetch JV, Cooper MD (1999a) Biochemical nature and cellular distribution of the paired immunoglobulin-like receptors, PIR-A and PIR-B. J Exp Med 189:309–317

Kubagawa H, Cooper MD, Chen CC, Ho LH, Alley TL, Hurez V, Tun T, Uehara T, Shimada T, Burrows PD (1999b) Paired immunoglobulin-like receptors of activating and inhibitory types. Curr Top Microbiol Immunol 244:137–149

Lanier LL (2001) Face off—the interplay between activating and inhibitory immune receptors. Curr Opin Immunol 13:326–331

Liang S, Baibakov B, Horuzsko A (2002) HLA-G inhibits the function of murine dendritic cells via the PIR-B immune inhibitory receptor. Eur J Immunol 32:2418–2426

Liu J, Liu Z, Witkowski P, Vlad G, Manavalan JS, Scotto L, Kim-Schulze S, Cortesini S, Hardy MA, Suciu-Foca N (2004) Rat $CD8^+$ $FOXP3^+$ T suppressor cells mediate tolerance to allogeneic heart transplants, including PIR-B in APC and rendering the graft invulnerable to rejection. Transplant Immunol 13:239–247

Long EO (1999) Regulation of immune responses through inhibitory receptors. Annu Rev Immunol 17:875–904

Maeda A, Kurosaki M, Ono M, Takai T, Kurosaki T (1998a) Requirement of SH2-containing protein tyrosine phosphatases SHP-1 and SHP-2 for paired immunogloblin-like receptor B (PIR-B)-mediated inhibitory signal. J Exp Med 187:1355–1360

Maeda A, Kurosaki M, Kurosaki T (1998b) Paired immunoglobulin-like receptor (PIR)-A is involved in activating mast cells through its association with Fc receptor γ chain. J Exp Med 188:991–995

Malbec O, Fridman WH, Daëron M (1999) Negative regulation of c-kit-mediated cell proliferation by FcγRIIB. J Immunol 162:4424–4429

Martin AM, Kulski JK, Witt C, Pontarotti P, Christiansen FT (2002) Leukocyte Ig-like receptor complex (LRC) in mice and men. Trends Immunol 23:81–88

Masuda K, Kubagawa H, Ikawa T, Chen CC, Kakugawa K, Hattori M, Kageyama R, Cooper MD, Minato N, Katsura Y, Kawamoto H (2005) Prethymic T-cell development defined by the expression of paired immunoglobulin-like receptors. EMBO J 24:4052–4060

Nakamura A, Kobayashi E, Takai T (2004) Exacerbated graft-versus-host disease in $Pirb^{-/-}$ mice. Nat Immunol 5:623–629

Ono M, Yuasa T, Ra C, Takai T (1999) Stimulatory function of paired immunoglobulin-like receptor-A in mast cell line by associating with subunits common to Fc receptors. J Biol Chem 274:30288–30296

Pereira S, Lowell C (2003) The Lyn tyrosine kinase negatively regulates neutrophil integrin signaling. J Immunol 171:1319–1327

Pereira S, Zhang H, Takai T, Lowell CA (2004) The inhibitory receptor PIR-B negative regulates neutrophil and macrophage integrin signaling. J Immunol 173:5757–5765

Ravetch JV, Lanier LL (2000) Immune inhibitory receptors. Science 290:84–89

Rudge EU, Cutler AJ, Pritchard NR, Smith KGC (2001) Interleukin 4 reduces expression of inhibitory receptors on B cells and abolishes CD22 and FcγRII-mediated B-cell suppression. J Exp Med 195:1079–1085

Shiroishi M, Tsumoto K, Amano K, Shirakihara Y, Colonna M, Braud VM, Allan DSJ, Makadzange A, Rowland-Jones S, Willcox B, Jones EY, van der Merwe PA, Kumagai I, Maenaka K (2003) Human inhibitory receptors Ig-like transcript 2 (ILT2) and ILT4 compete with CD8 for MHC class I binding and bind preferentially to HLA-G. Proc Natl Acad Sci USA 100:8856–8861

Sidorenko SP, Clark EA (2003) The dual-function CD150 receptor subfamily: the viral attraction. Nat Immunol 4:19–24

Takai T, Ono M (2001) Activating and inhibitory nature of the murine paired immunoglobulin-like receptor family. Immunol Rev 181:215–222

Takai T (2005a) Paired immunoglobulin-like receptors and their MHC class I recognition. Immunology 115:433–440

Takai T (2005b) A novel recognition system for MHC class I molecules contributed by PIR. Adv Immunol 88:161–192

Taylor LS, McVicar DW (1999) Functional association of FcεRIγ with arginine[632] of paired immunoglobulin-like receptor (PIR)-A3 in murine macrophages. Blood 94:1790–1796

Tun T, Kubagawa Y, Dennis G, Burrows PD, Cooper MD, Kubagawa H (2003) Genomic structure of mouse PIR-A6, an activating member of the paired immunoglobulin-like receptor gene family. Tissue Antigen 61:220–230

Uehara T, Bléry M, Kang DW, Chen CC, Ho LH, Gartland GL, Liu FT, Vivier E, Cooper MD, Kubagawa H (2001) Inhibition of IgE-mediated mast cell activation by the paired Ig-like receptor PIR-B. J Clin Invest 108:1041–1050

Ujike A, Takeda K, Nakamura A, Ebihara S, Akiyama K, Takai T (2002) Impaired dendritic cell maturation and increased TH2 responses in $PIR\text{-}B^{-/-}$ mice. Nat Immunol 3:542–548

Wagtmann N, Rojo S, Eichler E, Mohrenweiser H, Long EO (1997) A new human gene complex encoding the killer cell inhibitory receptors and related monocyte/macrophage receptors. Curr Biol 7:615–618

Yamashita Y, Fukuta D, Tsuji A, Nagabukuro A, Matsuda Y, Nishikawa Y, Ohyama Y, Ohmori H, Ono M, Takai T (1998a) Genomic structure and chromosomal location of p91, a novel murine regulatory receptor family. J Biochem (Tokyo) 123:358–368

Yamashita Y, Ono M, Takai T (1998b) Inhibitory and stimulatory function of paired Ig-like receptor (PIR) family in RBL-2H3 cells. J Immunol 161:4042–4047
Wang J, Koizumi T, Watanabe T (1996) Altered antigen receptor signaling and impaired Fas-mediated apoptosis of B cells in Lyn-deficient mice. J Exp Med 184:831–838
Wheadon H, Paling NRD, Welham MJ (2002) Molecular interactions of SHP1 and SHP2 in IL-3-signaling. Cell Signal 14:219–229
Willcox BE, Thomas LM, Bjorkman PJ (2003) Crystal structure of HLA-A2 bound to LIR-1, a host and viral major histocompatibility complex receptor. Nat Immunol 4:913–919
Zhang H, Meng F, Chu CC, Takai T, Lowell CA (2005) The Src family kinases Hck and Fgr negatively regulate neutrophil and dendritic cell chemokine signaling via PIR-B. Immunity 22:235–246

4
Self–nonself Recognition through B-Cell Antigen Receptor

Daisuke Kitamura

4.1 Introduction

Receptors on innate immune cells that recognize pathogens, such as Toll-like receptors, have been diversified and selected through evolution. In contrast, antigen receptors on B and T lymphocytes diversify enormously in each individual: genes for these receptors composed of multiple various parts are assembled randomly through gene recombination during development of these cells throughout the life of the individuals. While T-cell antigen receptor (TCR) recognizes antigenic peptides of proteins presented on the self MHC, B-cell antigen receptor (BCR) recognizes essentially any kinds of molecules. It is miraculous that only a few genes generate a repertoire of receptors that match essentially all of the molecular structures existing on earth including even newly generated chemical compounds, the mechanisms for which have mostly been clarified to date as briefly mentioned below. It is also amazing that each of the enormous number of receptors recognizes almost single specific structure. Another surprise is that such diverse receptors in an individual do not react with any structures contained in the self body (self-antigens) in principle, which is called self-tolerance.

B-cell antigen receptor and antibody, membrane-bound and secretary forms of Immunoglobulin (Ig), respectively, are produced from the same Ig genes through alternative splicing. The Ig genes are diversified during B-cell development through V(D)J recombination process in which D-J and then V-DJ recombination of heavy (H) chain gene locus successively take place in pro-B cells, and then V-J recombination of light (L) chain (κ or λ) genes in pre-B cells. Combination of multiple V, D, and J gene segments, and junctional nucleotide variations, makes enormous diversity among the assembled receptors. After successful recombination the of IgH gene, μH chain product is assembled with invariant surrogate light chains,

Department of Medicinal and Life Science, Faculty of Pharmaceutical Sciences, Division of Molecular Biology, Research Institute for Biological Sciences, Tokyo University of Science, Yamazaki 2669, Noda, Chiba 278-0022, Japan

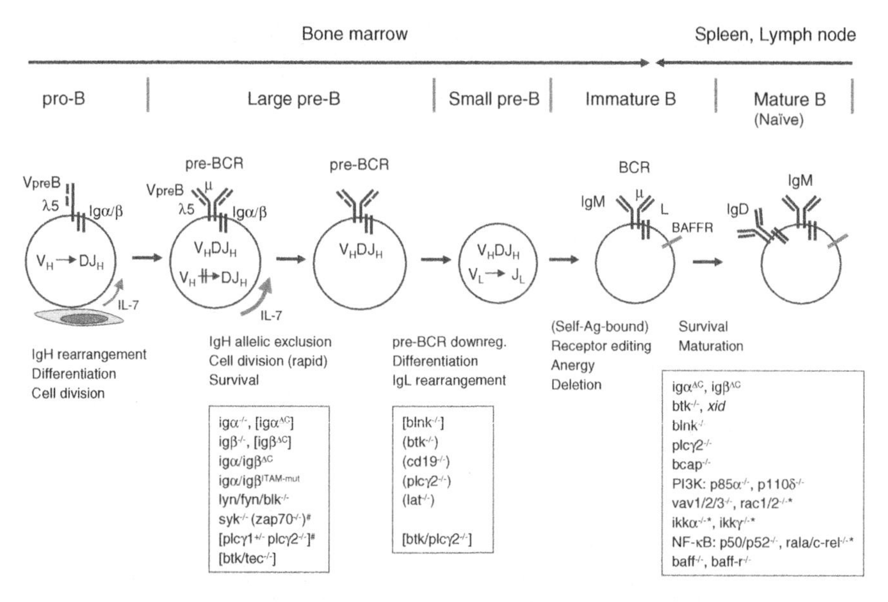

Fig. 1. Primary B-cell development regulated by pre-B/B-cell receptor signaling. Cellular events induced by pre-B-cell receptor (*BCR*)/BCR signals during the B-cell development are denoted below the symbolized cells of representative developmental stages. In *boxes* are listed the genotypes of the knockout mice showing defects (those with partial defects are in the *square brackets*) in the respective events. Gene symbols aligned with *slash* in between indicate double or triple gene knockout mice. In the *parentheses* are indicated the genotypes that significantly affect the phenotype only in combination with the adjoined genotype. *Hash mark*, defect of *IgH* allelic exclusion has been proven; *asterisk*, B-cell-specific conditional gene knockout mice or chimeric mice with the mutated hematopoietic cells have been analyzed

VpreB and λ5, forms a pre-B-cell receptor complex (pre-BCR), and is expressed on the surface of the cells (Fig. 1). Signals through pre-BCR, synergistically with interleukin-7, induce proliferation of what are then called large pre-B cells, followed by differentiation into the next stage called small pre-B cells where L chain gene rearrangements take place. The pre-BCR signal also prohibits rearrangement of another *IgH* allele, thus contributing to H chain allelic exclusion, the phenomenon that a single B-cell expresses H chain from only one allele of the two. The principle that one lymphocyte holds one receptor specificity is the basis of the clonal selection theory of the lymphoid system [reviewed by Rajewsky (1996) and Karasuyama et al. (1996)].

After successful V-J recombination of L chain gene, L chain products and μH chain are combined into IgM that is the first BCR class to emerge on B cells. The BCR signals the decision of the fate of each newly generated B cell. Depending on the affinity and local density of the self antigen bound to BCR, self-reactive B cells are inactivated (anergy) or eliminated (deletion) from the primary repertoire, or the self-reactive BCR is replaced with a non-harmful one (receptor editing), during the immature B-cell stage. This negative selection of self-reactivity of B cells is the basis for the self–nonself recognition (that is, tolerance for self) in the B-cell system and for avoidance of autoimmunity. After the negative selection, the remaining B cells emigrate into the peripheral lymphoid organs and further differentiate into so called mature B cells. The maturation and survival of B cells require antigen-independent signals from BCR. Upon encountering antigen through BCR and with T-cell help the mature B cells proliferate and undergo class switch recombination of H chain gene loci, then some of them differentiate into plasma cells that produce antibodies and others form germinal centers in which somatic hypermutation in the V region of H and L chains takes place. From the germinal centers develop memory B cells and long-lived plasma cells equipped with IgG or other classes of Ig containing the affinity-selected mutations in V regions. The random mutation of Ig genes should occasionally generate BCR that binds to self-antigen. However such self-binding B cells do not normally differentiate into memory or plasma cells, but are negatively selected by the mechanism that is largely unknown.

Both pre-BCR and BCR are bound with Igα/Igβ transmembrane proteins that serve as an intracellular signal transduction subunit. Despite the invariability of this subunit, signals from these receptors induce diverse cellular responses depending on the developmental stages of the cells, or possibly on the nature of the antigen binding. As mentioned above, pre-BCR, not by binding to antigen but probably by self-ligation (Ohnishi and Melchers 2003), regulates V(D)J recombination, cell proliferation, and differentiation. Antigen-bound BCR on newly generated B cells signals anergy, deletion, or receptor editing, whereas that on mature B cells signals proliferation, positive or negative selection, and differentiation into memory or plasma cells. The fact that the same Igα/Igβ signaling subunit induces such a diverse outcome suggests the presence of diversification and integration of the signaling pathways that eventually determine the cell response. This chapter gives an overview of our current knowledge on the molecular pathways of the pre-BCR and BCR signal transduction to understand the diversification and integration of signaling network, and to discuss from the signaling point of view on how B cells recognize self and nonself antigens.

4.2 Signal Transduction from BCR

4.2.1 Overview

B-cell antigen receptor signaling is primarily transduced from Igα/Igβ heterodimer, invariant membrane proteins non-covalently associated with H chain of IgM or other Ig classes. Igα and Igβ contain the immunoreceptor tyrosine-based activation motif (ITAM) in their cytoplasmic regions that is crucial for the signal transduction (Flaswinkel et al. 1995). Both Igα and Igβ are necessary for pre-BCR surface expression and thus for pro-B to pre-B-cell transition during the early B-cell development (Gong and Nussenzweig 1996; Pelanda et al. 2002). In mice with a targeted

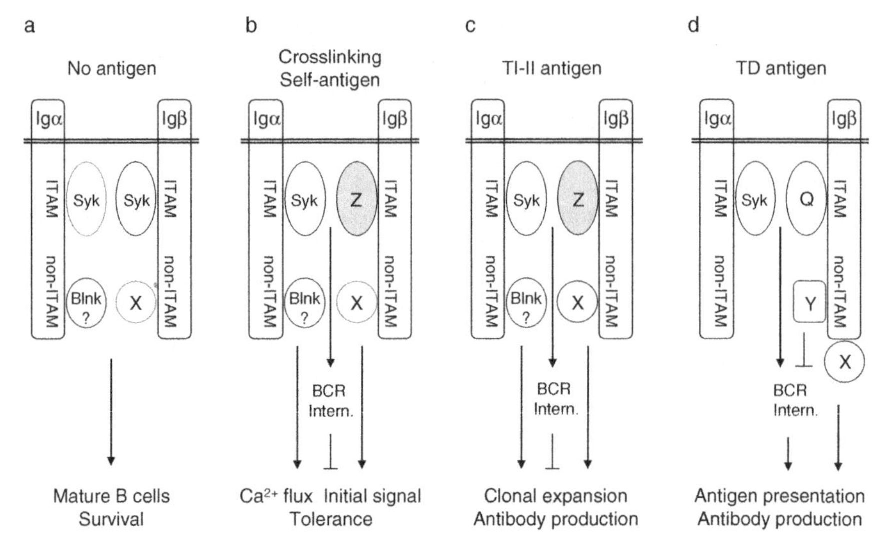

Fig. 2a–d. Differential function of ITAM and non-ITAM sequences in Igα and Igβ. *Each panel* indicates Igα and Igβ associated with one BCR (not shown). At least one ITAM of either Igα or Igβ likely functions through binding with Syk. There may be other ITAM-binding proteins (Z or Q). Although the essential non-ITAM sequences have not been identified, a BLNK-binding site in Igα may be the one. Putative Igβ non-ITAM-binding proteins are indicated as *X* or *Y*). (**a**) At least one ITAM of Igα or Igβ and non-ITAM part on the other (or both non-ITAM parts) are necessary for B-cell maturation and survival. (**b**) One ITAM mutation in either Igα or Igβ makes BCR-signal transduction stronger and induces more exaggerated tolerance, suggesting the presence of an inhibitory ITAM-binding protein (*Z*) and/or signal attenuation due to BCR-internalization promoted by the two ITAMs. ITAM and non-ITAM part in the same Igα (or Igβ) are enough for the initial signal and tolerance transduction. (**c**) One ITAM (particularly of Igβ) inhibits TI–II immune response as in (**b**) (Gazumyan et al. 2006). But non-ITAM parts of both Igα and Igβ are necessary for this response. (**d**) Both ITAMs contribute positively to BCR internalization and thus TD immune response. Non-ITAM part of Igβ may recruit a factor inhibiting BCR internalization (*Y*) as well as a factor activating B cells (*X*)

mutation that deletes a cytoplasmic domain of either Igα or Igβ (ΔC), this early development narrowly proceeds, but maintenance of peripheral B cells is impaired (Reichlin et al. 2001; Torres et al. 1996). Indeed, conditional deletion of Igα, of either whole molecule or a cytoplasmic domain, results in rapid loss of mature B cells (Kraus et al. 2004). In contrast, mutations of two tyrosine residues in an ITAM of either Igα or Igβ do not affect follicular B-cell development, suggesting the functional redundancy of the two ITAMs and a role for non-ITAM residues in B-cell maturation (Fig. 2a) However, at least one of these cytoplasmic domains and ITAMs therein is required for the pre-BCR-signaled B-cell development (Gazumyan et al. 2006; Kraus et al. 2001).

Two tyrosines in the ITAM are phosphorylated by cytoplasmic protein-tyrosine kinases (PTKs) of Src-family kinase (SFK) such as Lyn, Fyn, and Blk, and become docking sites for the two Src-homology (SH)2 domains of another PTK Syk. Recent fluorescence resonance energy transfer analysis of living cells demonstrated that ITAM phosphorylation by SFK induces reversible conformational changes of cytoplasmic domains of IgH, Igα and Igβ chains, which makes them open to Syk access (Tolar et al. 2005). Src-family kinase is potentiated by dephosphorylation of its C-terminal tyrosine, which is mediated by the transmembrane protein-tyrosine phosphatase CD45. Thus B cells from CD45-null mutant mice do not proliferate in response to BCR-ligation, although B-cell development is normal in these mice (Byth et al. 1996; Kishihara et al. 1993). Despite the central role of Lyn among the SFK for the initiator of BCR signaling, early B-cell development is unimpaired in Lyn-deficient mice, as in Fyn- or Blk-single knockout mice. However, Lyn/Fyn/Blk-triple knockout mice exhibit a developmental arrest at pro-B-cell stage, indicating the redundant function of the three kinases (Saijo et al. 2003). In the Lyn-single knockout mice mature B cells are moderately reduced in number but spontaneously activated and hyper-responsive to BCR stimulation in terms of Ca^{2+} flux, proliferation and antibody production, and the mice eventually develop autoimmune disease reminiscent of systemic lupus erythematosus (SLE). This is ascribed to a negative signaling role of Lyn: phosphorylation of immunoreceptor tyrosine-based inhibitory motifs (ITIM) in the cytoplasmic domains of inhibitory co-receptors such as CD22, CD72, and FcγRIIB, to which a protein-tyrosine phosphatase SHP-1 or an inositide phosphatase SHIP is recruited (see Chapter 7).

Upon binding to the phospho-ITAM, Syk becomes active possibly through its conformational change and auto-phosphorylation. Like Igα/Igβ, Syk is crucial for pre-BCR signaling because its ablation in mice results in a severe arrest of B-cell development at the pro-B-cell stage (Cheng et al. 1995; Turner et al. 1995), and a failure of H chain allelic exclusion (Schweighoffer et al. 2003). Yet another type of PTK, Btk, is also activated by Syk. The Btk gene is responsible for human X-linked agammaglobulinemia (XLA, also called Bruton's disease) and *xid* mice (spontaneous mutant mouse strain with B-cell deficiency), and Btk is essential for early B-cell development in humans. In mice, Btk is critical for B-cell maturation and activation: *xid* mice as well as Btk-knockout mice show the phenotype (referred to as *xid* phenotype hereafter) in which the number of mature B and peritoneal B1 cells are reduced, B-cell survival is shortened, serum IgM and IgG3 titer is low, BCR-mediated proliferation is abolished, and immune responses to thymus-independent type II (TI–II) antigens and primary, but not secondary, responses to thymus-dependent (TD) antigens are impaired (Khan et al. 1995; Kerner et al. 1995; Ridderstad et al. 1996). Activated PTKs phosphorylate and regulate the enzymatic activities of various signaling intermediates including phospholipase Cγ2 (PLCγ2), phosphoinositide 3-kinase (PI3K), and Vav, which in turn transmit the signals into distinct pathways, in part, leading to the activation of nuclear transcription factors such as AP-1, NFAT, and NF-κB (Kurosaki 2002; Reth and Wienands 1997). A cytoplasmic adaptor protein BLNK is phosphorylated by PTKs, recruits, and activates many of the signaling intermediates (Fig. 3).

CD19 is the coreceptor physically associated with BCR and also with complement receptor (CR) 2 and CD81, though it is unclear if these two complexes are

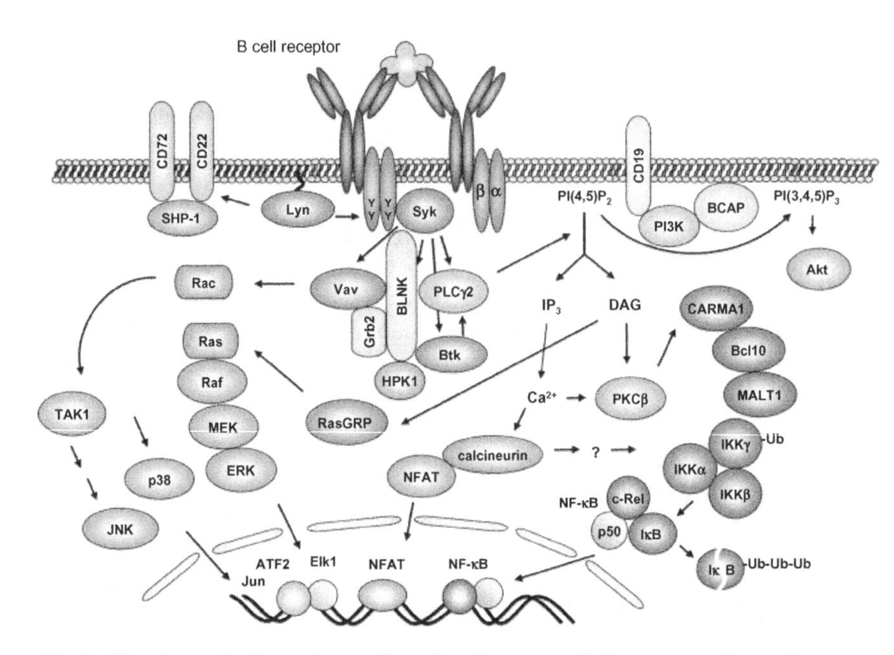

Fig. 3. Signal transduction pathways from B-cell receptor. Only representative pathways are shown. Note that all the depicted signaling events may not happen in the cells at the same time, but events may occur selectively depending on the developmental stages, activation or anergic states, nature of encountering antigen, etc. See text for more details. (See Color Plates)

independent of each other. Upon BCR ligation, tyrosine residues in the cytoplasmic region of CD19 are phosphorylated and bind phosphoinositide 3-kinase (PI3K) via its p85α regulatory subunit. Class I PI3K phosphorylates phosphatidylinositol (PI)-4,5-P_2 in the plasma membrane and generates PI-3,4,5-P_3, which recruit many enzymes required for B-cell survival and activation, such as phosphoinositide-dependent kinase 1 (PDK1), Akt (also known as PKB), Btk, Vav, and PLCγ2, through their PH domains. It follows that PDK1 phosphorylates and activates Akt, which leads to cell survival. Thus mice deficient for a regulatory subunit, p85α, or a catalytic subunit, p110δ, of PI3K show the *xid* phenotype with a defect in marginal-zone (MZ) B-cell development and a marked defect in BCR-mediated Akt activation (Clayton et al. 2002; Fruman et al. 1999; Jou et al. 2002; Okkenhaug et al. 2002; Suzuki et al. 1999, 2003). On the other hand, there is a report showing that BCR-induced activation of Btk is independent of p85α, and that Btk/p85α double-knockout mice had more severe defects than either single-knockout mouse in B-cell maturation and BCR-induced proliferation (Suzuki et al. 2003). It should be noted that redundant regulatory and catalytic subunits of class I PI3K may be involved in CD19, BCR, and other receptors in B and other cells, making the interpretation of the results difficult.

4.2.2 Adaptor Protein BLNK

A B-cell-specific member of the SLP-76 family of adaptor proteins, BLNK (also known as SLP-65 or BASH), has been proven to be critical in the signal transduction of BCR (Fu et al. 1998; Goitsuka et al. 1998; Ishiai et al. 1999; Wienands et al. 1998). BLNK possesses multiple phosphotyrosine-based SH2-binding motifs, proline-based SH3-binding motifs, and a C-terminal SH2 domain. Upon BCR-ligation, BLNK is recruited to the BCR through its SH2 domain binding to a phosphorylated non-ITAM tyrosine in Igα (Engels et al. 2001; Kabak et al. 2002). A leucine zipper motif in the N-terminal basic domain of BLNK has been shown to be necessary for its localization to the plasma membrane (Kohler et al. 2005). BLNK is primarily phosphorylated by Syk upon BCR stimulation and interacts with various signaling proteins. Simultaneous binding with Btk and PLCγ2, through their SH2 domains, allows Btk-phosphorylation and activation of PLCγ2 (Kurosaki and Tsukada 2000). In addition, BLNK binds another adaptor Grb2 and Vav, recruiting Vav to the plasma membrane and activating its GEF activity for a small GTPase Rac (Johmura et al. 2003). The SH2 domain of BLNK itself binds to Syk-phosphorylated HPK1, a serine/threonine kinase of the Ste20 family, and contributes to its activation that leads to NF-κB activation (Sauer et al. 2001; Tsuji et al. 2001). BLNK has been shown to be necessary for BCR-mediated calcium ion (Ca^{2+}) flux, the activation of mitogen-activated protein kinases (MAPKs) such as ERK, JNK, and p38, and transcription factors such as NFAT and NF-κB in DT40 chicken B cells (Fu et al. 1998; Ishiai et al. 1999).

BLNK knockout mice exhibit a severe *xid* phenotype, additionally with an incomplete block at large pre-B-cell stage of early B-cell development (Hayashi et al. 2000; Jumaa et al. 1999; Pappu et al. 1999; Xu et al. 2000). B-cell antigen receptor-mediated induction of cyclin D2, cdk4, and Bcl-x is impaired and B cells fail to enter the cell cycle (Tan et al. 2001). These defects in cell cycle entry and survival are common to Btk- or PI3K-deficient mice (Glassford et al. 2003; Suzuki et al. 2003). The similar peripheral B-cell phenotype in BLNK-, Btk-, PLCγ2-, and PI3K-deficient mice (Hashimoto et al. 2000; Wang et al. 2000) suggests their functional association and signifies the physical complex of these proteins, termed "BCR signalosome," in BCR signal transduction leading to survival, maturation, activation, and proliferation of peripheral B cells. In contrast to the result of DT40 cells, BCR-mediated activation of all MAPK is retained in primary B cells from BLNK-deficient mice (Tan et al. 2001).

BLNK is necessary, but not essential, for B-cell development since BLNK-deficient mice possess a small number of functional mature B cells that respond to T-cell-dependent antigen and conform memory B and long-lived plasma cells (Jumaa et al. 1999; Xu et al. 2000; Yamamoto et al. 2004). B-cell development is blocked nearly completely at the large pre-B-cell stage in the BLNK-deficient mice with combined deficiency of CD19, LAT, Btk, or PLCγ2, suggesting that the presence of BLNK-independent pre-BCR signaling pathway (Hayashi et al. 2003;

Kersseboom et al. 2003; Su and Jumaa 2003; Xu et al. 2006). A similar functional redundancy in the promotion of pre-BCR-mediated development is reported for Btk and Tec, or PLCγ2 and PLCγ1 (Ellmeier et al. 2000; Wen et al. 2004). Alternatively, but not exclusively, BLNK might function to enhance the signal by locally and temporally concentrating signaling molecules such as Btk and PLCγ2, since the early B-cell development and BCR signal transduction depend on the dosage of PLCγ2 in the absence of BLNK (Xu et al. 2006), and also combined Btk/PLCγ2 deficiencies cause the same developmental arrest at the large pre-B-cell stage as BLNK deficiency does (Xu et al. 2007). The local concentration of the signaling molecules might be negatively regulated by c-Cbl, a ubiquitin ligase targeting Igα, Syk, and Lyn, as the early B-cell development is partially rescued in BLNK/c-Cbl-double knockout mice (Song et al. 2007).

4.2.3 Calcium Signaling

After BCR-ligation, PLCγ2 is recruited to plasma membrane through its PH domain, and to BCR signalosome through binding to BLNK, as described above. PLCγ2 is phosphorylated and activated by Btk, and then hydrolyzes phosphatidylinositol-4,5-bisphosphate [PI(4,5)P_2] in the plasma membrane into inositol 1,4,5-triphosphate (IP_3) and diacylglycerol (DAG). Binding of IP_3 to its receptors on the ER membrane, which functions as Ca^{2+} channels, promotes flux of Ca^{2+} from ER into the cytoplasm. Immediately follows the Ca^{2+} influx through store-operated Ca^{2+} channels or other cation channels on the plasma membrane, making a transient increase of intracellular Ca^{2+} concentration. Accordingly, signalosome factors such as BLNK, Btk, and PLCγ2 are necessary for the BCR-induced Ca^{2+} flux response (Hashimoto et al. 2000; Kurosaki and Tsukada 2000; Wang et al. 2000). Interestingly, cytoplasmic tails or the tyrosines of ITAM in Igα or Igβ are not essential and rather inhibitory for the Ca^{2+} flux response and for tolerance induction, provided with one tail intact (Gazumyan et al. 2006; Kraus et al. 1999, 2001; Reichlin et al. 2001) (Fig. 2b) In contrast, a non-ITAM tyrosine, a binding site for BLNK, is necessary for full Ca^{2+} flux response as well as BLNK phosphorylation and its downstream events induced by BCR crosslinking (Patterson et al. 2006).

The increase of Ca^{2+} concentration induces activation of various calcium-dependent enzymes. One of such enzyme is Ca^{2+}/calmodulin-dependent serine-threonine phosphatase calcineurin. Calcineurin is known to dephosphorylate and activate the cytoplasmic components of NFAT transcription factors (NFATc). B cells express NFATc1, 2, and 3 of the four family members. Dephosphorylated NFATc shuttles into the nucleus and transcriptionally activates the target genes. A relatively low level of Ca^{2+} increase is enough to trigger the immediate NFATc translocation in B cells (Dolmetsch et al. 1997). Physiological role of the calcineurin/NFAT pathway in the B-cell system has been less studied compared with the T-cell system. Chimeric mice containing lymphocytes doubly deficient for NFATc1 and NFATc2 showed a hyperactivated B-cell phenotype, with markedly elevated serum IgG1 and IgE and plasma cell expansion, despite T-cell dysfunction, suggesting B-cell intrinsic function of these NFATs to negatively regulate terminal differentiation (Peng et al. 2001). B-cell-specific knockout of calcineurin (by deleting the B1 subunit gene) has recently revealed that this molecule is not necessary for follicular B-cell development but for the BCR-mediated activation and proliferation of B cells. The B-cell specific calcineurin B1-deficient mice showed a complex phenotype: higher serum IgM, reduced number of B1 cells, enhanced T-cell-independent (TI) type 1 but normal TI type 2 responses, moderately reduced T-cell-dependent immune responses and plasma cell differentiation, and normal B-cell tolerance (anergy), thus calcineurin may also be involved in other receptors (Winslow et al. 2006). Together with a marked immunosuppressive effect of the calcineurin inhibitors such as FK506, these results suggest the presence of unidentified calcineurin targets other than NFAT that promotes immune responses.

4.2.4 Protein Kinase C

The protein kinase C (PKC) family includes classical PKC (cPKC: α, β, and γ), novel PKC (nPKC: δ, ε, η, and θ) and atypical PKC (aPKC: ζ and λ/ι) isotypes. cPKC activation requires DAG (or its mimetic, phorbol esters such as PMA) and Ca^{2+} binding to its C1 and C2 domains, respectively, and nPKC requires DAG, whereas aPKC requires neither. Upon receptor stimulation, PKCs become phosphorylated, undergo conformational changes, and translocate to the plasma membrane to be activated. B-lineage cells express all PKC isoforms except PKCγ. Protein kinases C have been presumed to mediate key aspects of antigen receptor function, definite proofs for which have only recently been shown. PKCβ-knockout mice exhibit the *xid* phenotype, except that the number of follicular mature B cells is not reduced (Leitges et al. 1996) and that BCR-mediated induction of cyclin D2 expression is not impaired (Su et al. 2002). On the other hand, PKCβ is necessary for BCR-induced IKK-NF-κB activation, Bcl-xL expression and survival (Su et al. 2002; Saijo et al. 2002). These facts suggest that the BCR signal pathway diverges downstream of the signalosome into PKCβ-dependent and independent pathways.

In mice lacking PKCα, TCR-mediated T-cell proliferation is moderately impaired possibly due to low responsiveness to IL-2, but BCR-mediated B-cell proliferation is not. PKCα-deficient mice produced a reduced amount of antigen-specific IgG2a/2b antibodies upon immunization, suggesting a mild defect of Th1 cells (Pfeifhofer et al. 2006). PKCθ is known to be involved in TCR signal transduction, and the absence of PKCθ in mice leads to impaired T-cell activation, but B-cell function is normal (Sun et al. 2000).

Among other PKC isotypes, PKCζ has been demonstrated to be involved in B-cell function in vivo. In PKCζ-knockout mice, BCR-mediated survival, and proliferation in vitro are partially impaired and TD immune response in vivo is attenuated (Martin et al. 2002). It is indicated that PKCζ upregulates the transcriptional activity of NF-κB through direct phosphorylation of RelA (Leitges et al. 2001). In contrast to the role for PKCθ in T cells, another nPKC member PKCδ has a negative regulatory role in B-cell activation. PKCδ-knockout mice have increased numbers of B cells in the periphery and increased serum antibodies, and develop systemic autoimmune diseases being accounted for by a breakdown of peripheral B-cell tolerance (Mecklenbrauker et al. 2002; Miyamoto et al. 2002). It has been suggested that suppression of pro-apoptotic PKCδ function is integrated in BAFF-receptor-mediated, but not BCR-mediated, signaling for B-cell survival (Mecklenbrauker et al. 2004; Nojima et al. 2006). Expression of PKCη is reported to be readily detectable in pro-B cells but to markedly decrease along with the developmental transition into pre-B cells (Morrow et al. 1999). Although PKCη-knockout mice have not been reported so far, we have recently reported that retroviral expression of PKCη, as well as BLNK, in the leukemic pre-B-cell line derived from a BLNK-knockout mouse causes differentiation toward B cells, represented by κ-gene rearrangement (Yamamoto et al. 2006).

It has been shown that protein kinase D1 (PKD1, also known as PKCμ), a serine-threonine kinase containing a C1 domain, is a common substrate of cPKC and nPKC isoforms, and is activated by the PKCs downstream of BCR and TCR (Spitaler and Cantrell 2004). It has been proposed that PKD1 negatively regulate BCR signal transduction through inhibition of Syk (Sidorenko et al. 1996). PKCν (PKD3), another member of the PKD family, is abundant in B cells and activated upon BCR crosslinking in a BLNK/Btk/PLCγ2-dependent manner, through phosphorylation by nPKC, but not by cPKC (Matthews et al. 2003). DT40 B cells express PKD1 and PKD3, and both are redundantly necessary for BCR-induced phosphorylation and nuclear export of class II histone deacetylases, and the target promoter activation, but not for phosphorylation of ERK and Akt (Matthews et al. 2006). Physiological role for PKDs in B-cell development and activation remains to be determined.

4.2.5 Mitogen-Activated Protein Kinase Pathway

Upon crosslinking of BCR, small membrane-bound GTPases such as Ras and Rac are activated and constitutively bound GDP is rapidly replaced with GTP by the enzymes generally called a guanyl nucleotide exchange factor (GEF). The GTP-bound form recruits several cytoplasmic enzymes to the plasma membrane, including the serine threonine kinase Raf1 in case of Ras, which stimulates Raf1-MEK-ERK cascade. However, the active state of the GTPases is only transient since the bound GTP is rapidly metabolized into GDT with the aid of GTPase-activating proteins (GAP). Therefore the GTPases are regarded as rapid molecular switches. Although their requirement in BCR signaling has started to be shown recently, their role for activators of mitogen-activated protein kinase (MAPK) pathway has been well established in other cell lineages.

In the BCR signal transduction, Ras is mainly activated by RasGRP3 with a minor contribution of RasGRP1, both being Ras GEF that is activated upon binding with DAG (Coughlin et al. 2005; Oh-hora et al. 2003). It has recently been shown that PKC phosphorylates and activates RasGRP3, which is likely mediated through association of the two by binding to DAG at the plasma membrane (Aiba et al. 2004; Teixeira et al. 2003; Zheng et al. 2005). In B cells from RasGRP1/3 double null mutant mice, BCR ligation fails to induce activation of Ras and ERK as well as cell proliferation. Nevertheless, B-cell development is unaffected and T-dependent immune responses of only IgG1 and IgG2a isotypes are significantly impaired in the mutant mice (Coughlin et al. 2005). These results indicate that the Ras-ERK pathway of BCR signaling is essential for induction of cell proliferation but not for B-cell development and terminal differentiation. Ras-ERK pathway, possibly activated by other GEF such as Sos in cytokine-receptor signaling, may also play a role in B-cell development. For example, B cell-specific transgenic expression of a dominant-negative Ras mutant results in the developmental arrest at pro B-cell stage before the expression of pre-BCR (Iritani et al. 1997; Nagaoka et al. 2000). The same Ras mutant inhibits development or survival of a fraction of memory cells having high-affinity Ig mutations, and impairs differentiation of memory cells into antibody forming cells upon secondary antigen stimulation, which is rescued by overexpression of Bcl-2 (Takahashi et al. 2005). This suggests that Ras-mediated survival signal is critical for generation and response of memory B cells, although it is unknown for which receptor this signal is essential, BCR or cytokine-receptors. B cells express Raf-1 and B-Raf isoforms. It has been shown with DT40 cells that these two are necessary but mutually complementary in BCR-mediated ERK activation (Brummer et al. 2002). Contribution of MEK1 or MEK2 in BCR signaling is still unclear.

Rac is important for BCR-mediated reorganization of actin-cytoskeleton, and activation of JNK (SAPK) and p38. Rac is activated by Rac-GEF, Vav, in the BCR signal transduction. The B-cell adaptor molecule of 32 kDa (Bam32), containing SH2 and PH domains and acting downstream of PTK and PI3K, positively regulates Rac1 activation, actin remodeling and BCR internalization (Allam et al. 2004; Niiro et al. 2004). Various signaling cascades starting with GTP-bound Rac including PAK, MEKK, and SEK/MKKs that end with JNK/p38 activation have been reported in various receptor signaling, but little is known in BCR signaling. Loss of all Vav (Vav1–3) in mice results in a developmental arrest at immature B-cell stage, and at CD4/8 double-negative thymocytes as well. B-cell antigen receptor-ligation on the Vav-null B cells does not induce calcium flux and proliferation, but induces intact ERK1/2 activation. The numbers of immature and marginal zone B cells are moderately reduced. The Vav-null mice completely fail to mount both T-dependent and T-independent humoral responses, the latter being partly due to inability of MZ B-cell to differentiate into plasma cells (Fujikawa et al. 2003; Stephenson et al. 2006). Rac1/2-double knockout mice show the similar phenotype with more severe reduction of immature and MZ B cells (Walmsley et al. 2003). Recently, TAK1, a member of MAP3-kinase family, has been shown to be essential for BCR-mediated JNK, but not ERK or p38, activation (Sato et al. 2005; Shinohara et al. 2005). B-cell antigen receptor-induced strong calcium flux and calcineurin activation are also necessary for JNK activation (Dolmetsch et al. 1997; Healy et al.

1997). SEK1 (MKK4) is essential for the development of embryo, but B cells lacking SEK1 develop almost normally and produce antibody normally upon immunization in Rag-deficient recipient mice, and respond in vitro normally to BCR stimulation in terms of JNK activation and proliferation, indicating that SEK1 is dispensable for BCR signaling (Nishina et al. 1997; Swat et al. 1998).

4.2.6 Nuclear Factor (NF)-κB Pathway

Transcription factor NF-κB promotes transcription of various genes such as cytokines, cyclins, and anti-apoptotic genes. NF-κB is a protein family consisting of homo- or heterodimer of the subunits NF-κB1 (p50), NF-κB2 (p52), c-Rel, RelA (p65) and RelB. p50 and p52 are generated by proteolysis of their precursors p105 and p100, respectively, and lack activation domain that is required for transcriptional activation of the target gene. p50 and p52 act as transcriptional repressors in the forms of homodimers, but form transcriptionally active complexes with Rel proteins. In most resting cells, NF-κB is retained in the cytoplasm as an inactive complex with IκB proteins. Upon antigen-receptor stimulation, IκB kinase (IKK) complex consisting of two catalytic subunits, IKKα (IKK1) and IKKβ (IKK2), and a regulatory subunit, IKKγ (NEMO), is activated and phosphorylates IκB, which leads to polyubiquitination and proteasome-dependent degradation of IκB, and to the following nuclear accumulation of active NF-κB, mainly consisting of p50 and RelA or c-Rel. This signaling pathway is termed the "canonical" pathway (Hayden and Ghosh 2004). B-cell antigen receptor-ligation activates this pathway and expression of the target genes such as *bcl-2*, *bcl-x*, *A1*, *cdk4*, and *cyclinD2*, leading to proliferation of B cells. The BCR-induced B-cell proliferation and immune response are impaired in c-Rel-single knockout mice indicating a non-redundant role of c-Rel in the receptor-induced cell-cycle entry and survival (Grumont et al. 1998; Kontgen et al. 1995; Tumang et al. 1998).

In mature B cells, however, NF-κB is constitutively active to some extent, and this activity is required for development and survival of mature B cells including marginal zone and peritoneal B1 cells, as revealed by p50/p52-double knockout mice, bone-marrow chimeras transferred with IKKα- or RelA/c-Rel-deficient fetal liver cells, or B-cell specific IKKβ- or IKKγ-knockout mice (Franzoso et al. 1997; Grossmann et al. 2000; Kaisho et al. 2001; Pasparakis et al. 2002; Sasaki et al. 2006). The development and survival of mature B cells require "tonic" signal from unligated BCR (Kraus et al. 2004; Lam et al. 1997; Torres et al. 1996), as well as signal from BAFF receptor (BAFF-R) except for peritoneal B1 cells (Gross et al. 2001; Schiemann et al. 2001; Thompson et al. 2001). The BAFF-R signaling to NF-κB activation utilizes both "canonical" and "alternative" pathways, the latter involving NIK, IKKα, RelB, and p52, but not IKKγ (Claudio et al. 2002; Pasparakis et al. 2002; Sasaki et al. 2006). The defect of B-cell development in the NF-κB-deficient mice as mentioned above is, at least in part, ascribed to a defect in the canonical pathway from BAFF-R, since conditional expression of active IKKβ form recovers follicular mature and MZ B cells, possibly through PKCδ nuclear exclusion, in BAFF-R-deficient mice (Sasaki et al. 2006). RelB and p52 are essential for T-dependent humoral immune response, which is probably due to their critical role in CD40 signaling in B cells, and in TNF- and lymphotoxin-receptor signaling in splenic stromal cells (Franzoso et al. 1998; Weih et al. 2001). RelB in hematopoietic cells is necessary for MZ but not follicular B-cell development (Weih et al. 2001).

Btk, BLNK, PLCγ2, PI3K p85α, and PKCβ have been shown to be necessary for BCR signal leading to NF-κB activation (Bajpai et al. 2000; Hikida et al. 2003; Petro and Khan 2001; Petro et al. 2000; Saijo et al. 2002; Su et al. 2002; Suzuki et al. 2003; Tan et al. 2001). Interestingly, Src-family kinase, but not Syk, is required for Igβ-induced IKK-NF-κB activation in pro-B cells (Saijo et al. 2003), though Syk is critical for pro-B to pre-B-cell development. In this regard, there has been no evidence that B-cell intrinsic NF-κB activity is required for the early B-cell development. It has been shown that BCR-induced strong calcium flux and the following activation of calcineurin are necessary to trigger IκB degradation and NF-κB activation (Dolmetsch et al. 1997; Healy et al. 1997).

In peripheral B and T cells, three interacting proteins, CARMA1 (also known as CARD11), Bcl10, MALT1, are essential for IKK-NF-κB activation, cell activation and proliferation in response to stimulation through BCR, CD40, and TCR, but not for development of mature T cells and B cells except B1 and marginal zone B cells (Egawa et al. 2003; Hara et al. 2003; Jun et al. 2003; Newton and Dixit 2003; Ruefli-Brasse et al. 2003 Ruland et al. 2001, 2003; Xue et al. 2003). Therefore, the CARMA1/Bcl10/MALT1 complex, as well as PKCβ, are dispensable for the BCR "tonic" signal, but necessary for the "inductive" signal. It has been shown that BCR-induced PKCβ-binding and phosphorylation of CARMA1 leads to NF-κB activation (Sommer et al. 2005). Bcl10 and MALT1 have been shown to promote K63 polyubiquitination of IKKγ to activate IKK in other receptor signaling (Zhou et al. 2004). Thus, the CARMA1/Bcl10/MALT1 constitute an upstream part of the canonical pathway. Using DT40 cells, it has recently been shown that TAK1, a member of MAP3-kinase family, is essential for IKK and NF-κB activation in response to BCR stimulation, and that PKCβ-phosphorylated CARMA1 interacts with TAK1 and IKK, allowing TAK1-phosphorylation of IKK (Shinohara et al. 2005). On the other hand, BCR-mediated NF-κB activation is intact in primary B cells from mice with B cell-specific TAK1-deficiency, implying context-dependent TAK1 requirement (Sato et al. 2005). Taken together, it is proposed that BCR signalosome leads to PKCβ activation at plasma membrane, and that the CARMA1 modified by PKCβ serves as a docking site for IKKα/β/γ complex where IKKγ-ubiquitination is mediated by Bcl10 and MALT1. It remains to be determined how IKKα/β is then activated.

It has been demonstrated that PI3K is required for BCR-mediated activation of NF-κB, but not of Btk, implying a Btk-independent mechanism for NF-κB activation mediated by PI3K (Suzuki et al. 2003). In this regard, in TCR-signal transduction, PI3K-downstream PDK1 is shown to recruit PKCθ, IKK complex, and CARMA1/Bcl10/MALT1 complex together, resulting in IKK activation (Lee et al.

2005). In addition, BCAP, a BCR-associated adaptor protein, is necessary for maintenance of c-Rel protein level, implying another mechanism regulating NF-κB activation. Thus B cells from BCAP-knockout mice respond poorly to BCR cross-linking to express Bcl-xL, A1, Cyclin D2, and Cdk4, and to proliferate, and the mice exhibit the *xid* phenotype (Yamazaki et al. 2002; Yamazaki and Kurosaki 2003).

4.3 BCR Signaling Pathways for Immune Response

4.3.1 Thymus-Independent Response

Immune responses characterized by antibody production are classified as thymus-independent (TI) or thymus-dependent (TD) based on the requirement for T-cell help. Antigens that elicit TI responses are divided into type I and type II. The former (TI–I) antigens represented by lipopolysaccharide (LPS) are now known to stimulate polyclonal B cells through Toll-like receptors to proliferate and finally to differentiate into antibody-forming plasma cells, which is independent of BCR signaling. The latter (TI–II) antigens, represented by polysaccharide displaying repetitive epitopes, aggregate BCR and induce clonal expansion and differentiation of specific B cells into plasma cells in a short period (within a week). Typically, TI–II response is elicited by marginal zone B and B1 cells, which form short-lived plasma cells homing outside of the lymphoid follicles in the red pulp, and does not induce germinal centers (GCs) and affinity maturation of Ig (Fig. 4). In the TI–II response, B cells carrying BCRs with higher affinity to a given antigen respond more frequently to the antigen in vivo to proliferate and generate more plasma cells than those with lower-affinity BCRs, resulting in a selective production for high-affinity antibodies (Shih et al. 2002b). Recently it has been reported that TI–II response generates atypical memory B cells (Obukhanych and Nussenzweig 2006) and memory-type plasmablasts derived from B1b cells (Hsu et al. 2006).

4.3.2 Thymus-Dependent Response

In the primary response to TD antigens, antigen-specific B and T cells encounter and B cells are activated through antigen-bound BCR as well as by T-cell help through CD40 and cytokine receptors. After clonal expansion, some B cells form extrafollicular primary foci of plasmablasts secreting IgM, some undergo Ig class-switch and form the foci of IgG^{+} plasmablasts, and others aggressively proliferate in the B-cell follicles to form GCs (Fig. 4). Plasma cells from the foci are short-lived and produce antibodies for a couple of weeks after immunization. In GCs, proliferating B cells undergo Ig gene somatic hypermutation and selection for high-affinity BCR, and then the selected B cells differentiate into short-lived plasma,

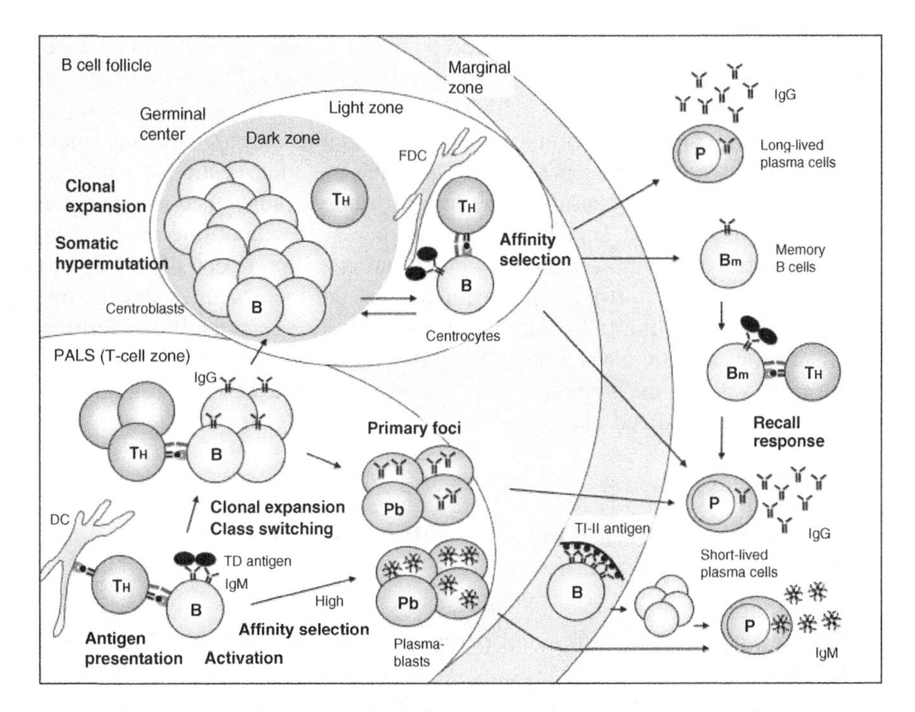

Fig. 4. Thymus-independent (*TI*)-II and thymus-dependent (*TD*) immune responses. See text for interpretation. Each event occurring in B cells during TD immune response is notated in bold. *B*, B cells; T_H, helper T cells; *Pb*, plasmablasts; *Bm*, memory B cells; *P*, plasma cells; *DC*, dendritic cells; *FDC*, follicular dendritic cells; *PALS*, periarteriolar lymphoid sheath. (See Color Plates)

memory B or long-lived plasma cells. Memory B cells are maintained for a long time without persisting antigen (Maruyama et al. 2000) and respond promptly to a secondary challenge of antigen (so-called recall response) to produce a large amount of high affinity antibodies of typically IgG class.

A conserved transmembrane/cytoplasmic tail of IgG was shown to be responsible for the efficient TD IgG response including Ig affinity maturation, memory formation, and higher antibody production partly due to enhanced survival of plasmablasts after clonal expansion compared to those mediated by IgM (Kaisho et al. 1997; Martin and Goodnow 2002). Molecular mechanism accounting for the IgG tail function is still unclear, besides an augmented Ca^{2+} response, and involvement of CD22 is currently under debate (Horikawa et al. 2007; Wakabayashi et al. 2002; Waisman et al. 2007). In addition to the class-switching of BCR itself, T-cell-mediated activation may drastically alter the BCR-downstream signaling pathway, as it was reported that prior B-cell exposure to CD40-ligand results in BCR-mediated activation of NF-κB and ERK pathways that are independent of Btk, PI3K or PLCγ2 [Mizuno and Rothstein (2005) and references therein].

In TD response, B cells with initially higher-affinity receptors for antigen are more preferentially selected for expansion in germinal centers and later for memory

B cells than those with lower-affinity receptors in a competitive condition, although both respond almost equally in non-competitive condition (Dal Porto et al. 2002; Shih et al. 2002a). The former cells in germinal centers accumulate less somatic hypermutation in Ig V region (albeit no less in the intron) than the latter in non-competitive condition, possibly because the mutation in the high-affinity V gene would more likely result in the loss of affinity, rather than the gain (Shih et al. 2002a). However, B cells initially carrying BCR with affinity higher than a certain threshold expand early upon immunization and preferentially differentiate into extrafollicular short-lived plasma cells, and are less recruited into germinal center reaction and the following long-lived plasma or memory B-cell formation (O'Connor et al. 2006; Paus et al. 2006). It is still unknown how B cells that acquire high affinity during the germinal center reaction are selected for differentiation into memory B or long-lived plasma cells.

4.3.3 Immune Responses in Mutant Mice Deficient for BCR-Signaling Molecules

Mice devoid of a cytoplasmic tail portion of either Igα or Igβ (ΔC), or one of BCR-proximal signaling molecules, such as Btk, PI3K (p85α or p110δ), PLCγ2, PKCβ, BLNK, or BCAP, are deficient for antibody production after immunization with TI–II antigens, indicating that the BCR signaling from both Igα and Igβ through these signaling molecules is essential for clonal expansion and/or plasma-cell differentiation upon receptor aggregation (Hashimoto et al. 2000; Jumaa et al. 1999; Khan et al. 1995; Leitges et al. 1996; Okkenhaug et al. 2002; Reichlin et al. 2001; Ridderstad et al. 1996; Suzuki et al. 1999; Torres et al. 1996; Wang et al. 2000; Xu et al. 2000; Yamamoto et al. 2004; Yamazaki et al. 2002). Involvement of SFK and Syk in the immune response cannot be assessed because of their functional redundancy and indispensability in pre-B-cell development, respectively. CD45 is essential for T-cell development and function, and B-cell-intrinsic function of CD45 in immune responses has not been reported.

The primary IgG response to TD antigen is severely decreased but detectable despite the greatly reduced number of mature B cells in mice with ΔC mutation in either Igα or Igβ, suggesting partial redundancy of the cytoplasmic tails in the TD response. Interestingly, in mice with replacement mutations at two tyrosine residues in ITAM of either Igα or Igβ, follicular mature B cells develop normally and fully (even better) respond to TI–II antigen, in contrast to the ΔC mutants as mentioned above, suggesting redundant function of the two ITAMs as well as the importance of non-ITAM residues in Igα/Igβ in the TI–II response (Gazumyan et al. 2006; Kraus et al. 2001) (Fig. 2c). In this regard, the mice with a mutation at a conserved non-ITAM tyrosine of Igα, which is necessary for BLNK binding and phosphorylation, respond several fold less to TI antigen but normally to TD antigen, which is similar to the response of BLNK-knockout mice (Patterson et al. 2006; Yamamoto et al. 2004). On the contrary, TD IgG1 response is significantly reduced in either

ITAM mutant mice, indicating that both ITAMs of Igα/Igβ are necessary for optimum TD response (Gazumyan et al. 2006; Kraus et al. 2001). In mice in which the cytoplasmic tail of Igβ has been replaced with that of Igα (β_c-α_c mutation), follicular mature B cells develop almost normally and are long-lived, but respond barely to TI–II antigen and several-fold less to TD antigen. Taken together with the dispensability (actually inhibitory role) of Igβ ITAM for the TI–II response, it indicates that Igβ non-ITAM sequence is necessary for the TI–II response (Reichlin et al. 2004) (Fig. 2c). In B cells of the β_c-α_c mutant mice, constitutive as well as crosslinking-induced BCR internalization is accelerated, surface BCR level reduced, and BCR signal transduction weakened (Reichlin et al. 2004), which is an almost opposite phenotype of B cells of either ITAM mutant (Gazumyan et al. 2006; Kraus et al. 2001). This suggests that two ITAMs of Igα/Igβ are necessary for the efficient BCR internalization, whereas the non-ITAM sequence in Igβ is involved in the negative regulation of the internalization (Fig. 2d).

The primary TD IgG response is delayed and reduced in Btk-, PKCβ-, PI3K p110δ-, or BLNK-deficient mice, with nearly absence of IgM production, which is not necessarily correlated with the number of mature B cells in each strain but may be due to less efficient activation and proliferation signaled through IgM BCR. It was reported to be normal in PLCγ2- or PI3K p85α-deficient mice, but kinetics data for them are missing. In addition the interpretation of the results requires some caution because the immune responses may be affected by the immunized antigens and conjugates or by genetic backgrounds of mouse strains. In contrast to the primary, the secondary IgG response to TD antigens is largely intact in these mutant mice (Hashimoto et al. 2000;Jumaa et al. 1999; Khan et al. 1995; Leitges et al. 1996; Okkenhaug et al. 2002; Ridderstad et al. 1996; Suzuki et al. 1999; Wang et al. 2000; Xu et al. 2000; Yamamoto et al. 2004). Finally, antibody affinity maturation, the recall response of memory B cells after a long-term interval, as well as long-term maintenance of specific antibodies, are intact in Btk- or BLNK-deficient mice (Ridderstad and Tarlinton 1997; Ridderstad et al. 1996; Yamamoto et al. 2004), while they remain unknown in other mutant mice.

In sharp contrast, CD19-deficient mice exhibited severely impaired TD primary and secondary responses including germinal center formation, antibody affinity maturation, memory B-cell formation and long-lasting antibody production, but rather enhanced TI-II response (Fehr et al. 1998; Rickert et al. 1995; Sato et al. 1995) (see also Chapter 7). The TD response is profoundly impaired, except for antibody affinity maturation, also in mice deficient for *Cr2* gene encoding CD21/CD35 complement receptors (Chen et al. 2000; Croix et al. 1996). These results are supported by the in vitro data showing that signaling through CD19/CD21 complex co-ligated with BCR by antigen-complement complex stabilizes the BCR and augments antigen processing/presentation (Cherukuri et al. 2001a,b), and in vivo/vitro data that CD21/CD35 signaling protects B cells from Fas-mediated apoptosis during primary TD-response (Barrington et al. 2005). Thus once the BCR class switches to IgG and T-cell help become available, BLNK/Btk signalosome is no more necessary, but the co-signaling from IgG and CD19 is still required for efficient induction of germinal center reaction and the following immune responses.

Among the downstream signaling pathways, CARMA1/Bcl10/MALT1 and cRel have been shown to be essential for B- and T-cell activation/proliferation through BCR, CD40, and TCR, and for both TI and TD immune responses (Hara et al. 2003; Jun et al. 2003; Kontgen et al. 1995; Newton and Dixit 2003; Ruefli-Brasse et al. 2003; Ruland et al. 2001, 2003). B-cell specific deletion of IKKβ in mice exhibit reduced TI and TD immune responses, but it is unclear which defect is responsible for it, reduced follicular, MZ, and B1 cells, lowered proliferative signals from BCR, TLR, or CD40, or reduced survival (Li et al. 2003). Thus contribution of the NF-κB pathway downstream of BCR for the TD immune response is still unclear. As mentioned above, Vav and Rac are also essential for both TI–II and TD responses, suggesting a signalosome-independent basic function of these molecules in BCR signal transduction. The significance of the Ras-MEK-ERK signaling pathway is currently unclear.

4.4 BCR Signaling Pathways for Self Tolerance

4.4.1 Clonal Deletion

Owing to the random nature of V(D)J recombination of Ig H and L genes and of H/L-chain pairing, many newly generated B cells express BCR that binds to self antigens. It has been proposed that such self-reactive B-cell clones are eliminated through BCR-signal-induced apoptosis in the bone marrow (clonal deletion), based on the in vitro model systems using ex vivo immature B cells or B lymphoma cell lines, and also in vivo model using transgenic mice expressing monoclonal BCR that reacts with membrane-bound (neo-) self-antigen or DNA (Fang et al. 1998; Lang et al. 1997; reviewed by Goodnow 1992; Nemazee et al. 1991). Although the mechanisms for the BCR-induced apoptosis of immature B cells have been extensively studied in the in vitro systems (reviewed by King and Monroe 2000), the specific mechanism that accounts for the BCR-mediated clonal deletion in vivo remains unclear. Detailed analysis of the anti-hen egg lysozyme (HEL) Ig/ membrane-bound-HEL (mHEL) transgenic mice revealed that the self-reactive immature B cells are not immediately deleted in the bone marrow but developmentally arrested at the immature B-cell stage, and ultimately die through apoptosis which can be rescued by artificial expression of bcl-2 (Hartley et al. 1993). This indicates that the signal from self-antigen bound IgM is likely to block maturation of self-reactive B cells but not to directly induce their apoptosis. Indeed the clonal deletion is a rare event in mice carrying self-reactive Ig transgenes targeted into the Ig gene loci so that they can be removed by de novo V(D)J rearrangements (receptor editing; see below). The arrested immature B cells expressing a low level of anti-self IgM in the mHEL/anti-HEL-Ig transgenic mice may represent the cells undergoing receptor editing. Accordingly BCR-ligation on immature B cells ex vivo induces Ig gene rearrangements but not apoptosis under appropriate culture conditions such as a use of feeder cells and/or monoclonal antibodies for the BCR ligation (Hertz and Nemazee 1997; Melamed and Nemazee 1997; Sandel and Monroe 1999). Thus it is likely that the immature B cells, having failed to edit their self-reactive IgM into a non-harmful one within the permissive period for the rearrangement, die through apoptosis (Melamed et al. 1998), possibly by a temporally controlled program.

It was shown that soluble antigen induces apoptosis of germinal-center B cells in vivo, which was thought to reflect the elimination of self-reactive B cells newly generated by hypermutation of Ig genes during the immune response (Pulendran et al. 1995; Shokat and Goodnow 1995). Even the resting mature B cells are rapidly eliminated in the periphery after forced replacement of their BCR into a self-reactive one by Cre/loxP-mediated recombination (Lam and Rajewsky 1998). Such self-antigen-induced elimination of mature B cells may be responsible, at least partly, for the peripheral tolerance. However the BCR-signaling mechanism specific for such elimination is unknown.

4.4.2 Receptor Editing

Receptor editing is a process by which the specificity of BCR is altered by successive V gene (mainly of L chain) rearrangements when the BCR on newly generated B cells is bound by self-antigen with repeated epitopes such as membrane proteins or DNA. The "Ig knock-in" mice, in which Ig gene loci are replaced through gene-targeting technology with rearranged V genes of an autoantibody, have revealed that the receptor editing plays a major role in eliminating the self-reactivity in the primary B-cell pool. In such mice, secondary rearrangement of the Ig gene loci (primarily at the *Igκ*, then at the *Igλ*, and less frequently at the *IgH* loci) replaces the self-reactive V gene with new ones, which efficiently rescues originally self-reactive B cells from clonal deletion (reviewed by Nemazee 2006). Cellular deletion occurs only when further rearrangement is precluded by the lack of the leftover of unrearranged J segments or of RAG proteins (Halverson et al. 2004; Xu et al. 1998). In accord with these in vivo data, BCR engagement on bone marrow immature B cells in vitro has been shown to induce expression of RAGs (when it is suppressed by transgenic BCR expression) and secondary L chain gene rearrangements, but not apoptosis (Hertz and Nemazee 1997; Melamed et al. 1998; Sandel and Monroe 1999).

B-cell antigen receptor signaling pathway to receptor editing has not been extensively studied, partly because the study requires genetically modified mice at multiple loci including the Ig genes as aforementioned and a gene of interest. It has been reported that tyrosine kinase Syk and Btk, tyrosine phosphatase CD45 and co-receptor CD19, all of which are known to positively regulate BCR signal transduction, have been reported to be dispensable for receptor editing in anti-self MHC or anti-HEL Ig-transgenic mouse systems (Dingjan et al. 2001; Meade et al. 2004; Shivtiel et al. 2002a,b). However, the apparent receptor editing in Syk- and CD19-knockout mice could be ascribed to the lack of allelic exclusion of light chains (Meade et al. 2004; Shivtiel et al. 2002b). We have recently demonstrated that receptor editing in anti-DNA Ig knock-in mice is impaired in the absence of BLNK. Accordingly BLNK-deficient mice with normal *Ig* loci possess less Igλ^+ B cells, have undergone less recombining sequence (RS) recombination, both of which are hallmarks of receptor editing, and respond more to DNA-immunization, than wild-type mice. Thus BLNK is necessary for the BCR signaling pathway leading to receptor editing that substantially contributes to the elimination of self-reactivity physiologically (Hayashi et al. 2004). It is currently unknown how BLNK transmits the signal for the receptor editing. Additional anti-self Ig knock-in mouse strains with genetic ablations of other BCR-signaling molecules would be necessary to solve this problem. Not only the membrane-bound but soluble self-antigens can induce the receptor editing to some extent (Hippen et al. 2005), indicating that the quantity of the BCR signaling plays a crucial role in determining the receptor editing. Signaling molecules such as BLNK may function to control the threshold of BCR signaling to induce the editing in an appropriate period of B-cell development. Bone-marrow microenvironment that has been shown to protect immature B cells from BCR-induced apoptosis and to allow receptor editing instead (Sandel and Monroe 1999) may also function to define a "time window" being permissive for the receptor editing during B-cell development.

4.4.3 Anergy

Self-reactive B-cell clones that have not been eliminated or edited in the bone marrow often become unresponsive to antigen, which is termed "anergy." This was first demonstrated and has been best characterized by the double transgenic mice that express anti-HEL Ig (Ig^{HEL}) and soluble HEL antigen (sHEL) (Goodnow 1992). Since the membrane-bound HEL antigen causes receptor editing or clonal deletion in the same Ig-transgenic mice as described above, it is believed that B cells are anergized when BCR is bound to self-antigen with relatively low avidity such as sHEL. This is supported by the fact that increased BCR signaling quantity rendered by the mutation of protein tyrosine phosphatase SHP-1 or Lyn, which constitute negative regulatory pathways, results in clonal deletion, rather than anergy, in the sHEL/Ig^{HEL} mouse system (Cyster and Goodnow 1995; Cornall et al. 1998). In the same system, continued binding of self-antigen is required for the B cells to be kept in anergic state (Gauld et al. 2005; Goodnow 1992). The anergic B cells show several features that partly differ depending on the experimental systems: cell surface expression level of BCR (mIgM) is reduced (Goodnow et al. 1988) and that of CD5 is increased (Hippen et al. 2000), lifespan is shortened, and entry into the lymphoid follicles is prohibited (Cyster et al. 1994). Their BCR are desensitized and therefore the B cells do not proliferate in response to antigens even in the presence of cognate T-cell help but are instead eliminated by Fas-induced apoptosis (Ho et al. 1994; Rathmell et al. 1995), which depends on repression of B7.2 on anergic B cells (Rathmell et al. 1998). The shortened lifespan and the follicular exclusion are evident when the anergic B cells and nonself-reactive B cells are coexist, which suggests that the anergic B cells fail to compete for a survival or chemotactic factor.

B-cell activation factor belonging to the tumor necrosis factor family (BAFF) is thought to be one of the survival factors that are critical to affect the tolerance, since overexpression of BAFF from transgene causes autoimmune disease featuring various autoantibodies in mice (Mackay et al. 1999). Although BAFF and its receptor (BAFF-R) are necessary for normal B-cell maturation and survival in the periphery (Mackay and Browning 2002), precise control of local concentration of BAFF appears to be important to maintain the follicular exclusion and the shortened lifespan of anergic B cells that are more dependent on BAFF (Lesley et al. 2004). BAFF negatively regulates the expression of pro-apoptotic protein Bim (Craxton et al. 2005) and the nuclear entry of death-promoting PKCδ (Mecklenbrauker et al. 2004). Loss of Bim or PKCδ has been shown to break B-cell anergy and to cause autoimmunity with autoantibodies in mice (Mecklenbrauker et al. 2002; Miyamoto et al. 2002; Oliver et al. 2006), indicating that Bim and PKCδ are crucial to shorten the life span of anergic B cells. In addition to the deregulated BAFF action, activation of anergic B cells through Toll-like receptors by pathogen-derived components such as unmethylated CpG-containing DNA or single-stranded RNA is proposed to break the tolerance and to induce autoimmune disease (Berland et al. 2006; Christensen et al. 2006; Leadbetter et al. 2002). Furthermore, overexpression of transgenic CD19 has been shown to break anergy in the sHEL/Ig^{HEL} system (Inaoki et al. 1997).

B-cell antigen receptor signaling pathway that induces and/or maintains anergy is poorly understood. B-cell development is blocked in Syk-deficient mice even in the presence of transgenic BCR expression, and therefore anergy induction cannot be tested (Cornall et al. 2000). Although Lyn-deficient mice develop autoimmune disease, they show normal tolerance induction in the sHEL/Ig^{HEL} system (Cornall et al. 1998). Recently it has been shown that the mice in which B cells are devoid of Cbl and Cbl-b, E3 ubiquitin ligases that target Syk and Igα in B cells, often suffer from SLE-like autoimmune disease and in the sHEL/Ig^{HEL} system B-cell anergy is impaired: B cells are mature, express a high level of Ig^{HEL}, respond to BCR-ligation by CD86 upregulation and Ca^{2+} influx. However the Cbl/Cbl-b-deficient B cells are not hyper-responsive in vitro as well as in vivo in terms of activation, proliferation, and antibody production, although they show enhanced tyrosine phosphorylation of Syk and its substrate proteins, except for BLNK, and Ca^{2+} response upon BCR-ligation, and impaired BCR downmodulation (Kitaura et al. 2007). These results suggest that Igα/Syk-ubiquitination by Cbl proteins and BCR downmodulation upon contact with self-antigen during the immature B-cell stage is necessary for the induction of B-cell anergy. PKCδ may also be necessary for the BCR signaling inducing anergy because PKCδ-deficient mice develop the similar autoimmune disease and in the sHEL/Ig^{HEL} system HEL-binding mature B cells and antibodies are not reduced, Ig^{HEL} is not downregulated, and BCR-induced CD86 upregulation, Ca^{2+} influx and proliferation are restored (Mecklenbrauker et al. 2002).

Using the sHEL/Ig^{HEL} system, it was shown that in anergic B cells basal calcium level is elevated because of repetitive calcium oscillations arising through continued BCR-stimulation by circulating self-antigen (sHEL), though BCR-ligation induces little initial calcium rise. This resulted in constitutive NFAT activation. ERK is also basically activated and fully activated by BCR ligation. Such activated phenotype of anergic B cells is dependent on the presence of CD45, suggesting the involvement of SFK-mediated signaling (Healy et al. 1997). Calcineurin-deficient B cells, however, have been shown to be normally tolerized in the same system, suggesting the basal NFAT activation is not necessary for the anergic state (Winslow et al. 2006). The anergic B cells do not respond by activation of JNK and NF-κB upon BCR-ligation, which requires large transient calcium rise and calcineurin activity (Dolmetsch et al. 1997). Since both JNK and NF-κB were fully activated in the anergic cells upon stimulation with phorbol ester plus calcium ionophore, the defect does not lie at CARMA1 or its downstream but probably at upstream of PKCβ (Healy et al. 1997). This chronic ERK activation prevents anergic B cells from producing autoantibody in response to CpG DNA or LPS through TLRs (Rui et al. 2003, 2006). Similarly it was reported that sustained activation of the Raf-MEK-ERK signaling pathway induces cytokine nonresponsiveness in T cells (Chen et al. 1999). In B cells, the active ERK inhibits LPS-mediated plasma-cell differentiation, which is relieved by IL-2/IL-5, T-cell cytokines known to promote plasma-cell differentiation, through induction of ERK-phosphatase DUSP5 (Rui et al. 2006). It remains to be clarified which each alteration in signaling is responsible for, induction of anergy, maintenance or both.

4.5 Concluding Remarks: To Respond or Not to Respond, That is the Question

Taking together all the evidence described so far, mainly from in vivo experiments, I have drawn a picture illustrating the signaling pathways from BCR that induce different B-cell responses depending on the nature of antigens and on the developmental stages of the cells (Fig. 5). These pathways are deduced from many pieces of evidence and of many inferences therefrom, and thus the picture is far from complete, but might be useful to clarify the unsolved problems. The difficulty in completing the picture stems from an uncertainty of cell type affected by a gene knockout, and of receptors affected even in the case of B cell-specific gene knockout, given the target signaling molecule may function under multiple receptors and in multiple types of cells. A new technology that enables targeting a molecule in a specific receptor signaling should be desired. In addition, the problem of molecular redundancy (functional compensation of one gene mutation by other genes) needs to be solved.

Despite the very incomplete understanding for the BCR signaling pathways, it may be meaningful at this point to deduce the strategy of how the B-cell system manages to discriminate self from nonself. Unlike toll-like receptors that have

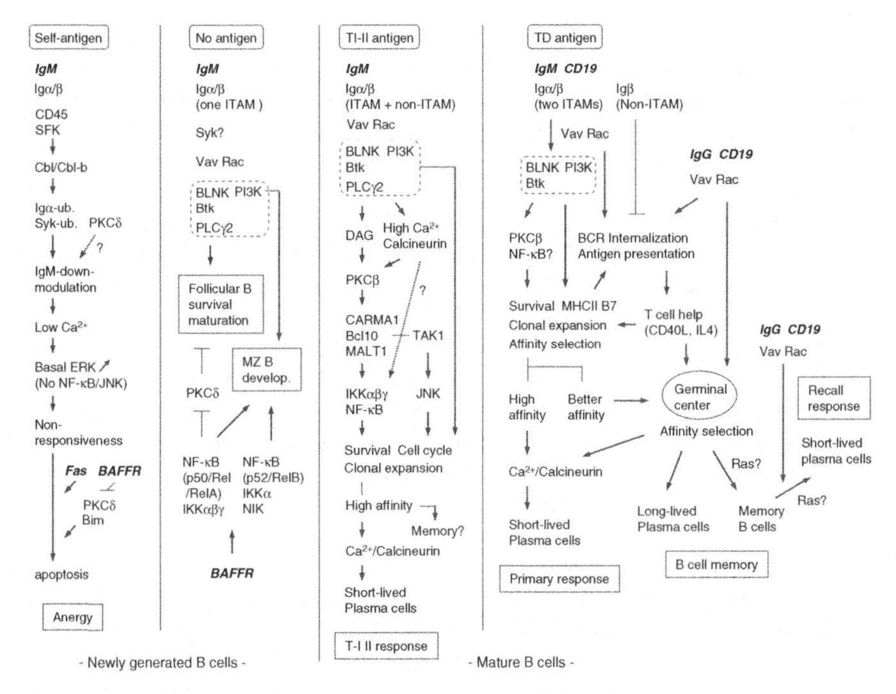

Fig. 5. B-cell receptor signaling pathways that induce B-cell anergy, maturation, TI–II, and TD immune responses. A hypothetical schema of B-cell receptor signaling pathways for each type of response (in *boxes*), which are deduced from many pieces of published data and inferences from them. Receptors inducing such responses are indicated in bold and italics. "BCR signalosome" is indicated by the *dotted box*. B1 cell development is not considered since it is substantially affected in a codominant fashion by the genetic background of 129 strains which many of the gene knock-out mice may have inherited (Corcoran and Metcalf 1999). See text for more details

evolved to recognize characteristic patterns of pathogens, BCR is primarily diversified in random fashion and has not evolved to recognize a particular structure. Therefore each B cell cannot discriminate self from nonself, but can respond to whatever the antigens in a manner depending on its developmental state and the valency of the antigens. Although it seems complicated how such a variety of B-cell responses are integrated into a whole B-cell system that performs humoral immune response to nonself and tolerance to self, I propose to take it rather simple as follows: the BCR signaling system may have evolved to let B cells respond to antigens as least toward antibody production. While a variety of inhibitory coreceptors, such as CD22, CD72, and FcγRIIB, obviously function to inhibit the activation signals from BCR, BCR signaling itself appears to be inhibitory. B-cell antigen receptor signal transduction machinery appears to be designed to transiently activate the cells but rapidly to endocytose any antigens that bind BCR, perhaps through Igα/Igβ ubiquitination by Cbl proteins, to ensure the cessation of the initial activation of the cells. In immature B cells that are not yet equipped for the antigen presentation (e.g., low MHC class II expression), this leads to low BCR expression and an unresponsive state, which results in the anergy to self antigens. In mature B cells, this endocytosis proceeds to antigen presentation, and the cells eventually die unless they are rescued by cognate helper T cells (TD response). Thus the response of peripheral self-reactive B cells, which may have escaped the central tolerance or to have been newly generated through somatic hypermutation, is prohibited by the T-cell tolerance system. In addition, the response to the TD antigens is restricted by the need of BCR-coligation with CD19/CD21 via complement C3d that marks nonself antigens like pathogens. Such passive tactics of BCR signaling appear to be a basis for the B-cell system that responds only to nonself.

Only when the antigen is so highly repetitive and large, such as bacterial capsular polysaccharides or TI–II antigens, that BCR is not able to endocytose, is BCR aggregated on the cell surface and continues to signal. In mature B cells (mostly MZ and B1 B cells) this results in clonal expansion and antibody production that is not checked by T cells. But this will not cause autoimmunity to similar repetitive self antigens because the same signal induces receptor editing in immature B cells still expressing RAG proteins and apoptosis of RAG-negative immature B cells, thus eliminating B cells carrying BCRs that react with such self antigens. It has not been established why immature B cells are destined to die upon BCR aggregation, but it was recently proposed that low DAG production unbalanced with Ca^{2+} flux upon BCR ligation is responsible for induction of apoptosis rather than maturation in immature B cells (Hoek et al. 2006). The remarkable feature of the TI–II response that every one of the BCR signalosome molecules is indispensable for it, in contrast to their dispensability for TD response, implies that the BCR signaling machinery for the TI–II response is an exceptional version, which might be adapted to tight self-control as no T-cell control is available.

References

Aiba Y, Oh-hora M, Kiyonaka S, Kimura Y, Hijikata A, Mori Y, Kurosaki T (2004) Activation of RasGRP3 by phosphorylation of Thr-133 is required for B-cell receptor-mediated Ras activation. Proc Natl Acad Sci USA 101:16612–16617

Allam A, Niiro H, Clark EA, Marshall AJ (2004) The adaptor protein Bam32 regulates Rac1 activation and actin remodeling through a phosphorylation-dependent mechanism. J Biol Chem 279:39775–39782

Bajpai UD, Zhang K, Teutsch M, Sen R, Wortis HH (2000) Bruton's tyrosine kinase links the B-cell receptor to nuclear factor kappaB activation. J Exp Med 191:1735–1744

Barrington R, Zhang M, Zhong X, Jonsson H, Holodick N, Cherukuri A, Pierce S, Rothstein T, Carroll M (2005) CD21/CD19 coreceptor signaling promotes B-cell survival during primary immune responses. J Immunol 175:2859–2867

Berland R, Fernandez L, Kari E, Han JH, Lomakin I, Akira S, Wortis HH, Kearney JF, Ucci AA, Imanishi-Kari T (2006) Toll-like receptor 7-dependent loss of B-cell tolerance in pathogenic autoantibody knockin mice. Immunity 25:429–440

Brummer T, Shaw PE, Reth M, Misawa Y (2002) Inducible gene deletion reveals different roles for B-Raf and Raf-1 in B-cell antigen receptor signalling. EMBO J 21:5611–5622

Byth KF, Conroy LA, Howlett S, Smith AJ, May J, Alexander DR, Holmes N (1996) CD45-null transgenic mice reveal a positive regulatory role for CD45 in early thymocyte development, in the selection of CD4+CD8+ thymocytes, and B-cell maturation. J Exp Med 183:1707–1718

Chen D, Heath V, O'Garra A, Johnston J, McMahon M (1999) Sustained activation of the raf-MEK-ERK pathway elicits cytokine unresponsiveness in T cells. J Immunol 163:5796–5805

Chen Z, Koralov SB, Gendelman M, Carroll MC, Kelsoe G (2000) Humoral immune responses in Cr2–/– mice: enhanced affinity maturation but impaired antibody persistence. J Immunol 164:4522–4532

Cheng AM, Rowley B, Pao W, Hayday A, Bolen JB, Pawson T (1995) Syk tyrosine kinase required for mouse viability and B-cell development. Nature 378:303–306

Cherukuri A, Cheng PC, Pierce SK (2001a) The role of the CD19/CD21 complex in B-cell processing and presentation of complement-tagged antigens. J Immunol 167:163–172

Cherukuri A, Cheng P, Sohn H, Pierce S (2001b) The CD19/CD21 complex functions to prolong B-cell antigen receptor signaling from lipid rafts. Immunity 14:169–179

Christensen S, Shupe J, Nickerson K, Kashgarian M, Flavell R, Shlomchik M (2006) Toll-like receptor 7 and TLR9 dictate autoantibody specificity and have opposing inflammatory and regulatory roles in a murine model of lupus. Immunity 25:417–428

Claudio E, Brown K, Park S, Wang H, Siebenlist U (2002) BAFF-induced NEMO-independent processing of NF-kappa B2 in maturing B cells. Nat Immunol 3:958–965

Clayton E, Bardi G, Bell S, Chantry D, Downes C, Gray A, Humphries L, Rawlings D, Reynolds H, Vigorito E, Turner M (2002) A crucial role for the p110delta subunit of phosphatidylinositol 3-kinase in B-cell development and activation. J Exp Med 196:753–763

Corcoran LM, Metcalf D (1999) IL-5 and Rp105 signaling defects in B cells from commonly used 129 mouse substrains. J Immunol 163:5836–5842

Cornall RJ, Cyster JG, Hibbs ML, Dunn AR, Otipoby KL, Clark EA, Goodnow CC (1998) Polygenic autoimmune traits: Lyn, CD22, and SHP-1 are limiting elements of a biochemical pathway regulating BCR signaling and selection. Immunity 8:497–508

Cornall R, Cheng A, Pawson T, Goodnow C (2000) Role of Syk in B-cell development and antigen-receptor signaling. Proc Natl Acad Sci USA 97:1713–1718

Coughlin JJ, Stang SL, Dower NA, Stone JC (2005) RasGRP1 and RasGRP3 regulate B-cell proliferation by facilitating B-cell receptor-Ras signaling. J Immunol 175:7179–7184

Craxton A, Draves KE, Gruppi A, Clark EA (2005) BAFF regulates B-cell survival by downregulating the BH3-only family member Bim via the ERK pathway. J Exp Med 202:1363–1374

Croix DA, Ahearn JM, Rosengard AM, Han S, Kelsoe G, Ma M, Carroll MC (1996) Antibody response to a T-dependent antigen requires B-cell expression of complement receptors. J Exp Med 183:1857–1864

Cyster JG, Goodnow CC (1995) Protein tyrosine phosphatase 1C negatively regulates antigen receptor signaling in B lymphocytes and determines thresholds for negative selection. Immunity 2:13–24

Cyster JG, Hartley SB, Goodnow CC (1994) Competition for follicular niches excludes self-reactive cells from the recirculating B-cell repertoire. Nature 371:389–395

Dal Porto JM, Haberman AM, Kelsoe G, Shlomchik MJ (2002) Very low affinity B cells form germinal centers, become memory B cells, and participate in secondary immune responses when higher affinity competition is reduced. J Exp Med 195:1215–1221

Dingjan GM, Middendorp S, Dahlenborg K, Maas A, Grosveld F, Hendriks RW (2001) Bruton's tyrosine kinase regulates the activation of gene rearrangements at the lambda light chain locus in precursor B cells in the mouse. J Exp Med 193:1169–1178

Dolmetsch RE, Lewis RS, Goodnow CC, Healy JI (1997) Differential activation of transcription factors induced by Ca2+ response amplitude and duration. Nature 386:855–858

Egawa T, Albrecht B, Favier B, Sunshine M, Mirchandani K, O'Brien W, Thome M, Littman D (2003) Requirement for CARMA1 in antigen receptor-induced NF-kappa B activation and lymphocyte proliferation. Curr Biol 13:1252–1258

Ellmeier W, Jung S, Sunshine MJ, Hatam F, Xu Y, Baltimore D, Mano H, Littman DR (2000) Severe B-cell deficiency in mice lacking the tec kinase family members Tec and Btk. J Exp Med 192:1611–1624

Engels N, Wollscheid B, Wienands J (2001) Association of SLP-65/BLNK with the B-cell antigen receptor through a non-ITAM tyrosine of Ig-alpha. Eur J Immunol 31:2126–2134

Fang W, Weintraub BC, Dunlap B, Garside P, Pape KA, Jenkins MK, Goodnow CC, Mueller DL, Behrens TW (1998) Self-reactive B lymphocytes overexpressing Bcl-xL escape negative selection and are tolerized by clonal anergy and receptor editing. Immunity 9:35–45

Fehr T, Rickert R, Odermatt B, Roes J, Rajewsky K, Hengartner H, Zinkernagel R (1998) Antiviral protection and germinal center formation, but impaired B-cell memory in the absence of CD19. J Exp Med 188:145–155

Flaswinkel H, Barner M, Reth M (1995) The tyrosine activation motif as a target of protein tyrosine kinases and SH2 domains. Semin Immunol 7:21–27

Franzoso G, Carlson L, Poljak L, Shores E, Epstein S, Leonardi A, Grinberg A, Tran T, Scharton-Kersten T, Anver M, et al (1998) Mice deficient in nuclear factor (NF)-kappa B/p52 present with defects in humoral responses, germinal center reactions, and splenic microarchitecture. J Exp Med 187:147–159

Franzoso G, Carlson L, Xing L, Poljak L, Shores EW, Brown KD, Leonardi A, Tran T, Boyce BF, Siebenlist U (1997) Requirement for NF-kappaB in osteoclast and B-cell development. Genes Dev 11:3482–3496

Fruman D, Snapper S, Yballe C, Davidson L, Yu J, Alt F, Cantley L (1999) Impaired B-cell development and proliferation in absence of phosphoinositide 3-kinase p85alpha. Science 283:393–397

Fu C, Turck C, Kurosaki T, Chan A (1998) BLNK: a central linker protein in B-cell activation. Immunity 9:93–103

Fujikawa K, Miletic AV, Alt FW, Faccio R, Brown T, Hoog J, Fredericks J, Nishi S, Mildiner S, Moores SL, et al (2003) Vav1/2/3-null mice define an essential role for Vav family proteins in lymphocyte development and activation but a differential requirement in MAPK signaling in T and B cells. J Exp Med 198:1595–1608

Gauld SB, Benschop RJ, Merrell KT, Cambier JC (2005) Maintenance of B-cell anergy requires constant antigen receptor occupancy and signaling. Nat Immunol 6:1160–1167

Gazumyan A, Reichlin A, Nussenzweig MC (2006) Ig beta tyrosine residues contribute to the control of B-cell receptor signaling by regulating receptor internalization. J Exp Med 203:1785–1794

Glassford J, Soeiro I, Skarell SM, Banerji L, Holman M, Klaus GG, Kadowaki T, Koyasu S, Lam EW (2003) BCR targets cyclin D2 via Btk and the p85alpha subunit of PI3-K to induce cell cycle progression in primary mouse B cells. Oncogene 22:2248–2259

Goitsuka R, Fujimura Y, Mamada H, Umeda A, Morimura T, Uetsuka K, Doi K, Tsuji S, Kitamura D (1998) BASH, a novel signaling molecule preferentially expressed in B cells of the bursa of Fabricius. J Immunol 161:5804–5808

Gong S, Nussenzweig MC (1996) Regulation of an early developmental checkpoint in the B-cell pathway by Ig beta. Science 272:411–414

Goodnow CC (1992) Transgenic mice and analysis of B-cell tolerance. Annu Rev Immunol 10:489–518

Goodnow CC, Crosbie J, Adelstein S, Lavoie TB, Smith-Gill SJ, Brink RA, Pritchard-Briscoe H, Wotherspoon JS, Loblay RH, Raphael K, et al (1988) Altered immunoglobulin expression and functional silencing of self-reactive B lymphocytes in transgenic mice. Nature 334:676–682

Gross JA, Dillon SR, Mudri S, Johnston J, Littau A, Roque R, Rixon M, Schou O, Foley KP, Haugen H, et al (2001) TACI–Ig neutralizes molecules critical for B-cell development and autoimmune disease. impaired B-cell maturation in mice lacking BLyS. Immunity 15:289–302

Grossmann M, O'Reilly LA, Gugasyan R, Strasser A, Adams JM, Gerondakis S (2000) The anti-apoptotic activities of Rel and RelA required during B-cell maturation involve the regulation of Bcl-2 expression. EMBO J 19:6351–6360
Grumont RJ, Rourke IJ, O'Reilly LA, Strasser A, Miyake K, Sha W, Gerondakis S (1998) B lymphocytes differentially use the Rel and nuclear factor kappaB1 (NF-kappaB1) transcription factors to regulate cell cycle progression and apoptosis in quiescent and mitogen-activated cells. J Exp Med 187:663–674
Halverson R, Torres RM, Pelanda R (2004) Receptor editing is the main mechanism of B-cell tolerance toward membrane antigens. Nat Immunol 5:645–650
Hara H, Wada T, Bakal C, Kozieradzki I, Suzuki S, Suzuki N, Nghiem M, Griffiths E, Krawczyk C, Bauer B, et al (2003) The MAGUK family protein CARD11 is essential for lymphocyte activation. Immunity 18:763–775
Hartley S, Cooke M, Fulcher D, Harris A, Cory S, Basten A, Goodnow C (1993) Elimination of self-reactive B lymphocytes proceeds in two stages: arrested development and cell death. Cell 72:325–335
Hashimoto A, Takeda K, Inaba M, Sekimata M, Kaisho T, Ikehara S, Homma Y, Akira S, Kurosaki T (2000) Cutting edge: essential role of phospholipase C-gamma 2 in B-cell development and function. J Immunol 165:1738–1742
Hayashi K, Nittono R, Okamoto N, Tsuji S, Hara Y, Goitsuka R, Kitamura D (2000) The B cell-restricted adaptor BASH is required for normal development and antigen receptor-mediated activation of B cells. Proc Natl Acad Sci USA 97:2755–2760
Hayashi K, Yamamoto M, Nojima T, Goitsuka R, Kitamura D (2003) Distinct signaling requirements for Dmu selection IgH allelic exclusion, pre-B-cell transition, and tumor suppression in B cell progenitors. Immunity 18:825–836
Hayashi K, Nojima T, Goitsuka R, Kitamura D (2004) Impaired receptor editing in the primary B-cell repertoire of BASH-deficient mice. J Immunol 173:5980–5988
Hayden MS, Ghosh S (2004) Signaling to NF-kappaB. Genes Dev 18:2195–2224
Healy J, Dolmetsch R, Timmerman L, Cyster J, Thomas M, Crabtree G, Lewis R, Goodnow C (1997) Different nuclear signals are activated by the B-cell receptor during positive versus negative signaling. Immunity 6:419–428
Hertz M, Nemazee D (1997) BCR ligation induces receptor editing in IgM + IgD- bone marrow B cells in vitro. Immunity 6:429–436
Hikida M, Johmura S, Hashimoto A, Takezaki M, Kurosaki T (2003) Coupling between B-cell receptor and phospholipase C-gamma2 is essential for mature B-cell development. J Exp Med 198:581–589
Hippen KL, Tze LE, Behrens TW (2000) CD5 maintains tolerance in anergic B cells. J Exp Med 191:883–890
Hippen KL, Schram BR, Tze LE, Pape KA, Jenkins MK, Behrens TW (2005) In vivo assessment of the relative contributions of deletion, anergy, and editing to B-cell self-tolerance. J Immunol 175:909–916
Hoek K, Antony P, Lowe J, Shinners N, Sarmah B, Wente S, Wang D, Gerstein R, Khan W (2006) Transitional B-cell fate is associated with developmental stage-specific regulation of diacylglycerol and calcium signaling upon B-cell receptor engagement. J Immunol 177:5405–5413
Horikawa K, Martin SW, Pogue SL, Silver K, Peng K, Takatsu K, Goodnow CC (2007) Enhancement and suppression of signaling by the conserved tail of IgG memory-type B-cell antigen receptors. J Exp Med
Ho WY, Cooke MP, Goodnow CC, Davis MM (1994) Resting and anergic B cells are defective in CD28-dependent costimulation of naive CD4+ T cells. J Exp Med 179:1539–1549
Hsu M, Toellner K, Vinuesa C, Maclennan I (2006) B-cell clones that sustain long-term plasmablast growth in T-independent extrafollicular antibody responses. Proc Natl Acad Sci USA 103:5905–5910

Inaoki M, Sato S, Weintraub BC, Goodnow CC, Tedder TF (1997) CD19-regulated signaling thresholds control peripheral tolerance and autoantibody production in B lymphocytes. J Exp Med 186:1923–1931

Iritani BM, Forbush KA, Farrar MA, Perlmutter RM (1997) Control of B-cell development by Ras-mediated activation of Raf. EMBO J 16:7019–7031

Ishiai M, Kurosaki M, Pappu R, Okawa K, Ronko I, Fu C, Shibata M, Iwamatsu A, Chan AC, Kurosaki T (1999) BLNK required for coupling Syk to PLC gamma 2 and Rac1-JNK in B cells. Immunity 10:117–125

Johmura S, Oh-hora M, Inabe K, Nishikawa Y, Hayashi K, Vigorito E, Kitamura D, Turner M, Shingu K, Hikida M, Kurosaki T (2003) Regulation of Vav localization in membrane rafts by adaptor molecules Grb2 and BLNK. Immunity 18:777–787

Jou ST, Carpino N, Takahashi Y, Piekorz R, Chao JR, Carpino N, Wang D, Ihle JN (2002) Essential, nonredundant role for the phosphoinositide 3-kinase p110delta in signaling by the B-cell receptor complex. Mol Cell Biol 22:8580–8591

Jumaa H, Wollscheid B, Mitterer M, Wienands J, Reth M, Nielsen PJ (1999) Abnormal development and function of B lymphocytes in mice deficient for the signaling adaptor protein SLP-65. Immunity 11:547–554

Jun JE, Wilson LE, Vinuesa CG, Lesage S, Blery M, Miosge LA, Cook MC, Kucharska EM, Hara H, Penninger JM, et al (2003) Identifying the MAGUK protein Carma-1 as a central regulator of humoral immune responses and atopy by genome-wide mouse mutagenesis. Immunity 18:751–762

Kabak S, Skaggs BJ, Gold MR, Affolter M, West KL, Foster MS, Siemasko K, Chan AC, Aebersold R, Clark MR (2002) The direct recruitment of BLNK to immunoglobulin alpha couples the B-cell antigen receptor to distal signaling pathways. Mol Cell Biol 22:2524–2535

Kaisho T, Schwenk F, Rajewsky K (1997) The roles of gamma 1 heavy chain membrane expression and cytoplasmic tail in IgG1 responses. Science 276:412–415

Kaisho T, Takeda K, Tsujimura T, Kawai T, Nomura F, Terada N, Akira S (2001) IkappaB kinase alpha is essential for mature B-cell development and function. J Exp Med 193:417–426

Karasuyama H, Rolink A, Melchers F (1996) Surrogate light chain in B-cell development. Adv Immunol 63:1–41

Kerner J, Appleby M, Mohr R, Chien S, Rawlings D, Maliszewski C, Witte O, Perlmutter R (1995) Impaired expansion of mouse B-cell progenitors lacking Btk. Immunity 3:301–312

Kersseboom R, Middendorp S, Dingjan GM, Dahlenborg K, Reth M, Jumaa H, Hendriks RW (2003) Bruton's tyrosine kinase cooperates with the B-cell linker protein SLP-65 as a tumor suppressor in Pre-B cells. J Exp Med 198:91–98

Khan WN, Alt FW, Gerstein RM, Malynn BA, Larsson I, Rathbun G, Davidson L, Muller S, Kantor AB, Herzenberg LA, et al (1995) Defective B-cell development and function in Btk-deficient mice. Immunity 3:283–299

King LB, Monroe JG (2000) Immunobiology of the immature B cell: plasticity in the B-cell antigen receptor-induced response fine tunes negative selection. Immunol Rev 176:86–104

Kishihara K, Penninger J, Wallace V, Kündig T, Kawai K, Wakeham A, Timms E, Pfeffer K, Ohashi P, Thomas M (1993) Normal B lymphocyte development but impaired T-cell maturation in CD45-exon6 protein tyrosine phosphatase-deficient mice. Cell 74:143–156

Kitaura Y, Jang IK, Wang Y, Han YC, Inazu T, Cadera EJ, Schlissel M, Hardy RR, Gu H (2007) Control of the B cell-intrinsic tolerance programs by ubiquitin ligases Cbl and Cbl-b. Immunity 26:567–578

Kohler F, Storch B, Kulathu Y, Herzog S, Kuppig S, Reth M, Jumaa H (2005) A leucine zipper in the N terminus confers membrane association to SLP-65. Nat Immunol 6:204–210

Kontgen F, Grumont RJ, Strasser A, Metcalf D, Li R, Tarlinton D, Gerondakis S (1995) Mice lacking the c-rel proto-oncogene exhibit defects in lymphocyte proliferation, humoral immunity, and interleukin-2 expression. Genes Dev 9:1965–1977

Kraus M, Saijo K, Torres RM, Rajewsky K (1999) Ig-alpha cytoplasmic truncation renders immature B cells more sensitive to antigen contact. Immunity 11:537–545

Kraus M, Pao L, Reichlin A, Hu Y, Canono B, Cambier J, Nussenzweig M, Rajewsky K (2001) Interference with immunoglobulin (Ig)alpha immunoreceptor tyrosine-based activation motif (ITAM) phosphorylation modulates or blocks B-cell development, depending on the availability of an Igbeta cytoplasmic tail. J Exp Med 194:455–469

Kraus M, Alimzhanov MB, Rajewsky N, Rajewsky K (2004) Survival of resting mature B lymphocytes depends on BCR signaling via the Igalpha/beta heterodimer. Cell 117:787–800

Kurosaki T (2002) Regulation of B-cell signal transduction by adaptor proteins. Nat Rev Immunol 2:354–363

Kurosaki T, Tsukada S (2000) BLNK: connecting Syk and Btk to calcium signals. Immunity 12:1–5

Lam K, Rajewsky K (1998) Rapid elimination of mature autoreactive B cells demonstrated by Cre-induced change in B-cell antigen receptor specificity in vivo. Proc Natl Acad Sci USA 95:13171–13175

Lam KP, Kuhn R, Rajewsky K (1997) In vivo ablation of surface immunoglobulin on mature B cells by inducible gene targeting results in rapid cell death. Cell 90:1073–1083

Lang J, Arnold B, Hammerling G, Harris AW, Korsmeyer S, Russell D, Strasser A, Nemazee D (1997) Enforced Bcl-2 expression inhibits antigen-mediated clonal elimination of peripheral B cells in an antigen dose-dependent manner and promotes receptor editing in autoreactive, immature B cells. J Exp Med 186:1513–1522

Leadbetter EA, Rifkin IR, Hohlbaum AM, Beaudette BC, Shlomchik MJ, Marshak-Rothstein A (2002) Chromatin-IgG complexes activate B cells by dual engagement of IgM and Toll-like receptors. Nature 416:603–607

Leitges M, Schmedt C, Guinamard R, Davoust J, Schaal S, Stabel S, Tarakhovsky A (1996) Immunodeficiency in protein kinase cbeta-deficient mice. Science 273:788–791

Leitges M, Sanz L, Martin P, Duran A, Braun U, Garcia JF, Camacho F, Diaz-Meco MT, Rennert PD, Moscat J (2001) Targeted disruption of the zetaPKC gene results in the impairment of the NF-kappaB pathway. Mol Cell 8:771–780

Lesley R, Xu Y, Kalled SL, Hess DM, Schwab SR, Shu HB, Cyster JG (2004) Reduced competitiveness of autoantigen-engaged B cells due to increased dependence on BAFF. Immunity 20:441–453

Li ZW, Omori SA, Labuda T, Karin M, Rickert RC (2003) IKK beta is required for peripheral B-cell survival and proliferation. J Immunol 170:4630–4637

Mackay F, Browning JL (2002) BAFF: a fundamental survival factor for B cells. Nat Rev Immunol 2:465–475

Mackay F, Woodcock SA, Lawton P, Ambrose C, Baetscher M, Schneider P, Tschopp J, Browning JL (1999) Mice transgenic for BAFF develop lymphocytic disorders along with autoimmune manifestations. J Exp Med 190:1697–1710

Martin P, Duran A, Minguet S, Gaspar ML, Diaz-Meco MT, Rennert P, Leitges M, Moscat J (2002) Role of zeta PKC in B-cell signaling and function. EMBO J 21:4049–4057

Martin SW, Goodnow CC (2002) Burst-enhancing role of the IgG membrane tail as a molecular determinant of memory. Nat Immunol 3:182–188

Maruyama M, Lam K, Rajewsky K (2000) Memory B-cell persistence is independent of persisting immunizing antigen. Nature 407:636–642

Matthews SA, Dayalu R, Thompson LJ, Scharenberg AM (2003) Regulation of protein kinase Cnu by the B-cell antigen receptor. J Biol Chem 278:9086–9091

Matthews SA, Liu P, Spitaler M, Olson EN, McKinsey TA, Cantrell DA, Scharenberg AM (2006) Essential role for protein kinase D family kinases in the regulation of class II histone deacetylases in B lymphocytes. Mol Cell Biol 26:1569–1577

Meade J, Tybulewicz VL, Turner M (2004) The tyrosine kinase Syk is required for light chain isotype exclusion but dispensable for the negative selection of B cells. Eur J Immunol 34:1102–1110

Mecklenbrauker I, Saijo K, Zheng NY, Leitges M, Tarakhovsky A (2002) Protein kinase Cdelta controls self-antigen-induced B-cell tolerance. Nature 416:860–865

Mecklenbrauker I, Kalled SL, Leitges M, Mackay F, Tarakhovsky A (2004) Regulation of B-cell survival by BAFF-dependent PKCdelta-mediated nuclear signalling. Nature 431:456–461

Melamed D, Nemazee D (1997) Self-antigen does not accelerate immature B-cell apoptosis, but stimulates receptor editing as a consequence of developmental arrest. Proc Natl Acad Sci USA 94:9267–9272

Melamed D, Benschop R, Cambier J, Nemazee D (1998) Developmental regulation of B lymphocyte immune tolerance compartmentalizes clonal selection from receptor selection. Cell 92:173–182

Miyamoto A, Nakayama K, Imaki H, Hirose S, Jiang Y, Abe M, Tsukiyama T, Nagahama H, Ohno S, Hatakeyama S, Nakayama KI (2002) Increased proliferation of B cells and auto-immunity in mice lacking protein kinase Cdelta. Nature 416:865–869

Mizuno T, Rothstein TL (2005) B-cell receptor (BCR) cross-talk: CD40 engagement creates an alternate pathway for BCR signaling that activates I kappa B kinase/I kappa B alpha/NF-kappa B without the need for PI3K and phospholipase C gamma. J Immunol 174:6062–6070

Morrow TA, Muljo SA, Zhang J, Hardwick JM, Schlissel MS (1999) Pro-B-cell-specific transcription and proapoptotic function of protein kinase Ceta. Mol Cell Biol 19:5608–5618

Nagaoka H, Takahashi Y, Hayashi R, Nakamura T, Ishii K, Matsuda J, Ogura A, Shirakata Y, Karasuyama H, Sudo T, et al(2000) Ras mediates effector pathways responsible for pre-B-cell survival, which is essential for the developmental progression to the late pre-B-cell stage. J Exp Med 192:171–182

Nemazee D (2006) Receptor editing in lymphocyte development and central tolerance. Nat Rev Immunol 6:728–740

Nemazee D, Russell D, Arnold B, Haemmerling G, Allison J, Miller JF, Morahan G, Buerki K (1991) Clonal deletion of autospecific B lymphocytes. Immunol Rev 122:117–132

Newton K, Dixit V (2003) Mice lacking the CARD of CARMA1 exhibit defective B lymphocyte development and impaired proliferation of their B and T lymphocytes. Curr Biol 13: 1247–1251

Niiro H, Allam A, Stoddart A, Brodsky FM, Marshall AJ, Clark EA (2004) The B lymphocyte adaptor molecule of 32 kilodaltons (Bam32) regulates B-cell antigen receptor internalization. J Immunol 173:5601–5609

Nishina H, Bachmann M, Oliveira-dos-Santos AJ, Kozieradzki I, Fischer KD, Odermatt B, Wakeham A, Shahinian A, Takimoto H, Bernstein A, et al (1997) Impaired CD28-mediated interleukin 2 production and proliferation in stress kinase SAPK/ERK1 kinase (SEK1)/ mitogen-activated protein kinase kinase 4 (MKK4)-deficient T lymphocytes. J Exp Med 186:941–953

Nojima T, Hayashi K, Goitsuka R, Nakayama K, Nakayama K, Kitamura D (2006) Double knockout mice show BASH and PKCdelta have different epistatic relationships in B-cell maturation and CD40-mediated activation. Immunol Lett 105:48–54

Obukhanych T, Nussenzweig M (2006) T-independent type II immune responses generate memory B cells. J Exp Med 203:305–310

O'Connor BP, Vogel LA, Zhang W, Loo W, Shnider D, Lind EF, Ratliff M, Noelle RJ, Erickson LD (2006) Imprinting the fate of antigen-reactive B cells through the affinity of the B-cell receptor. J Immunol 177:7723–7732

Oh-hora M, Johmura S, Hashimoto A, Hikida M, Kurosaki T (2003) Requirement for Ras guanine nucleotide releasing protein 3 in coupling phospholipase C-gamma2 to Ras in B-cell receptor signaling. J Exp Med 198:1841–1851

Ohnishi K, Melchers F (2003) The nonimmunoglobulin portion of lambda5 mediates cell-autonomous pre-B-cell receptor signaling. Nat Immunol 4:849–856

Okkenhaug K, Bilancio A, Farjot G, Priddle H, Sancho S, Peskett E, Pearce W, Meek S, Salpekar A, Waterfield M, et al (2002) Impaired B and T-cell antigen receptor signaling in p110delta PI 3-kinase mutant mice. Science 297:1031–1034

Oliver PM, Vass T, Kappler J, Marrack P (2006) Loss of the proapoptotic protein, Bim, breaks B-cell anergy. J Exp Med 203:731–741

Pappu R, Cheng AM, Li B, Gong Q, Chiu C, Griffin N, White M, Sleckman BP, Chan AC (1999) Requirement for B-cell linker protein (BLNK) in B-cell development. Science 286: 1949–1954

Pasparakis M, Schmidt-Supprian M, Rajewsky K (2002) IkappaB kinase signaling is essential for maintenance of mature B cells. J Exp Med 196:743–752

Patterson HC, Kraus M, Kim YM, Ploegh H, Rajewsky K (2006) The B-cell receptor promotes B-cell activation and proliferation through a non-ITAM tyrosine in the Igalpha cytoplasmic domain. Immunity 25:55–65

Paus D, Phan TG, Chan TD, Gardam S, Basten A, Brink R (2006) Antigen recognition strength regulates the choice between extrafollicular plasma cell and germinal center B-cell differentiation. J Exp Med 203:1081–1091

Pelanda R, Braun U, Hobeika E, Nussenzweig MC, Reth M (2002) B-cell progenitors are arrested in maturation but have intact VDJ recombination in the absence of Ig-alpha and Ig-beta. J Immunol 169:865–872

Peng SL, Gerth AJ, Ranger AM, Glimcher LH (2001) NFATc1 and NFATc2 together control both T and B-cell activation and differentiation. Immunity 14:13–20

Petro JB, Khan WN (2001) Phospholipase C-gamma 2 couples Bruton's tyrosine kinase to the NF-kappaB signaling pathway in B lymphocytes. J Biol Chem 276:1715–1719

Petro JB, Rahman SM, Ballard DW, Khan WN (2000) Bruton's tyrosine kinase is required for activation of IkappaB kinase and nuclear factor kappaB in response to B-cell receptor engagement. J Exp Med 191:1745–1754

Pfeifhofer C, Gruber T, Letschka T, Thuille N, Lutz-Nicoladoni C, Hermann-Kleiter N, Braun U, Leitges M, Baier G (2006) Defective IgG2a/2b class switching in PKC alpha–/– mice. J Immunol 176:6004–6011

Pulendran B, Kannourakis G, Nouri S, Smith K, Nossal G (1995) Soluble antigen can cause enhanced apoptosis of germinal-centre B cells. Nature 375:331–334

Rajewsky K (1996) Clonal selection and learning in the antibody system. Nature 381:751–758

Rathmell JC, Cooke MP, Ho WY, Grein J, Townsend SE, Davis MM, Goodnow CC (1995) CD95 (Fas)-dependent elimination of self-reactive B cells upon interaction with CD4+ T cells. Nature 376:181–184

Rathmell JC, Fournier S, Weintraub BC, Allison JP, Goodnow CC (1998) Repression of B7.2 on self-reactive B cells is essential to prevent proliferation and allow Fas-mediated deletion by CD4(+) T cells. J Exp Med 188:651–659

Reichlin A, Hu Y, Meffre E, Nagaoka H, Gong S, Kraus M, Rajewsky K, Nussenzweig M (2001) B-cell development is arrested at the immature B-cell stage in mice carrying a mutation in the cytoplasmic domain of immunoglobulin beta. J Exp Med 193:13–23

Reichlin A, Gazumyan A, Nagaoka H, Kirsch K, Kraus M, Rajewsky K, Nussenzweig M (2004) A B-cell receptor with two Igalpha cytoplasmic domains supports development of mature but anergic B cells. J Exp Med 199:855–865

Reth M, Wienands J (1997) Initiation and processing of signals from the B-cell antigen receptor. Annu Rev Immunol 15:453–479

Rickert R, Rajewsky K, Roes J (1995) Impairment of T-cell-dependent B-cell responses and B-1 cell development in CD19-deficient mice. Nature 376:352–355

Ridderstad A, Nossal GJ, Tarlinton DM (1996) The xid mutation diminishes memory B-cell generation but does not affect somatic hypermutation and selection. J Immunol 157:3357–3365

Ridderstad A, Tarlinton D (1997) B-cell memory in xid mice is long-lived despite reduced memory B-cell frequency. Scand J Immunol 45:655–659

Ruefli-Brasse A, French D, Dixit V (2003) Regulation of NF-kappaB-dependent lymphocyte activation and development by paracaspase. Science 302:1581–1584

Rui L, Vinuesa CG, Blasioli J, Goodnow CC (2003) Resistance to CpG DNA-induced autoimmunity through tolerogenic B-cell antigen receptor ERK signaling. Nat Immunol 4:594–600

Rui L, Healy JI, Blasioli J, Goodnow CC (2006) ERK signaling is a molecular switch integrating opposing inputs from B-cell receptor and T-cell cytokines to control TLR4-driven plasma cell differentiation. J Immunol 177:5337–5346

Ruland J, Duncan GS, Elia A, del Barco Barrantes I, Nguyen L, Plyte S, Millar DG, Bouchard D, Wakeham A, Ohashi PS, Mak TW (2001) Bcl10 is a positive regulator of antigen receptor-induced activation of NF-kappaB and neural tube closure. Cell 104:33–42

Ruland J, Duncan G, Wakeham A, Mak T (2003) Differential requirement for Malt1 in T and B-cell antigen receptor signaling. Immunity 19:749–758

Saijo K, Mecklenbrauker I, Santana A, Leitger M, Schmedt C, Tarakhovsky A (2002) Protein kinase C beta controls nuclear factor kappaB activation in B cells through selective regulation of the IkappaB kinase alpha. J Exp Med 195:1647–1652

Saijo K, Schmedt C, Su IH, Karasuyama H, Lowell CA, Reth M, Adachi T, Patke A, Santana A, Tarakhovsky A (2003) Essential role of Src-family protein tyrosine kinases in NF-kappaB activation during B-cell development. Nat Immunol 4:274–279

Sandel P, Monroe J (1999) Negative selection of immature B cells by receptor editing or deletion is determined by site of antigen encounter. Immunity 10:289–299

Sasaki Y, Derudder E, Hobeika E, Pelanda R, Reth M, Rajewsky K, Schmidt-Supprian M (2006) Canonical NF-kappaB activity, dispensable for B-cell development, replaces BAFF-receptor signals and promotes B-cell proliferation upon activation. Immunity 24:729–739

Sato S, Steeber D, Tedder T (1995) The CD19 signal transduction molecule is a response regulator of B-lymphocyte differentiation. Proc Natl Acad Sci USA 92:11558–11562

Sato S, Sanjo H, Takeda K, Ninomiya-Tsuji J, Yamamoto M, Kawai T, Matsumoto K, Takeuchi O, Akira S (2005) Essential function for the kinase TAK1 in innate and adaptive immune responses. Nat Immunol 6:1087–1095

Sauer K, Liou J, Singh SB, Yablonski D, Weiss A, Perlmutter RM (2001) Hematopoietic progenitor kinase 1 associates physically and functionally with the adaptor proteins B-cell linker protein and SLP-76 in lymphocytes. J Biol Chem 276:45207–45216

Schiemann B, Gommerman JL, Vora K, Cachero TG, Shulga-Morskaya S, Dobles M, Frew E, Scott ML (2001) An essential role for BAFF in the normal development of B cells through a BCMA-independent pathway. Science 293:2111–2114

Schweighoffer E, Vanes L, Mathiot A, Nakamura T, Tybulewicz VL (2003) Unexpected requirement for ZAP-70 in pre-B-cell development and allelic exclusion. Immunity 18:523–533

Shih TA, Meffre E, Roederer M, Nussenzweig MC (2002a) Role of BCR affinity in T-cell dependent antibody responses in vivo. Nat Immunol 3:570–575

Shih TA, Roederer M, Nussenzweig MC (2002b) Role of antigen receptor affinity in T cell-independent antibody responses in vivo. Nat Immunol 3:399–406

Shinohara H, Yasuda T, Aiba Y, Sanjo H, Hamadate M, Watarai H, Sakurai H, Kurosaki T (2005) PKC beta regulates BCR-mediated IKK activation by facilitating the interaction between TAK1 and CARMA1. J Exp Med 202:1423–1431

Shivtiel S, Leider N, Melamed D (2002a) Receptor editing in CD45-deficient immature B cells. Eur J Immunol 32:2264–2273

Shivtiel S, Leider N, Sadeh O, Kraiem Z, Melamed D (2002b) Impaired light chain allelic exclusion and lack of positive selection in immature B cells expressing incompetent receptor deficient of CD19. J Immunol 168:5596–5604

Shokat K, Goodnow C (1995) Antigen-induced B-cell death and elimination during germinal-centre immune responses. Nature 375:334–338

Sidorenko SP, Law CL, Klaus SJ, Chandran KA, Takata M, Kurosaki T, Clark EA (1996) Protein kinase C mu (PKC mu) associates with the B-cell antigen receptor complex and regulates lymphocyte signaling. Immunity 5:353–363

Sommer K, Guo B, Pomerantz J, Bandaranayake A, Moreno-García M, Ovechkina Y, Rawlings D (2005) Phosphorylation of the CARMA1 linker controls NF-kappaB activation. Immunity 23:561–574

Song H, Zhang J, Chiang YJ, Siraganian RP, Hodes RJ (2007) Redundancy in B-cell developmental pathways: c-Cbl inactivation rescues early B-cell development through a B-cell linker protein-independent pathway. J Immunol 178:926–935

Spitaler M, Cantrell DA (2004) Protein kinase C and beyond. Nat Immunol 5:785–790

Stephenson LM, Miletic AV, Kloeppel T, Kusin S, Swat W (2006) Vav proteins regulate the plasma cell program and secretory Ig production. J Immunol 177:8620–8625

Su YW, Jumaa H (2003) LAT links the pre-BCR to calcium signaling. Immunity 19:295–305

Su TT, Guo B, Kawakami Y, Sommer K, Chae K, Humphries LA, Kato RM, Kang S, Patrone L, Wall R, et al (2002) PKC-beta controls I kappa B kinase lipid raft recruitment and activation in response to BCR signaling. Nat Immunol 3:780–786

Sun Z, Arendt CW, Ellmeier W, Schaeffer EM, Sunshine MJ, Gandhi L, Annes J, Petrzilka D, Kupfer A, Schwartzberg PL, Littman DR (2000) PKC-theta is required for TCR-induced NF-kappaB activation in mature but not immature T lymphocytes. Nature 404:402–407

Suzuki H, Matsuda S, Terauchi Y, Fujiwara M, Ohteki T, Asano T, Behrens TW, Kouro T, Takatsu K, Kadowaki T, Koyasu S (2003) PI3K and Btk differentially regulate B-cell antigen receptor-mediated signal transduction. Nat Immunol 4:280–286

Suzuki H, Terauchi Y, Fujiwara M, Aizawa S, Yazaki Y, Kadowaki T, Koyasu S (1999) Xid-like immunodeficiency in mice with disruption of the p85alpha subunit of phosphoinositide 3-kinase. Science 283:390–392

Swat W, Fujikawa K, Ganiatsas S, Yang D, Xavier RJ, Harris NL, Davidson L, Ferrini R, Davis RJ, Labow MA, et al (1998) SEK1/MKK4 is required for maintenance of a normal peripheral lymphoid compartment but not for lymphocyte development. Immunity 8:625–634

Takahashi Y, Inaminc A, Hashimoto S, Haraguchi S, Yoshioka E, Kojima N, Abe R, Takemori T (2005) Novel role of the Ras cascade in memory B-cell response. Immunity 23:127–138

Tan JE, Wong SC, Gan SK, Xu S, Lam KP (2001) The adaptor protein BLNK is required for b cell antigen receptor-induced activation of nuclear factor-kappa B and cell cycle entry and survival of B lymphocytes. J Biol Chem 276:20055–20063

Teixeira C, Stang SL, Zheng Y, Beswick NS, Stone JC (2003) Integration of DAG signaling systems mediated by PKC-dependent phosphorylation of RasGRP3. Blood 102:1414–1420

Thompson JS, Bixler SA, Qian F, Vora K, Scott ML, Cachero TG, Hession C, Schneider P, Sizing ID, Mullen C, et al (2001) BAFF-R, a newly identified TNF receptor that specifically interacts with BAFF. Science 293:2108–2111

Tolar P, Sohn HW, Pierce SK (2005) The initiation of antigen-induced B-cell antigen receptor signaling viewed in living cells by fluorescence resonance energy transfer. Nat Immunol 6:1168–1176

Torres RM, Flaswinkel H, Reth M, Rajewsky K (1996) Aberrant B-cell development and immune response in mice with a compromised BCR complex. Science 272:1804–1808

Tsuji S, Okamoto M, Yamada K, Okamoto N, Goitsuka R, Arnold R, Kiefer F, Kitamura D (2001) B-cell adaptor containing src homology 2 domain (BASH) links B-cell receptor signaling to the activation of hematopoietic progenitor kinase 1. J Exp Med 194:529–539

Tumang JR, Owyang A, Andjelic S, Jin Z, Hardy RR, Liou ML, Liou HC (1998) c-Rel is essential for B lymphocyte survival and cell cycle progression. Eur J Immunol 28:4299–4312

Turner M, Mee PJ, Costello PS, Williams O, Price AA, Duddy LP, Furlong MT, Geahlen RL, Tybulewicz VL (1995) Perinatal lethality and blocked B-cell development in mice lacking the tyrosine kinase Syk. Nature 378:298–302

Waisman A, Kraus M, Seagal J, Ghosh S, Melamed D, Song J, Sasaki Y, Classen S, Lutz C, Brombacher F, et al (2007) IgG1 B-cell receptor signaling is inhibited by CD22 and promotes the development of B cells whose survival is less dependent on Ig{alpha}/{beta}. J Exp Med 204:747–758

Wakabayashi C, Adachi T, Wienands J, Tsubata T (2002) A distinct signaling pathway used by the IgG-containing B-cell antigen receptor. Science 298:2392–2395

Walmsley MJ, Ooi SK, Reynolds LF, Smith SH, Ruf S, Mathiot A, Vanes L, Williams DA, Cancro MP, Tybulewicz VL (2003) Critical roles for Rac1 and Rac2 GTPases in B-cell development and signaling. Science 302:459–462

Wang D, Feng J, Wen R, Marine JC, Sangster MY, Parganas E, Hoffmeyer A, Jackson CW, Cleveland JL, Murray PJ, Ihle JN (2000) Phospholipase Cgamma2 is essential in the functions of B-cell and several Fc receptors. Immunity 13:25–35

Weih DS, Yilmaz ZB, Weih F (2001) Essential role of RelB in germinal center and marginal zone formation and proper expression of homing chemokines. J Immunol 167:1909–1919

Wen R, Chen Y, Schuman J, Fu G, Yang S, Zhang W, Newman DK, Wang D (2004) An important role of phospholipase Cgamma1 in pre-B-cell development and allelic exclusion. EMBO J 23:4007–4017

Wienands J, Schweikert J, Wollscheid B, Jumaa H, Nielsen PJ, Reth M (1998) SLP-65: A new signaling component in B lymphocytes which requires expression of the antigen receptor for phosphorylation. J Exp Med 188:791–795

Winslow MM, Gallo EM, Neilson JR, Crabtree GR (2006) The calcineurin phosphatase complex modulates immunogenic B-cell responses. Immunity 24:141–152

Xu H, Li H, Suri-Payer E, Hardy R, Weigert M (1998) Regulation of anti-DNA B cells in recombination-activating gene-deficient mice. J Exp Med 188:1247–1254

Xu S, Tan JE, Wong EP, Manickam A, Ponniah S, Lam KP (2000) B-cell development and activation defects resulting in xid-like immunodeficiency in BLNK/SLP-65-deficient mice. Int Immunol 12:397–404

Xu S, Huo J, Chew WK, Hikida M, Kurosaki T, Lam KP (2006) Phospholipase Cgamma2 dosage is critical for B-cell development in the absence of adaptor protein BLNK. J Immunol 176:4690–4698

Xu S, Lee KG, Huo J, Kurosaki T, Lam KP (2007) Combined deficiencies in Bruton tyrosine kinase and phospholipase Cgamma2 arrest B-cell development at a pre-BCR+ stage. Blood 109:3377–3384

Xue L, Morris S, Orihuela C, Tuomanen E, Cui X, Wen R, Wang D (2003) Defective development and function of Bcl10-deficient follicular, marginal zone and B1 B cells. Nat Immunol 4:857–865

Yamamoto M, Hayashi K, Nojima T, Matsuzaki Y, Kawano Y, Karasuyama H, Goitsuka R, Kitamura D (2006) BASH-novel PKC-Raf-1 pathway of pre-BCR signaling induces kappa gene rearrangement. Blood 108:2703–2711

Yamamoto M, Nojima T, Hayashi K, Goitsuka R, Furukawa K, Azuma T, Kitamura D (2004) BASH-deficient mice: limited primary repertoire and antibody formation, but sufficient affinity maturation and memory B-cell generation, in anti-NP response. Int Immunol 16:1161–1171

Yamazaki T, Kurosaki T (2003) Contribution of BCAP to maintenance of mature B cells through c-Rel. Nat Immunol 4:780–786

Yamazaki T, Takeda K, Gotoh K, Takeshima H, Akira S, Kurosaki T (2002) Essential immunoregulatory role for BCAP in B-cell development and function. J Exp Med 195:535–545

Zheng Y, Liu H, Coughlin J, Zheng J, Li L, Stone JC (2005) Phosphorylation of RasGRP3 on threonine 133 provides a mechanistic link between PKC and Ras signaling systems in B cells. Blood 105:3648–3654

Zhou H, Wertz I, O'Rourke K, Ultsch M, Seshagiri S, Eby M, Xiao W, Dixit V (2004) Bcl10 activates the NF-kappaB pathway through ubiquitination of NEMO. Nature 427:167–171

5
How Do T Cells Discriminate Self from Nonself?

Catherine Mazza and Bernard Malissen

5.1 Introduction

There are two forms of immune responses, innate and adaptive. Individuals from species capable of innate immune responses possess limited repertoires of receptors dedicated to this task. Innate receptors recognizes either stress-induced self-molecules (as exemplified by the NKG2D receptor, see Chapter 2), or bacterial, viral, or protozoan components that are difficult to mutate without an impact on pathogen replicative capacity (as exemplified by the Toll-like receptors, see Chapter 1). The specificity of these receptors is encoded in the germline, and although their expression may be restricted to a certain cell type, they are not clonally distributed (reviewed in Beutler 2003). Most individuals within a species capable of innate immune responses share very similar repertoires of microbial sensors (Pisitkun et al. 2006). Although the ligands of the Toll-like receptors were initially defined as "pathogen-associated molecular patterns," it should be stressed that in the case of bacteria these ligands are not exclusively derived from pathogens. Therefore, sensors of innate immunity such as Toll-like receptors do not distinguish microbial commensals from pathogens.

Individuals from species capable of adaptive immune responses assemble large repertoires of T-cell antigen receptors (TCRs) and B-cell antigen receptors (BCRs) to anticipate encounter with any possible antigen. TCR and BCR diversity is generated somatically through site-specific DNA recombinations, and each receptor of a particular specificity is expressed in a clone of lymphocytes. The potential repertoire of αβ TCRs that can be created by this process is exceedingly large and outnumber the number of T cells that are present in an individual at a given moment. The recognition by TCRs of antigenic peptides (p) bound to major histocompatibility complex (MHC)-encoded molecules is at the basis of adaptive

Centre d'Immunologie de Marseille-Luminy, INSERM U631, CNRS UMR6102, Université de la Méditerranée, Case 906, 13288 Marseille Cedex 9, France

Correspondence to: B. Malissen

immune responses. The fact that TCRs are blind to antigenic peptides that are not delivered to the cell surface through MHC molecules accounts for the MHC-restricted nature of antigen recognition by T cells. MHC class Ia and class II genes are highly polymorphic. This extraordinary diversity primarily involves the nucleotide sequences coding for the peptide-binding domain of the MHC molecules and is thought to reflect evolutionary pressure for recognition of a repertoire of antigenic peptides as broadly inclusive as possible. In contrast to the somatic site-specific DNA recombination that generates a vast TCR repertoire in each individual, the evolutionary forces operating on MHC molecules generate diversity at the population level. Accordingly, within a given species, the existence of multiple alleles for each MHC isoform increases the chance that at least a few individuals will express an allele capable of binding to at least one peptide derived from any pathogen encountered. MHC molecules do not discriminate self from non-self peptides. This ability fully relies on T cells and is in part acquired during T-cell differentiation in the thymus (central self-tolerance). Tolerance to self pMHC complexes that are not expressed in the thymus is further achieved after naïve T cells leave the thymus and circulate throughout secondary lymphoid organs (peripheral self-tolerance). Therefore, for T cells, the definition of the "immunologic self" is specified de novo in each individual and varies among individuals based on the inherited MHC class Ia and class II alleles, and on the repertoire of self-peptides capable of being presented by the products of these alleles (reviewed in Vivier and Malissen 2005). Considering, that some T cells normally reside in barrier organs that are laden with microbial symbionts (e.g., the gut), it is likely that a comprehensive definition of the "immunologic self" should also includes the peptides derived from those obligate microbial symbionts or postulate tolerance rules proper to organ-specific immune responses (Raz 2007). Among the peptides that are sampled by MHC molecules and recognized by TCRs during immune responses to highly variable viruses (as exemplified by HIV), a few derive from conserved viral regions that, if mutated, had a substantial impact on viral replicative fitness. Because they are not prone to evade T-cell responses through escape mutations, these rare peptides might be used to induce highly focused immune responses resulting in a slower disease progression (Altfeld and Allen 2006).

The selection of the binding-specificity of TCRs and of innate sensors such as Toll-like receptors occurs at two different time scales since innate sensors are selected within a species over evolutionary time, whereas TCRs are selected during the life of an individual. Most receptors encountered in Biology are encoded in the germline and committed to bind a predetermined physiological ligand with invariant thermodynamic parameters. In the case of stochastically generated TCRs, there is no a priori physiological ligand, and a given pMHC complex is qualified as "the cognate ligand" only after it successfully activates a given T-cell clone. Because the manifestations of T-cell responses (proliferation, cytotoxicity, cytokine production, etc.) have distinct activation thresholds, quantitative differences in TCR ligand binding translates into qualitatively different signals and leads to responses that can range from maximal activation to desensitization (reviewed in Bongrand and Malissen 1998). Among the pMHC ligands that interact with a given TCR, those denoted as agonists elicit the entire range of responses from expression of activation markers to stimulation of cell division and production of cytokines. Other pMHC ligands, termed partial agonists elicit some, but not all, of these responses. Finally, others are termed antagonists because they elicit no obvious response, except that they specifically block T-cell responses to agonist ligands. The ability of a given TCR to elicit a diverse range of biological responses according to its affinity for pMHC ligands plays a critical role in self–nonself discrimination. For instance, the developmental choices (deletion or differentiation into mature T cells) followed by immature T cells are primarily determined by the affinity of their TCRs for the self pMHC ligands expressed in the thymus (reviewed in von Boehmer et al. 2003). In this chapter, we provide an overview on the structure of the TCR complex, and on the way it recognizes foreign pMHC complexes and triggers T-cell activation. We also revisit the phenomenon of TCR binding degeneracy, a property with important biological consequences since there are far more potentially antigenic peptides in the environment of a mouse than it has T cells. Finally, we discuss whether the globally conserved mode of TCR docking on MHC molecules is due to coevolutionary pressure that shapes TCR and MHC gene products, or to the action of the CD4 and CD8 coreceptors.

5.2 Structure of the TCR Complex

5.2.1 Subunit Composition

The recognition of antigens by T cells and the ensuing transduction of intracellular signals are accomplished by a multisubunit transmembrane complex denoted as the TCR–CD3 complex. Mature T cells can be divided into two lineages on the basis of their antigen-binding TCR module. In adult mice, most T cells express TCR heterodimers consisting of covalently associated α and β chains, whereas a minor population expresses an alternative TCR isoform consisting of covalently associated γ and δ chains. The TCR α, β, γ, and δ chains each comprise an amino-terminal, clonally variable (V) region, and a carboxy-terminal constant (C) region. Their type I, single-pass transmembrane segment is followed by a very short cytoplasmic tail. Peptide loops homologous to immunoglobulin (Ig) complementarity-determining regions (CDRs) protrude at the membrane-distal end of TCR V domains. They exhibit interclonal sequence variation and constitute the binding site for pMHC ligands (see Fig. 1).

Transport of TCR heterodimers to the cell surface is dependent on their prior assembly with the invariant CD3 polypeptides. As shown in Fig. 2, the CD3γ, CD3δ, and CD3ε subunits are expressed as noncovalently associated CD3γε and CD3δε pairs, whereas CD3ζ polypeptides combine to form disulfide linked CD3ζζ homodimers. The stoichiometry of the TCR–CD3 complex has been the object of intense debate (Hayes and Love 2006). The αβ TCR–CD3 complex likely contains a single αβ TCR heterodimer that assembles with single CD3γε, CD3δε, and

Fig. 1. Structure of a TCR-peptide–major histocompatibility complex class I (pMHCI) complex (LC13, PDB accession number: 1mi5). The TCR is on the top, the α chain is colored *light blue*, and the β chain is colored *light pink*. The MHCI molecule is on the bottom, colored *yellow* with the noncovalently associated β2-microglobulin in *magenta*. The antigenic peptide is in *pink*. The complementarity-determining region (CDR) loops are colored in *green* (CDR3α and β), *blue* (CDR1α and β), and *red* (CDRα and β). (See Color Plates)

CD3ζζ pairs (Fig. 2). Aside from their role in allowing expression of the TCR at the cell surface, CD3 subunits are also responsible for coupling the antigen-binding TCR heterodimers to intracellular signaling pathways. None of the CD3 subunits possesses a cytoplasmic domain endowed with recognizable enzymatic activity. However, each contains one or several copies of a conserved sequence that is referred to as an immunoreceptor tyrosine-based activation motif (ITAM). ITAMs are also found in the transducing subunits of the BCR, the receptors for the Fc domain of IgE and IgG, activating natural killer (NK) cell receptors, and glycoprotein VI (GPVI), a receptor supporting platelet adhesion to collagen. The probable evolutionary relationships existing between the ITAM-containing signaling subunits associated with immunoreceptors is mirrored by their capacity to activate similar signaling pathways (Wegener et al. 1992). In the case of the

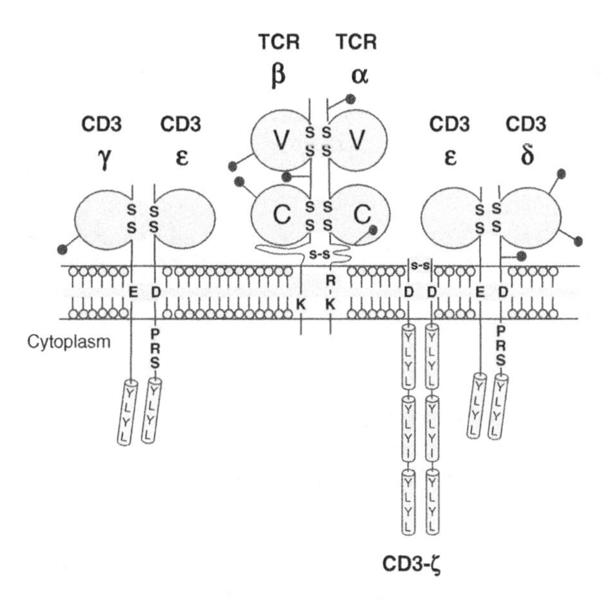

Fig. 2. Putative subunit composition of αβ T-cell receptor (*TCR*) complexes. Each of the three basic residues (*K* and *R* in the single-letter amino acid code) in the transmembrane segment of the TCRα and β polypeptides serves as critical contact for one of the three pairs of acidic residues (D/D in the case of CD3ζζ and D/E in the case of CD3γε and CD3δε) found in the transmembrane segments of the CD3 signaling dimers. The immunoreceptor tyrosine-based activation motifs (ITAMs) found in each of the CD3 subunits are shown as cylinders containing tyrosine-based docking motifs (*YLYL*). Sites with N-linked carbohydrates are indicated by *black dots*. The *semi-circles* depict sequence segments that fold as C or V immunoglobulin domain. The CD3ζ chain forms a disulfide-linked homodimer, and has an extracellular portion only nine amino acids long. *PRS* corresponds to the proline-rich sequence found in the cytosolic juxtamembrane segment of CD3ε. *S–S* corresponds to an intrachain or interchain disulfide bond

TCR, the ITAMs found in the CD3 subunits are phosphorylated by the Src-family protein tyrosine kinases (PTK) Lck and Fyn. Phosphorylated CD3 ITAMs subsequently recruit ZAP-70, a cytosolic tyrosine kinase belonging to the Syk-family. As a result, ZAP-70 becomes activated and phosphorylates a number of downstream molecules, among which the adaptor molecule Linker for Activation of T cells (LAT) plays a cardinal role in that it coordinates the assembly of one of the signalosomes linking the TCR–CD3 complex to a wealth of signaling pathways (reviewed in Malissen et al. 2005).

5.2.2 *Quaternary Structure*

The four pairs of polypeptides (αβ, CD3γε, CD3δε, and CD3ζζ) that compose αβ TCR–CD3 complexes are brought together via non-covalent interactions involving residues found in their helical transmembrane segments (Call et al. 2002). The

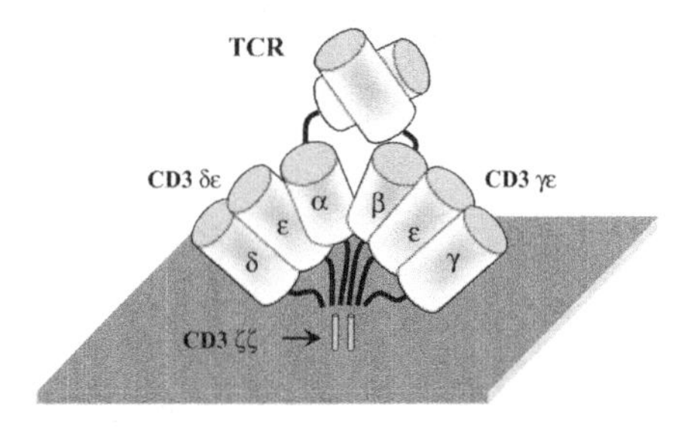

Fig. 3. A model depicting the association of the CD3 and αβ TCR ectodomains

TCRα chain transmembrane segment contains both a lysine and an arginine residue, whereas that of the TCRβ chain contains a single lysine residue. Each of the basic residues found in the TCR transmembrane segments interacts with a pair of acidic residues donated by the transmembrane segments of a CD3 dimer to constitute a tight three-helix association. The formation of three distinct three-helix associations (TCRα-CD3δε, TCRα-CD3ζζ, and TCRβ-CD3γε) constitutes the driving force for TCR–CD3 complex assembly and contributes to specifying its composition (Call et al. 2002).

Analysis of the extracellular domains of CD3γε and CD3δε dimers showed that the stalks that connect them to the transmembrane segments are shorter than those found in the TCR αβ heterodimer (Sun et al. 2004). When considered together with the tight association existing between the transmembrane segments of the TCR and of the CD3 dimers, this suggests that the extracellular domains of the CD3γε and CD3δε heterodimers lie just below the TCR C domains. As depicted in Fig. 3, the TCR αβ heterodimer with a vertical dimension of 80 Å likely projects from the cell membrane, and is flanked on either side by the shorter (40 Å) CD3 heterodimers (Sun et al. 2004). Epitope mapping analysis and structural modeling further suggested that the CD3γε and CD3δε dimers map to the TCRβ- and TCRα-side, respectively, and that the glycosylated CD3γ and CD3δ subunits lie away from the TCR heterodimer (Fig. 2). When the widths of the CD3δε and CD3γε dimers, 50 Å and 55 Å, respectively, are added to that of the αβ TCR heterodimer (58 Å), the full TCR–CD3 complex spans approximately 160 Å (excluding glycans).

As stated by Garcia and colleagues (Garcia and Adams 2005), "the extracellular contacts between the TCR and the CD3γε and CD3δε dimers may simply be imposed by the transmembrane interactions, like balloons that bump into each other because their strings are held together in a bunch, with no specificity whatsoever." However, existing data suggest that specific extracellular interactions occur between the αβ TCR and the CD3γε and CD3δε dimers. γδ TCR heterodimers have an overall shape distinct from αβ TCR heterodimers, and their C domains present unique molecular surfaces for the docking of CD3 subunits (Allison et al. 2001). As a consequence, in the mouse, CD3δε dimers fail to associate with γδ TCR heterodimers, and two CD3γε dimers are incorporated per γδ TCR complex (Hayes and Love 2002, 2006). The absence of the CD3δε dimer does not impede, however, signaling through γδ TCRs. Following cross-linking with anti-CD3 antibodies, γδ TCR complexes induce phosphorylation of appropriate substrates, mobilization of Ca^{2+}, and activation of MAP-Kinases with a greater amplitude than αβ TCR complexes. It is thus likely that the CD3δε pair additionally found in αβ TCRs constitutes a device needed for coupling αβ TCRs to the CD8 or CD4 coreceptors rather than to a unique signaling cassette (Doucey et al. 2003a). Another example of the influence exerted by the ligand-binding module on the subunit composition of the TCR–CD3 complex can be found in developing T cells. In contrast to mature T cells, immature T cells lack TCRα chain and express a substitute known as the pre-TCRα chain (pTα). pTα associates with TCRβ and gives rise to pre-TCR complexes which are composed of CD3γε and CD3ζζ dimers, and lack or are loosely associated with CD3δε dimers (von Boehmer 2005). Therefore, the nature of the ligand-binding module found in TCR complexes (αβ TCR, γδ TCR, and pre-TCR) clearly influences the spectrum of associated CD3 transducing subunits.

ITAM-containing signaling subunits that are evolutionary related to CD3 subunits have been also co-opted by other immunoreceptors such as Fc receptors and activating NK cell receptors (Malissen 2003a). The latter contain ligand-binding polypeptides with extracellular domains that belong to the Ig or to the C-type lectin families. The amino terminus is located in the extracellular space for Ig family polypeptides and in the cytoplasm for C-type lectin polypeptides. Despite this distinct topology, the transmembrane segment of the ligand-binding polypeptides of both families of activating NK cell receptors contain a basic residue (lysine or arginine) that associates according to a three-helix mode with signaling homodimers (DAP-10, DAP-12, etc.) that contain an aspartic acid in their transmembrane segments. Interestingly, activating NK receptors have inhibitory counterparts that lack a transmembrane lysine, and contain immunoreceptor tyrosine-based inhibitory (ITIM) in their longer cytoplasmic tail. Provided that inhibitory NK receptors are evolutionary more ancient than activating NK receptors, it has been suggested that, following gene duplication, one of the ancestral daughter gene gained an activating function through introduction of a positive charge into its transmembrane segment, and lost its inhibitory function by introduction of a stop codon prior to the sequence coding for the ITIM (Abi-Rached and Parham 2005). The acquisition of such "activating" functionality occurred independently in NK cell receptors belonging to the Ig or C-type lectin families, and might constitute an example of convergent evolution. Therefore, the noncovalent assembly mechanism that holds together the ligand-binding and transducing subunits and is based on membrane-embedded polar interactions constitutes a "signature" of activating immunoreceptors (Feng et al. 2005). During evolution, it likely permitted the "cheap" capture of some predating signaling pathways and also allowed some combinatorial diversification. For instance, the mouse NKG2D receptor can alternatively associate with either DAP-10 or DAP-12 (Diefenbach et al. 2002), and the αβ TCR complexes expressed on $CD8\alpha\alpha^{+}$-intestinal intraepithelial lymphocytes with CD3ζ or FcεRIγ dimers (Guy-Grand et al. 1994).

5.3 How Does the αβ TCR Convey Signals Across the Membrane?

How information is passed from the αβ TCR antigen-binding site to the CD3 ITAMs, a process termed TCR triggering, remains highly controversial (Malissen 2003a). None of the existing models satisfactorily accounts for the sensitivity of TCR recognition and the fact that the TCR embodies not only a switch that turns on signals in response to pMHC ligand, but also a timer that determines how long the signal will stay on. This "timer" dimension was originally revealed by the correlation existing between the bioactivity of a given pMHC ligand and the lifetime ($t_{1/2}$) of its association with the TCR. A recent study showed, however, that if changes in heat capacity of the TCR–pMHC interaction (ΔCp) were combined with the $t_{1/2}$ of the interaction, a better correlation was found with pMHC bioactivity (Krogsgaard et al. 2003). In contrast to the situation observed for the αβ TCR, for most germline-encoded receptors encountered in biology, the "timer" dimension is only revealed through the synthesis of weak agonist analogs of an agonist physiological ligand.

Apart from a model developed by Gil and colleagues (Gil et al. 2002, 2005) that suggests that the unmasking of a proline-rich sequence (PRS) found in the CD3ε cytoplasmic tail is the seminal event in TCR triggering, most other models posit that tyrosine phosphorylation of CD3 ITAMs constitutes the initial trigger. All these models of TCR triggering can be organized into three categories. The first one postulates the occurrence of pMHC-induced changes in the quaternary structure of the TCR–CD3 complex, whereas the second one considers that the whole TCR–CD3 complex behaves as a rigid body and stresses the importance of ligand-induced homo-oligomerization or hetero-oligomerization with the CD4 or CD8 coreceptors. The third model considers that the seminal event in TCR triggering is the exclusion of the protein tyrosine phosphatases (PTPases) from the close- contact zones that form between a T-cell and an antigen-presenting cell (APC). Before discussing each of these models, it should be emphasized that in the absence of structural data on the organization of the extracellular domains of the TCR–CD3 complex under pMHC-liganded and -unliganded conditions, these models remain largely speculative (Mitra et al. 2004).

5.3.1 Models Implying Architectural Changes in TCR–CD3 Quaternary Structure

The elegance of the concept of allosteric control where binding of a ligand at one site affects a distant functional site through a conformational change has stimulated efforts to document whether allosteric changes occur during TCR triggering. Some studies suggested that after TCR engagement, a conformational change in the quaternary structure of the TCR–CD3 complex demasks the evolutionary conserved

PRS sequence found in the cytosolic juxtamembrane segment of CD3ε, and permits the docking of the Nck adaptor through its first Src-homology (SH) 3 domain (Gil et al. 2002, 2005). Ligand-induced PRS exposure appears to occur prior to and independent of ITAM phosphorylation. Considering that overexpression of a dominant negative form of Nck inhibits early T-cell activation events, it has been suggested that PRS exposure constitutes a crucial "trigger" in the causality chain initiated by TCR engagement. Note that among ITAM-containing subunits, the PRS is unique to the CD3ε polypeptide. Therefore, considering that αβ and γδ TCRs are the only immunoreceptors that contain CD3ε polypeptides, they may function via a mode distinct from that of all other ITAM-associated immunoreceptors.

In mutant T cells expressing a CD3ε polypeptide where all the proline residues of the PRS sequence have been converted into alanine, the interaction between CD3ε and Nck was abolished, confirming that it depends on the CD3ε PRS. This interaction was, however, dispensable for T-cell development and function (Szymczak et al. 2005). The demonstration that the Nck–PRS interaction is biologically irrelevant does not invalidate the hypothesis that the TCR–CD3 complex undergoes conformational change after pMHC binding. Exposure of the PRS may be only one among several changes that occur in the architecture of the TCR–CD3 complex as a result of TCR engagement. For instance, it has been shown that ligation of the TCR leads to exposure of the amino terminus of CD3ζ, which is ordinarily buried within the complex, and to the dissociation of CD3ζ from the remainder of the TCR–CD3 complex (La Gruta et al. 2004). Note that the strong association observed between the two transmembrane helices of the CD3ζζ transmembrane likely precludes them from undergoing rotational motion (Call et al. 2006). In contrast to PRS unmasking, the architectural changes involving CD3ζ are dependent on Src family PTKs and thus presumably occur after or concurrently with kinase activation. The biological relevance of the changes postulated to occur in the TCR–CD3 quaternary structure remains however to be validated.

The view that conformational changes in the TCR–CD3 complex mediate critical "data transfer" from the external environment to the intracellular compartment is inconsistent with comparison of crystal structures of soluble αβ TCRs in their liganded and unliganded states. Although these structural data contribute only part of the information required to improve our understanding of TCR triggering in the context of the cell surface, they clearly show that TCR engagement does not demask a "trigger" in the TCR ectodomain resembling the dimerization motif found in the epidermal growth factor receptor (EGFR) ectodomain (Garrett et al. 2002; Ogiso et al. 2002). Moreover, no positional shift occurs at the membrane-proximal ends of the TCR following binding to pMHC, although conformational changes in the TCR antigen-binding site can be considerable (reviewed in Rudolph et al. 2006). Likewise, the TCR structure is essentially the same whether bound to agonist, superagonist, or antagonist pMHC ligands. However, one exception reported by Kjer-Nielsen is the subtle difference in the conformation of the A-B loop of the Cα domain in the crystal structure of the human TCR LC13 in isolation and in complex with its cognate pMHC ligand (Kjer-Nielsen et al. 2003). Considering that the Cα domain shows both some flexibility and an unusual Ig fold due to the absence of a typical outer β-sheet, it has been speculated that it can switch to a distinct conformation upon pMHC binding. This switch may affect the organization of CD3 components and trigger the activation cascade. Interestingly, the structure of the TCR Cγ and Cδ domains is distinct from that of Cα (Allison et al. 2001). Both the Cγ and Cδ domains are composed of a "classical" Ig domain with an outer β-sheet made of three strands. This suggests that in the case Cα plays an important function in data transfer during TCR triggering, the corresponding mechanism does not extend to γδ TCR–CD3 complex.

The view that TCR–CD3 quaternary changes are implicated in critical "data transfer" is also inconsistent with the observation that the ectodomains of the TCR–CD3 complex are dispensable for T-cell activation. For instance, following their artificial attachment to the inner leaflet of the membrane and their fusion to the FK506-binding protein 12 (FKBP12), monomeric CD3ζ chains can be oligomerized by the small synthetic molecule FKBP12, and trigger activation events that faithfully mimic those induced by antibody-mediated cross-linking of the TCR–CD3 ectodomains (Spencer et al. 1993). Moreover, in many engineered TCRs, extracellular domains and ITAM-containing intracellular domains have been mismatched via different intervening transmembrane segments. Given that the majority of these hybrid receptors were capable of activating T cells, little selectivity and structural constraint seems required at the level of the segments connecting extracellular and intracellular domains (Choudhuri et al. 2005b). These rather loose constraints should be compared to the tight constraints needed for the activation of seven-transmembrane-domain receptors, a class of receptors where the transmembrane segments act as a switch that transmit ligand-induced conformational changes across the membrane to large heterotrimeric G proteins.

5.3.2 Models Not Relying on TCR–CD3 Quaternary Changes

The fact that ITAM-containing transducing subunits can indiscriminately cooperate with ligand binding units belonging to the Ig superfamily or to the C-type lectin family raises the question of how ligand binding to such diverse extracellular domains can be communicated to the common architecture of the cytoplasmic domains of the transducing subunits. It has been suggested that all the ITAM-associated immunoreceptors use a common strategy for conveying signals through the membrane that consists in changing the local, steady-state balance of intracellular PTKs and PTPases through ligand-induced oligomerization (reviewed in Davis and van der Merwe 2006; Malissen 2003a; Strong and McFarland 2004). Once phosphorylated by the Src family PTK Lyn, the ITAM-associated with the transducing subunits of the receptor for the Fc domain of IgE (FcεRI) recruit Syk, a PTK that constitutes the founding member of the Syk family. Using the FcεRI receptor, a more tractable receptor than the TCR, it has been possible to demonstrate that when the relative Lyn/FcεRI molecular ratio is low, the Lyn-mediated phosphorylation of the FcεRI ITAMs is absolutely dependent on prior ligand-induced receptor oligomerization. In contrast, at a high Lyn/FcεRI ratio, the rate of spontaneous ITAM phosphorylation is faster than the rate of spontaneous ITAM dephosphorylation. Accumulation of tyrosine-phosphorylated FcεRI subunits, even in the absence of ligand-induced oligomerization, is the result (Scharenberg et al. 1995; Vonakis et al. 2001). Therefore, for at least some ITAM-based receptors, changing the intracellular balance of PTKs and PTPs can replace external stimuli and convert a ligand-activated switch into a constitutively active device. The ability to manipulate the ground-state PTK/PTP ratio in a ligand-dependent and focal manner also lies at the heart of the three models of TCR triggering to be described next.

5.3.2.1 "pMHC-Mediated Heterodimerization" Model

The TCR binds to the top of the MHC peptide-binding groove, whereas the CD4 and CD8 molecules bind to an invariant surface found at the bases of the MHC class II (MHCII) and MHC class I (MHCI) peptide-binding domain, respectively. Because the two surfaces recognized by the TCR and the CD4/C8 molecules are non-overlapping, it is thus sterically possible for a single pMHCI/II complex to be bound simultaneously by a specific TCR and a CD4/CD8 molecule (Chang et al. 2005). Early work suggested that this simultaneous engagement is required for optimal TCR triggering, and led CD4 and CD8 to be denoted as "coreceptors". The divalent nature of a pMHC monomer is the basis of the "pMHC-mediated heterodimerization" model of TCR triggering (Malissen 1996). According to this model, pMHC-dependent aggregation of a monomeric TCR–CD3 complex and of a Lck-associated coreceptor molecule suffices to modify in the TCR vicinity the steady state PTK/PTP balance in favor of PTK, and to trigger T-cell activation. This model applies equally well to CD8-dependent, MHCI-restricted TCRs as to CD4-dependent, MHCII-restricted TCRs. We will discuss it below for the former case.

The activation of ZAP-70, the "effector" PTK in charge of propagating TCR-elicited signals requires a "priming module" made of Lck and of ITAM-containing CD3 subunits (reviewed in Malissen 2003a). This module is primarily intended to create high affinity, phosphotyrosine-based docking sites for the tandem SH2 domains found in ZAP-70, and to destabilize the inhibited conformation assumed by ZAP-70 under resting conditions (Brdicka et al. 2005). The "priming" Lck kinase is anchored in the membrane via lipid modification, and also associates to the cytoplasmic tail of CD8 through a zinc chelate complex. Thereby, CD8 constitutes a specialized device that can bring, via translational diffusion, Lck into contact with a pMHCI-occupied TCR–CD3 complex and increase its local concentration. Following CD8–MHCI association, the probability of an encounter between a CD8-associated Lck and the CD3 cytoplasmic tails of pMHC-engaged TCR–CD3 complex increases. This results in strong phosphorylation of the CD3 ITAMs and subsequent recruitment and activation of ZAP-70. Although TCR–CD8-pMHCI ternary complexes have a greater stability (longer half-life) than TCR–pMHCI binary complexes (Luescher et al. 1995), the primary function of CD8 is probably not to reinforce the adhesiveness between T cells and antigen-presenting cells, but rather to increase the time an Lck kinase will reside in contiguity with CD3 ITAMs and ZAP-70.

What physiological benefit might result from the fragmentation of the TCR into receptor and coreceptor subcomplexes recognizing two distinct sites on the same pMHC ligand? It has been hypothesized that this structural fragmentation lies at the heart of T-cell physiology in that it permits a kinetic discrimination between short- and longer-lived TCR–pMHC interactions (Germain and Stefanova 1999). Provided that in the resting state most coreceptors are physically separated from the pool of TCR–CD3 complexes, some time must elapse between the encounter of a pMHCI-occupied TCR–CD3 complex and of a coreceptor (Yachi et al. 2005). This time delay in coreceptor recruitment depends primarily on the surface density of TCRs and of coreceptors, and thus fixes a kinetic threshold to TCR triggering. A long lived TCR–pMHCI complex would then have a high probability of recruiting a coreceptor and of leading to TCR triggering, whereas signaling proficient TCR–CD8–pMHCI ternary complex would have a low probability of assembling if a low affinity pMHCI ligand dissociates almost immediately from the TCR. Consistent with the above views, it should be mentioned that the contribution of CD8 to TCR triggering becomes particularly important when TCR–pMHCI interactions exhibit K_D values above 3 μM. Therefore, since the majority of TCR–pMHC interactions occur in the micromolar range, they likely require CD8 to result in TCR triggering (Holler and Kranz 2003).

According to the "pMHC-mediated heterodimerization" model, the CD4 and CD8 coreceptors constitute "kinetic gatekeepers" allowing the length of occupancy of the TCR binding site to be measured. Therefore, they constitute "intrinsic components" of the TCR–CD3 complex rather than "plain" co-inhibitors and co-stimulators. The function of the latter is more likely to be to relay information on the cellular environment in which antigen recognition occurs and to act as "context" detectors capable of tuning activation thresholds. In support of the "pMHC-mediated heterodimerization" model, monomers of pMHCI have been shown capable of activating adherent T cells (Delon et al. 1998). This observation is, however, controversial and other studies have suggested that soluble pMHC ligands need to be at least dimeric to trigger T-cell activation (Stone and Stern 2006). It has been suggested that these inconsistencies in the nature of the basic signaling unit that results in TCR triggering reflect differences in the state of activation of the T cells under study. For instance, in the case of adherent T cells, coreceptor-associated Lck kinases might be in a "high" state of activation, whereas when T cells are kept in solution, coreceptor-associated Lck kinases might be in a "low" state of activation (Doucey et al. 2003b; Randriamampita et al. 2003). In the latter instance, dimerization of TCR–CD3-CD8 complexes would result in two Lck kinases being brought together, and thereby convert them through trans-autophosphorylation into a "high" state of activation (Malissen 2003a). Importantly, however, the "pMHC-mediated heterodimerization" model fails to explain the fact that γδ T-cell activation does not rely on the presence of the CD4/CD8 coreceptors. Elucidating the nature and the valency of the ligands recognized by γδ T cells will certainly help our understanding of the mechanisms leading to coreceptor-independent γδ TCR firing, and will probably also shed light on the fact that for some αβ TCRs expressed on T-cell clones and hybridomas, coreceptors are clearly dispensable when agonist ligands are expressed at supra-physiological concentrations.

5.3.2.2 "Pseudodimer" Model

Because the intracellular machinery that generates pMHC ligands does not sort out foreign peptides from self peptides, the surface of a cell that is being scanned by TCRs is a mosaic of self- and foreign-pMHC ligands. In contrast to other models of TCR triggering that focus on foreign, agonist pMHC ligands, the "pseudodimer" model confers an important role to MHCII molecules bearing peptides derived from self-proteins, and that are referred to here as endogenous pMHCII complexes (Krogsgaard and Davis 2005; Krogsgaard et al. 2005). The "pseudodimer" model was originally developed for $CD4^+$ T cells. It stems from the observation that endogenous pMHCII complexes accumulate in immunological synapses when agonist pMHCII complexes are also present to drive activation. Although they are unable to induce TCR triggering on their own, endogenous pMHCII complexes are capable of facilitating TCR triggering by very low densities of agonist pMHCII complexes. Based on the assumption that in the resting state some CD4 coreceptors are physically associated with TCR–CD3 complexes and that, in contrast to CD8, CD4 could not bind the same pMHCII as the TCR, a "pseudodimer" model was proposed in which the basic signaling unit involves two TCRs, one CD4 molecule, one MHCII molecule bound to an agonist peptide, and a second MHCII molecule bound to a self-peptide. According to that model, a cooperative effect results from the complexation of one TCR binding to an agonist pMHCII ligand and another binding to an endogenous pMHCII, with CD4 acting to link the two TCR–CD3 complexes. The "pseudodimer" model appears only applicable to some $CD8^+$ T-cell clones (Cebecauer et al. 2005; Sporri and Reis e Sousa 2002).

Considering that the affinity of CD4 for MHCII is much lower than the affinity of CD8 for MHCI, Choudhuri and colleagues (Choudhuri et al. 2005a) suggested that the "pseudodimer" and "pMHC-mediated heterodimerization" models can be integrated into a single one. They proposed that the affinity of the CD8–MHCI interaction is sufficient to recruit CD8 to pMHCI-engaged TCR–CD3 monomeric complex and initiate triggering. In contrast, the CD4–MHCII interaction is too weak to drive the assembly of TCR–CD3–CD4 complexes, and an additional weak interaction between a CD4-preassociated TCR–CD3 complex and a MHCII molecule bearing an endogenous peptide is simultaneously required to stabilize CD4–MHCII interaction and initiate triggering. Note that according to the "pseudodimer" model, the Lck molecule associated to CD4 can phosphorylate CD3 subunits belonging to either of the two TCR–CD3 complexes present in the "pseudodimer".

It has been previously suspected that a weak reactivity toward self pMHC complexes expressed on the APCs found in secondary lymphoid organs permits the survival of naïve T cells in the periphery and keeps them in a state of heightened antigen sensitivity (Stefanova et al. 2002). The observation that suboptimal TCR signals might prime naïve T cells and allow them to respond strongly to a subsequent agonistic challenge is reminiscent of the situation documented for the IFN-α/β signaling pathway (Taniguchi and Takaoka 2001). Interestingly, the "pseudodimer" model extends the role attributed to self pMHC complexes by demonstrating that they also assist the TCR during agonist stimulation, and thus constitute bona fide "accessory self-peptides". Therefore, the thymus is not the only anatomical site where the TCR makes biologically relevant, low-affinity contacts with self pMHC complexes. Therefore, the purpose of TCR αβ positive selection is probably to select a peripheral TCR repertoire that can use self pMHC complexes to achieve maximal sensitivity toward the same self MHC molecules bound to foreign peptides.

5.3.2.3 "Kinetic-Segregation" Model

The cumulative length of the ectodomains of a pair of TCR–pMHC molecules is relatively short, spanning approximately 14 nm, whereas receptor PTPases known to inhibit TCR triggering, including CD45 and CD148, have much larger ectodomains. This observation suggested that the passive, size-based exclusion of the PTPases from the close- contact zones that form between a T-cell and an APC probably increases the half-lives of phosphorylated species in that region and thus favors tyrosine phosphorylation of TCR–CD3 complexes (Springer 1990). By holding the T-cell and APC membranes the appropriate distance apart, small adhesion molecules such as CD2 likely facilitate size-based exclusion of bulky PTPases and the encounter between the TCR and its pMHC ligand. It should be stressed that prior to the encounter of an APC, intermolecular collisions are likely to occur between Lck and CD3 subunits in the plane of the T-cell membrane. These transient interactions probably result in some constitutive phosphorylation of the CD3 subunits of the TCR complex. However, due to the action of PTPases, the net level of CD3 phosphorylation is kept too low for activation, although it may be sufficient to heighten T-cell sensitivity and promote T-cell survival (Stefanova et al. 2002). Size-based exclusion of PTPases constitutes the basis of the model referred to as the "kinetic-segregation model" (Choudhuri et al. 2005b). Importantly, this model allows for weak, ligand-independent TCR triggering in the PTPase-free, close-contact zones. The presence of agonist pMHC ligands in the close-contact zone essentially prevents ligand-engaged TCRs to diffuse outside of the close-contact zone and thus increases the half-life of the phosphorylated CD3 ITAMs they associate with. This increase is commensurable to the half-life of the TCR–pMHC interaction. It is important to note that the "kinetic-segregation model" avoids an absolute requirement for coreceptor by postulating that free Lck molecules associated with the membrane inner leaflet suffice to phosphorylate the CD3 subunits present in the PTPase-free close-contact zones. According to this model, the function of coreceptors is thus limited to signal amplification by recruiting additional Lck in the vicinity of ligand-engaged TCRs and by stabilizing their interaction with their substrates (CD3 ITAMs and ZAP-70) (Davis and van der Merwe 2006).

5.3.2.4 Other TCR Triggering Models

Provided that the CD3 subunits are tightly associated to the TCR, the whole TCR–CD3 complex would likely behave as a rigid body, and it has been suggested that pMHC binding pushes the whole TCR–CD3 complex into the cell membrane. This "piston-like" movement could expose CD3 ITAMs to the constitutive action of Lck and trigger the signaling cascade (reviewed in Krogsgaard and Davis 2005). Finally, it should be mentioned that TCR–CD3 complexes have been postulated to exist in the membrane as higher-order structures (TCR oligomers) even in the absence of cognate pMHC ligands (Schamel et al. 2005). Binding of pMHC to these postulated oligomers could alter their distance or the relative orientation of their individual components and result in TCR triggering.

5.4 TCR Assembly During Intrathymic Development

The α and β chains of the TCR control both the physiology of mature T cells and the unfolding of the intrathymic T-cell development program. During the latter process, TCR chains constitute key components of molecular sensors that counteract the stochastic nature of V(D)J recombinations and prevent the possible emergence of cells bearing strongly autoreactive TCRs (von Boehmer et al. 2003). Genetic studies have greatly contributed to the definition of the two consecutive developmental checkpoints that are controlled by these sensors and at which T cells progressing along the αβ-lineage undergo programmed cell death if they fail to rearrange TCR genes productively or if they express a TCR with inappropriate specificities.

5.4.1 TCR β selection

Transition through the earliest developmental checkpoint requires the operation of a sensor known as the pre-TCR. Because TCRβ gene rearrangements precede rearrangements at the TCRα locus, the pre-TCR complex lacks a TCRα chain. It is composed instead of a pTαTCRβ heterodimer that is non-covalently associated to CD3γε and CD3ζζ dimers (reviewed in Malissen et al. 1999)). At this stage of development, the pTα and CD3 components of the pre-TCR are already available and it is the TCRβ polypeptide that constitutes the rate-limiting factor in the assembly of the pre-TCR complex. As stressed by Harald von Boehmer, the invariant pTα subunit of the pre-TCR is more than just a structural substitute for the missing TCRα chain (von Boehmer 2005). It contains a unique proline-rich motif in its cytoplasmic tail that appears important for proper pre-TCR signaling. It has been further suggested that once assembled, the pre-TCR signals without binding to exogenous ligands. This cell-autonomous and ligand-independent mode of action is likely due to the fact that, once incorporated in the pre-TCR complex, pTα ectodomains have the ability to spontaneously form oligomers (Yamasaki et al. 2006). The phenotypic transition induced by the pre-TCR is generally referred to as β-selection. There is no selection for particular Vβ gene products at this checkpoint; it ensures only that double-negative (DN) $CD4^-CD8^-$ cells with productive TCRβ gene rearrangements become double-positive (DP) $CD4^+CD8^+$ cells (Wilson et al. 2001). The pre-TCR also contributes to establish clonal expression of TCR αβ heterodimers in that it prohibits further TCRβ chain rearrangements on the second TCRβ allele through the process of allelic exclusion.

5.4.2 *TCR αβ selection*

Signals emanating from the pre-TCR also trigger the induction of a high rate of TCRα rearrangements among DP cells. Following productive TCRα rearrangements and substitution of pTα by TCRα, DP cells express at their surface clonally distributed mature αβ TCR–CD3 complexes. Based on the specificity of their TCR for self-pMHC ligands, a small percentage (3 to 5%) of DPs cells are rescued from programmed cell death and induced to differentiate into $CD4^+CD8^-$ and $CD4^-CD8^+$ single positive (SP) cells. Such second developmental checkpoint that marks the boundary between the DP and the SP stage of thymic differentiation is referred to as TCR αβ selection, and is dependent on TCR binding specificity (see below). It is ablated by a number of genetic defects affecting (1) the assembly and proper cellular display of self pMHC ligands at the surface of thymic stromal cells (2) the synthesis of TCR–CD3 complexes (3) the CD4 and CD8 coreceptors, and (4) components of the signaling cassette activated by αβ TCRs (reviewed in (Malissen et al. 1999). SP cells correspond to the end products of the intrathymic αβ T-cell differentiation sequence, and gradually exit from the thymus to reach peripheral secondary lymphoid organs.

Because the site-specific DNA recombination reactions that affect the TCR genes result in the random reassortment of a limited set of germ-line encoded CDR1 and CDR2 loops and of a large set of somatically encoded CDR3 loops, the population of DP T lymphocytes present in the thymus shows a diverse repertoire of clonally distributed αβ TCRs. To function appropriately, TCRs have to match the combination of self-MHC gene products that are coincidently inherited and available from conception. Consequently, after the generation of an MHC allele-neutral TCR repertoire in DP thymocytes, the phase of epigenetic molecular matching known as positive TCR αβ selection is intended to favor the development of those clones expressing TCRs capable of weakly interacting with self-pMHC. Most DP thymocytes express TCRs with insufficient self-reactivity to mediate positive TCR αβ selection and die in situ through programmed cell death. A subset of DP thymocytes bearing TCRs with excess self-reactivity is also eliminated by apoptosis. This last mechanism of clonal deletion is known as negative TCR αβ selection and results in central tolerance. Therefore, whereas negative thymic selection prevents the emergence of strongly autoreactive T-cell clones, positive selection prevents the overloading of the periphery with T cells expressing TCR unable to fit structurally and cooperate with the inherited MHC alleles. It should be stressed that the nature and complexity of the self-peptides involved in positive selection and their degree of structural relatedness to the antigenic peptides that are encountered in the periphery remains controversial (Hogquist et al. 2005). The antigenic specificity of a given T-cell clone is fixed by the irreversible shut-down of the V(D)J recombinase machinery that is associated with positive TCR αβ selection. Moreover, in contrast to Ig genes, TCR genes are not subjected to somatic hypermutation in the periphery.

5.4.3 Secondary TCRα Rearrangements and Editing

The structure of TCRα locus is well suited to secondary rearrangements, providing a given DP cell with the possibility to express consecutively several TCRs and increasing its probability of being positively selected and thus of escaping "death by neglect". It is not clear, however, whether there is enough time between consecutive rounds of TCRα rearrangements to test each novel TCRα chain for binding to self pMHC ligands (Davodeau et al. 2001). It has also been suggested that DP cells can escape death by negative selection using secondary rearrangements as well (Hogquist et al. 2005). The physiological role of this phenomenon, referred to as receptor editing, remains however to be determined.

5.5 How TCRs Bind pMHC

5.5.1 General Features

The first TCR–pMHCI crystal structures have permitted to define a few "general" features of TCR–pMHCI interactions (Bankovich and Garcia 2003; Housset and Malissen 2003; Krogsgaard and Davis 2005; Rudolph and Wilson 2002; Rudolph et al. 2006). For instance, the docking of the TCR on the pMHC is such that the Vα and Vβ domains are closest to the amino- and carboxy-terminal residues of the antigenic peptide, respectively. Moreover, the extremely variable CDR3s are generally located over the center of the pMHCI surface and make contacts with the antigenic peptide as well as with the MHC α helices, whereas the less variable CDR1–CDR2 loops contact the termini of the bound peptide and residues of the MHC α helices (CDR1s), or exclusively the central part of the MHC α-helices (CDR2s). This "diagonal" TCR docking topology allows the relatively flat TCR binding surface to "slot" between the two high "peaks" that are found near the amino termini of the two MHC α helical regions, and thus maximizes the readout of the few peptide residues that are exposed to the TCR.

MHCII molecules bind processed peptides as long as 20 amino acids such that the peptide termini extend out of the open ends of the MHC groove (Batalia and Collins 1997). The peptide backbone in MHCII molecules is bound in an extended polyproline type II conformation and adopts a rather deep course in the binding groove. This contrasts with the situation observed in MHCI alleles where conserved hydrogen bonds fix the position of the amino and carboxy termini of the bound peptides, forcing the central part of peptides longer than eight amino acids to bulge out of the groove and become more accessible for TCR inspection. The structure of several TCR–pMHCII complexes has been recently determined (Hahn et al. 2005; Hennecke and Wiley 2002; Li et al. 2005a; Maynard et al. 2005; Reinherz et al. 1999). As previously observed for MHCI-restricted TCRs, the orientation of the TCR over the pMHCII surface maximizes contacts with the bound peptide by avoidance of the high points of the MHCII helices and of the ridge created by the amino terminus of the bound peptide. Analysis of the collection of available TCR–pMHCI and TCR–pMHCII complexes reveals the extent of variation afforded around the standard "diagonal" docking mode, and shows that there is no docking mode proper to MHC class I or class II molecules. A single TCR, denoted Ob.1A12 and originating from a patient with multiple sclerosis, showed a totally unusual binding topology (Hahn et al. 2005). Ob.1A12 recognizes with low affinity an immunodominant peptide of myelin basic protein (MBP residues 85–99) bound to HLA-DR2. Rather than being centered over the pMHCII surface it is instead positioned over the P2 peptide residue. This binding mode centers Ob.1A12 over the high point of the DRα helix and on the ridge created by the N-terminal extension of the peptide, indicating that these topological features do not preclude TCR binding as previously postulated. Moreover, the two CDR3 loops of Ob.1A12 create a dome-shaped cavity that is large enough to accommodate both an MHC residue and the P2 side chain. This feature is unprecedented, as the corresponding pocket found in some other TCRs accommodates a single peptide side chain but no MHC residue.

5.5.2 Flexibility of the TCR Antigen-Binding Site

Several TCRs have been crystallized in their unliganded and liganded states. Comparisons of these structures showed important changes that are restricted to the antigen-binding site, and that unevenly affected the various CDRs. For instance in the case of the KB5-C20 and LC13 TCRs, large-scale conformational reorganization of CDR3 loops have been observed upon binding to the pMHC surface, whereas CDR1 and CDR2 loops were the object of "en bloc" movements or of moderate conformational changes (Kjer-Nielsen et al. 2003; Reiser et al. 2003). In contrast, the 1G4 TCR showed very little conformational change upon recognition of HLA-A2 loaded with a peptide denoted ESO 9C and derived from a tumor-specific antigen (Chen et al. 2005). The ESO 9C-HLA-A2 ligand displays a prominent methionine-tryptophan "peg" in the central portion of the ESO 9C peptide, which forms a large protrusion from the surface of the pMHC. The cavity formed between the CDR1α, CDR3α, and CDR3β loops, and used by 1G4 to accommodate the methionine-tryptophan peg was preformed in the unliganded state. Therefore, in contrast to the conformational flexibility observed during the formation of TCR–pMHC complexes involving the KB5-C20 and LC13 TCRs, the 1G4 TCR showed that a well optimized pMHC-binding surface can be achieved through relatively small structural changes.

Thermodynamic parameters have been measured for a number of TCR–pMHC interactions establishing a recurrent thermodynamic profile whereby TCR ligation is governed by favorable enthalpic forces and unfavorable entropy (Boniface et al. 1999). The entropic penalty probably results from the stabilization of flexible CDR3 loops upon docking to the pMHC, whereas the favorable enthalpic forces result from the extensive network of bonds formed upon ligation. However, this thermodynamic signature may only apply to a subset of TCRs. For instance, it has been shown that the interaction of the LC13 TCR with the FLR–HLA-B8 ligand is entropically and enthalpically driven (Ely et al. 2006). Structural analyses of the LC13–FLR–HLA-B8 complex revealed that the CDR3 loops are well ordered in the unliganded state and undergo a conformational isomerization to adopt the liganded state, with the simultaneous expulsion of water molecules from the TCR–pMHC interface. The thermodynamics of LC13–FLR–HLA-B8 ligation is not the only exception. A similar thermodynamic profile is also found in the case of the formation of other TCR–pMHC complexes (Davis-Harrison et al. 2005; Mazza et al. 2007).

5.5.3 TCR Flexibility Facilitates TCR Binding Degeneracy

Binding specificity and degeneracy are imprecise but widely used concepts in immunology. Degeneracy (also called polyspecificity or molecular promiscuity) can be defined as the ability of a given receptor to interact with structurally distinct ligands. Conversely, specificity (also called monospecificity or molecular monogamy) can be defined as the ability of a given receptor to interact with one or a few ligands that are closely related in structure. Note that drug action is based on binding degeneracy since most drugs bind to sites that evolved to interact with physiological ligands. Degeneracy in protein-mediated recognition can be achieved through structural flexibility of the binding site, allowing the receptor to adapt to distinct ligands through conformational changes. Alternatively, studies of the interaction between the human growth hormone and the growth hormone receptor and of the NKG2D receptor with MHCI-like ligands have shown that an identical and rather rigid surface can be used to contact chemically different ligand surfaces (McFarland et al. 2003). It has also been shown that a given antibody can recognize several 14-mer peptides unrelated in sequences via differential usage of residues located in the binding site, the constellation of interacting residues being unique for each peptide. Finally, binding degeneracy can also be achieved via the recognition of sequence-independent features. For instance in the case of the MHC peptide-binding groove, interactions with the peptide main chain provides a generic and largely sequence-independent set of interactions.

A given TCR can recognize other pMHC ligands in addition to its cognate pMHC ligand. When the cross-recognized pMHC ligand fortuitously displays the same constellation of TCR contact residues, a phenomenon referred to as molecular mimicry, cross-recognition may be achieved without flexibility in the TCR antigen-binding site (Lang et al. 2002). In contrast, the BM3.3 TCR provided the first example of bona fide TCR degeneracy, in that the flexibility of the TCR antigen-binding site is exploited to facilitate its adaptation to distinct pMHC surfaces (Reiser et al. 2003). Comparative analysis of the BM3.3 TCR in complex with two distinct peptides, pBM1 and VSV8, bound to H-2K^b, showed that BM3.3 recognition focused on peptide position 6 (P6) of pBM1 and of VSV8. P6 is the only position contacted by the TCR that shows a homologous asparagine to glutamine replacement in the cross-recognized VSV8 peptide. All the other positions contacted by the TCR in VSV8 corresponded to non-conservative replacements, and affected TCR binding in a positive (replacement at P4) or a negative (replacement at P7) way. The structurally different replacement observed at P4 of VSV8 was exploited through CDR3α flexibility, which enables the residues at the apex of the CDR3α loop to be used fully and to compensate in part for the loosened interactions between the CDR3β loop and the residue found at P7 of VSV8. Altogether, these results show how structurally different replacements at P4 and P7 were accommodated through CDR3α loop flexibility, and demonstrate that the BM3.3 TCR adopts unique structural solution to adapt to each peptide. Therefore, TCR cross-reactivity does not always require a high degree of pMHC structural mimicry, and TCR degeneracy is primarily permitted by the malleability of the TCR antigen-binding site.

5.5.4 Affinity, Degeneracy, and Bioactivity

It has been originally suggested that the absence of mature T cells with high affinity TCRs towards foreign pMHC ligands is due to the fact that those TCRs are detrimental to T-cell physiology (as postulated by the "serial engagement model" of T-cell activation [Gonzalez et al. 2005]). Conversely, high-affinity TCRs have been postulated to have a higher probability of cross-reacting with several self-pMHC complexes, leading to intrathymic deletion of the corresponding T cells (Donermeyer et al. 2006; Huseby et al. 2006). However, in the case of antibodies, no obvious link has been found between affinity maturation during secondary B-cell responses and the magnitude of crossreactivity (James and Tawfik 2003). In some instances, affinity maturation of antibodies might even result in the generation of more rigid antigen-binding sites with a narrower specificity. Moreover, in the case of TCRs engineered in vitro to yield variants capable of binding the cognate pMHC ligand with higher affinity, the resulting increase in affinity (up to 26 pM) was not systematically associated with a broadening of specificity (Holler et al. 2003; Laugel et al. 2005; Li et al. 2005b). Therefore, TCR affinity and cross-reactivity appear to constitute two independent variables, and the absence of mature T cells with high-affinity TCRs is likely due to the fact that long TCR–pMHC bond lifetimes are detrimental to T-cell activation when pMHC ligands are present in limited amounts (Gonzalez et al. 2005).

5.5.5 Raison d'Être of TCR Degeneracy

Although the notion of degeneracy tends to be linked with deterioration in quality, it has been argued that TCR degeneracy has been selected to allow TCRs to recognize a universe of structurally diverse peptide antigens much larger in number than its own combinatorial diversity and than the number of T cells contained at a given moment in each individual (Mason 1998). Landsteiner and Pauling similarly hypothesized in the 1930s, that provided that a given antibody exists as an ensemble of isomers, each with a different antigen-binding site structure capable of accommodating a different ligand, then functional diversity could go far beyond primary sequence diversity. T-cell antigen receptor degeneracy is also at play during intrathymic T-cell positive selection and antigenic responses in the periphery. In these two instances, the cross-recognized ligands consist of structurally distinct peptides presented by the same self-MHC molecule. T-cell antigen receptor degeneracy is also directly involved in heterologous immunity (Selin et al. 2004), and allows adaptive immunity to cope with antigenic variation during antiviral immunity (Goulder and Watkins 2004). Considering that much of the burden of infectious diseases today is caused by antigenically variable pathogens that can escape from adaptive immunity induced by prior infection or vaccination, TCR, as well as antibody, degeneracy is of immense importance for vaccine design. Finally, considering that the TCR can translates small quantitative differences in ligand binding into qualitatively different signals, TCR degeneracy does permit the occurrence of T-cell responses that range from maximal activation to desensitization. In the absence of TCR-binding degeneracy, no variation in TCR–pMHC bond lifetimes would be allowed for a given TCR, and its signaling properties would be limited to a mere on-off switch.

5.5.6 Alloreactivity and Xenoreactivity

Effector T cells preferentially interact with target cells bearing MHC alleles that the T cells had experienced in the thymic environment during development, and physiological cross-recognized ligands consist primarily of distinct peptides presented by the same self-MHC molecule. However under experimental or clinical conditions, TCRs can even react against MHC molecules not seen during thymic selection. For instance, many TCRs selected to respond to foreign peptides bound to a self-MHC molecule display a concomitant cross-reactivity for intraspecies allelic variants of self-MHC molecules that can differ by up to 20 amino acids. This property, termed alloreactivity, causes graft rejection and graft-*versus*-host disease. Most of these polymorphic residues line the peptide-binding groove where they determine peptide-binding specificity, while a few of them are located on the top of the α-helices that form the groove and are thus available for TCR contact. Based on crystal structures of TCR–pMHC allocomplexes, it appears that during allorecognition, TCRs neither avoid contacting the bound peptide, nor focus on the polymorphic residues that are exposed on the top of the allo-MHC α helices (Housset and Malissen 2003). Although structures corresponding to the same TCR in complex with a self- and an allo-MHC are not available yet, a parsimonious interpretation of the available data suggests that during allorecognition, TCRs exploit the similarities rather than the differences between the top of the helices of self- and allo-MHC molecules, and that the high precursor frequency of alloreactive T cells is accounted for by the fact that allo-MHC molecules display a totally new constellation of endogenous peptides against which the repertoire of mature T cells has not been negatively selected in the thymus (Housset and Malissen 2003).

Some TCRs can even find compatible landmarks on MHC molecules belonging to other species. Such cross-reactivity for MHC across species is termed xenoreactivity. For instance, the mouse TCR repertoire is capable of recognizing transgenic human HLA molecules, although with a greatly reduced efficiency (Firat et al. 2002). This trans-species cross-reactivity is likely due to the shared amino acids that occur at positions through the entire spectrum of mammalian MHCI molecules. By elucidating the structure of a mouse TCR bound to a human MHCI molecule Buslepp and colleagues (Buslepp et al. 2003) provided the first hints to the structural basis of TCR xenoreactivity. They showed that xenoreactive recognition follows the same conserved docking orientation as self-pMHC and allo-pMHC recognition. Moreover, the bound peptide is engaged as a specificity element, and the TCR does not appear to focus on the species-specific residues that are exposed on the outer surface of xeno-MHC helices. Therefore, TCRs might exploit the similarities rather than the differences between the top of the helices of self- and xeno-MHC molecules, and avoid a global repositioning on binding to a xeno-pMHC surface. T-cell antigen receptor trans-species cross-reactivity appears thus permitted by the shared amino acids that fortuitously occur at TCR contact positions through the entire spectrum of mammalian MHC molecules (Malissen 2003b). A direct consequence of these observations is that in a clinical allotransplantation set-up, selecting donor-recipient combinations that maximize the differences in the set of residues that are sensed by the TCRs should dampen the strength of T-cell alloresponses (Housset and Malissen 2003). Conversely, MHC alleles differing by only a minimal mismatch might induce alloresponses with as great or even with a greater magnitude than observed across more disparate MHC allelic differences. Consistent with these views, Macdonald and colleagues (Macdonald et al. 2003) have recently demonstrated that powerful alloreactivity occurs across a "minimal" mismatch that involves one residue on the α2 helix that is inaccessible to direct TCR recognition and only alters the selection of the peptides.

5.5.7 *Revisiting the Extent of TCR Degeneracy and the Specificity of Cross-Reactivity*

Recent data suggest that TCRs are probably much less degenerate than assumed on the basis of theoretical considerations (Mason 1998), or on the utilization of combinatorial peptide libraries (Maynard et al. 2005). For instance, in the case of the BM3.3 TCR system, the upward pointing peptide residue P6 serves as a focal point for the BM3.3 TCR (Reiser et al. 2000; Reiser et al. 2003). P6 tolerates very few if any substitution and thus constitutes a "primary TCR contact residue" according to the nomenclature introduced by Allen and colleagues (Shih and Allen 2004). Residue P4 and P7 constitute "secondary TCR contact residues" and some mutations in these residues are less detrimental to the recognition process. The existence of "primary TCR contact residues" explains how some T cells can discriminate between peptide structures that differ by a single oxygen atom in a peptide side chain (for instance as the result of a phenylalanine to tyrosine substitution), and achieve specific recognition under conditions where there is generally a 10^3–10^4 excess of peptides that derive from self-proteins and may differ from the foreign peptide by only a single amino acid. Conversely, the malleability of the CDR3α loop allows the BM3.3 TCR to cope with some non-conservative substitutions in the P4 and P7 secondary TCR contact residues, and account to its degeneracy. Importantly, each alternative docking solution achieved by BM3.3 involves a highly specific and fortuitous bond network (Reiser et al. 2003). As vividly stated by Garcia and colleagues, "CDR3 can assume different conformations to recognize a limited number of alternative peptides with a high degree of specificity, but they are not easily accommodating limp noodles" (Garcia and Adams 2005). Therefore, the BM3.3 TCR illustrates how a single TCR antigen-binding site can display both high specificity and degeneracy, two structural properties that appear a priori conflicting. It also demonstrates how the tight constraints imposed by a primary TCR contact residue and the limited number of permissible changes allowed in the secondary TCR contact residues significantly restrict the number of potential ligands that are recognizable as agonist by a given TCR.

5.5.8 *Receptor Versus Cellular Specificity*

Receptor specificity and cross-reactivity constitute empirical and assay-dependent properties. Accordingly, the sensitivity of some in vitro assays is sometimes much too high and reveals cross-reactions that are not biologically relevant (Zinkernagel 2002). For instance, the observation that cytotoxic T cells stained with pMHCI multimeric probes does not always mean that they are capable of productive responses in vivo. Importantly, the range of cross-reactivity exhibited by a given TCR is not a property intrinsic to the TCR. It also depends on the developmental stage of the T-cell clone that expresses it, and on the architecture of the signaling network that acts downstream of the TCR. For instance, it has been shown that the protein tyrosine phosphatase SHP-1 can increase the specificity of TCR signaling by preventing spurious T-cell activation with large quantities of low-affinity pMHC ligands, while permitting sensitive responses towards small quantities of more strongly binding pMHC complexes (Altan-Bonnet and Germain 2005; Stefanova et al. 2003). Importantly, the point where ligand discrimination is set is not "hard-wired" into the TCR but can be modulated by changing the concentration of key components of the feedback loops that control the TCR signaling cassette. A state of TCR-signaling hypersensitivity can also be induced during lymphopenia (Goldrath and Bevan 1999), following adhesion-induced T-cell priming (Doucey et al. 2003a), or in the presence of the CD4 or CD8 coreceptors (Donermeyer et al. 2006; Holler and Kranz 2003; Renard et al. 1996), resulting in broadened TCR specificity.

T-cell antigen receptor (or more appropriately T cell) specificity is thus a property of the whole T cell that can be either sharpened by negative feedback loops that blunt signals emanating from weak pMHC ligands, or conversely relaxed by enhancing the output of the TCR signaling cassette (Altan-Bonnet and Germain 2005; Lin et al. 1997). Therefore, factors not directly related to the affinity of the TCR–pMHC interaction can determine the outcome of a T-cell response. This is due to the fact that the TCR is only one part of a complex cellular sensor, and that the state of activation of APCs and even the quality of the neighboring T cells can dramatically influence the response of a given T-cell clone. For instance, a study by Kedl and colleagues (Kedl et al. 2003) showed that high-affinity T cells can induce a selective loss of peptide-MHC complexes from the surface of antigen-presenting cells (APCs). By reducing the antigenic determinants on APCs, high-affinity T cells likely outcompete low-affinity ones. Along the same line, regulatory T cells can "deactivate," in a direct or indirect (via APCs) manner, signaling proficient T cells (Fontenot and Rudensky 2005). Therefore, regardless of the fact that functional specificity originates from the V domains of TCRs, operational end points for assessing immunological specificity must be higher-level cellular functions or even organismal functions. As stated by R. Zinkernagel, observing whether a mouse ends up "legs up or legs down" following a viral challenge remains the ultimate cut-off for adaptive immunity.

5.6 What Causes the Restriction in Orientation Imposed on TCR–pMHC Interactions?

All but one of the TCRs analyzed to date bind to pMHC with roughly similar diagonal geometry. However, the shapes and chemical properties of the interacting surfaces found in these complexes are so diverse that no fixed contact exists between conserved TCR residues and a conserved area of the MHC α-helices (Gagnon et al. 2005; Garcia and Adams 2005; Housset and Malissen 2003; Rudolph et al. 2006). This absence of recognizable landmarks or pivot points within the

TCR–MHC interface suggested that the molecular cues responsible for enforcing the globally conserved mode of TCR docking on MHC molecules are encoded outside of the TCR/MHC pair, and might, for instance, be sought in the action of the CD4 and CD8 coreceptors. It has been argued that the conserved orientation of the TCR over the pMHC corresponds to the need to interact maximally with the peptide. However, these considerations would also be compatible with TCR engaging the pMHC with a 180° rotation relative to the observed orientation. Therefore, prior to the phase of intrathymic TCR αβ selection, the repertoire of assembled TCR binding sites might contain TCRs capable of engaging pMHC with a 180° rotation relative to the "canonical" orientation. However, considering that CD8 likely interacts specifically with the CD3δε dimer of the TCR–CD3 complex (Doucey et al. 2003a; Yachi et al. 2005), it is plausible that in the case of a TCR–pMHCI docking geometry rotated by 180°, the Lck kinase associated with the CD8 coreceptor will be prevented from phosphorylating the CD3 ITAMs and ZAP-70, thereby impeding selection of the clones bearing those TCRs. Therefore, the need for coreceptor function during TCR αβ selection may eliminate all the DP cells that are present in the pre-selected TCR repertoire and that express TCR antigen-binding site unable to engage the pMHC surface according to a fixed "signaling-proficient" geometry (Buslepp et al. 2003). The non-canonical docking observed for the coreceptor-dependent Ob.1A12 TCR (see section 5.5.1) further questioned the view that MHC restriction is encoded in the CDR1–CDR2 loops of the TCR V domains.

Although TCR–pMHC crystal structures have not revealed a straightforward "TCR–MHC recognition code," loose guidance cues in charge of steering TCR–pMHC interactions can still be encoded within the germline encoded CDR1–CDR2 loops of the TCR V domains and in the MHC α helices themselves. A wide range of TCR docking modes can be deduced from all the TCR–pMHC structures solved so far (Rudolph et al. 2006). In some TCR–pMHCI complexes, the CDR2 loops of both Vα and Vβ domains contact the central part of the MHC α2 and α1 helices, respectively. However, in other TCR–pMHC complexes, the two MHC helices are not contacted in a concerted fashion. For instance, the occurrence of contacts between the CDR2β loop and the MHC α1 helix is clearly optional and contingent on the size of the CDR3β loop (Ding et al. 1998). Conversely, in the JM22-MP(58–66)-HLA-A2 complex, the JM22 TCR footprint is translated such that the CDR2α loop makes minimal contact with the MHC α2 helix (Stewart-Jones et al. 2003). Considering that the range of footprints adopted by the CDR1–CDR2 loops encompasses several turns of the MHC α helices and that the detailed atomic features of these interactions are very different, it has been suggested that the apex of the CDR1 and CDR2 loops and the top of the MHC α helix function as a kind of molecular "Velcro" offering multiple "continuous" docking possibilities (Housset and Malissen 2003). Along the same line, Huseby and colleagues have suggested that the CDR1–CDR2 loops primarily contact the main chain atoms of the MHC α helices, and have thus an intrinsic predisposition to adapt to their top regardless of their amino acid side chain content (Huseby et al. 2005). This postulated propensity may account for the fact that several mouse TCRs are capable of recognizing both pMHCI and pMHCII ligands (see for instance Ge et al. 2006). Whether these unique TCRs uses a largely sequence-independent set of interactions with the MHC surface as postulated by Huseby and colleagues (Huseby et al. 2005), or exploit the existence of fortuitous, side-chain encoded sequence similarities on the top of MHCI and MHCII helices remains, however, to be demonstrated.

Recent data, however, are not consistent with the view that the top of the MHC α helices constitutes "sticky" matrices promoting TCR binding in different registers. For instance, comparison of two TCRs (172.10 and scD10) that use the same Vβ gene segment (Vβ8.2), showed that a set of interactions dependent on amino acid side chains might play a key role in steering TCR–pMHC interactions. Comparison of the 172.10-MBP-1-11-I-A^u and scD10-CA-I-A^k complexes showed that the Vβ8.2 CDR1–CDR2 loops establish the same multi-point contact patch with the I-A^u and I-A^k MHC α1 helix (Maynard et al. 2005; Reinherz et al. 1999). This superimposable region comprises five shared TCR residues interacting with five shared MHC residues, forming five hydrogen bonds and ten van der Waals contacts. The correspondence of the CDR1–CDR2 contacts with the MHC α1 helix in the two structures is highly suggestive of a conserved anchor point. In contrast, comparison of the KB5-C20-pKB1-H-$2K^b$ and BM3.3-pBM1-H-$2K^b$ complexes revealed that the CDR1–CDR2 loops found in the Vβ2 domain used by both TCRs adopt different binding registers when docking to the same MHC α helix (Reiser et al. 2002). Therefore, the CDR1–CDR2 loops of a given TCR V domain might alternatively adopt a small set of discrete docking geometries on a given MHC allele (Maynard et al. 2005). The "docking subsite" to be used by a given V would then depend primarily on the V partner it associates with. In support of this view, it should be noted that in each of the KB5-C20-pKB1-H-$2K^b$ and BM3.3-pBM1-H-$2K^b$ complexes, the Vβ2 domain was associated with a distinct Vα, whereas the Vα found in 172.10 and D10 TCRs belonged to the same Vα subfamily. It is thus possible that the unique combination of CDR1–CDR2 loops found in a given VαVβ pair fixes the overall geometry of the TCR–pMHC interaction, the associated CDR3 loops being able to adapt to this imposed geometry for some pMHC ligands.

The present day TCR V gene segments have evolved through several rounds of duplication. It is thus possible that the position and nature of the primordial set of CDR1–CDR2 residues involved in contacting a fixed site on MHC helices and in steering the binding orientation of a primordial TCR–MHC pair have drifted during the process of V gene diversification, making their recognition difficult. Accordingly, the globally conserved diagonal mode of interaction adopted by all the present day TCR–pMHC pairs could result from a founder-like effect, and constitute variations on the binding geometry adopted by an ancestral TCR–pMHC pair (Housset and Malissen 2003).

The HLA-B*3508 allele binds a 13-amino acid peptide (LPEP) derived from the BZLF1 antigen of Epstein–Barr virus. Considering that the MHC class I antigen-processing pathway and antigen-presentation pathway is biased toward 8- to 10-amino acid peptides, LPEP constitutes thus an unusually long peptide (Burrows et al. 2006). When bound to HLA-B*3508, LPEP adopts a bulged conformation in its center while maintaining conserved networks of hydrogen bonds at its amino and carboxy termini. In addition, LPEP assumes a rigid conformation and has thus the potential to block sterically TCR engagement. Indeed, determination of the structure of LPEP–HLA-B*3508 in complex with the SB27 TCR showed that the bulged peptide prevented a full engagement of the TCR with the HLA molecule. The interface with the TCR was dominated by peptide-mediated interactions rather than by MHC-mediated contacts; in other TCR–pMHC complexes, up to two third of atomic contacts are with the MHC itself (Tynan et al. 2005). However, the SB27 TCR was still MHC-restricted in that it contacted both residues 65 and 69 of the α1 helix and residues 150–158 of the α2 helix. Further structural studies of TCRs specific for other longer MHCI-bound peptides should confirm the need for minimal TCR–MHC contacts, or conversely reveal the occurrence of MHC-restricted T-cell recognition without significant TCR–MHC contacts, thereby indicating that MHC restriction is a consequence of extrinsic forces such as coreceptor interactions (Burrows et al. 2006).

It has been suggested that a large part of the binding energy of TCRs for pMHC ligands stems from direct contacts between the TCR and the MHC, the CDR3-peptide interaction modulating a preexisting affinity of the TCR CDR1–CDR2 loops for the MHC and raising the binding energy above the threshold required for productive engagement (Baker et al. 2001; Wu et al. 2002). In contrast to that view, a recent study using the LC13–FLR–HLA-B8 system showed that the interaction between the CDR1–CDR2 loops and the MHC helices contributed very little energy to the contact interface, whereas almost all of the binding energetics derived from peptide-CDR3 contacts (Borg et al. 2005). Likewise, in the case of intrathymic positive selection, the underlying energetic principles may vary according to the TCR and to the MHC class and allele under study.

Given the difficulty, so far, in seeing a TCR–MHC recognition code from crystal structures, Garcia and colleagues have recently stressed the limits of interpretations that are solely based on TCR–pMHC crystals (Garcia and Adams 2005). They suggested that the "TCR–MHC recognition code" may become only manifest at the level of the initial TCR–pMHC encounter complex. When the putative encounter complex relaxes into the most energetically stable final complex, the original contacts adopted by the CDR1–CDR2 and "signing" MHC restriction may undergo some reorganization that will depend on CDR3-peptide interactions and will thus differ for each CDR3 combination. In support of these views, recent studies of the kinetics and thermodynamics of TCR–pMHC binding suggest that conformational changes at the TCR–pMHC interface occur after an initial permissive encounter (Lee et al. 2004; Miley et al. 2004).

5.7 Rationalizing the Purpose of TCR αβ Positive Selection

The proportion of de novo assembled TCRs capable of interacting with self-pMHC complexes is difficult to estimate due to the possible existence of sequential TCRα locus rearrangements (see section 5.4.3), and to the efficient removal of apoptotic thymocytes. It has been, however, suggested that prior to TCR αβ selection, up to 30% of de novo assembled TCRs can interact with the self-pMHC complexes encountered in the thymus (Merkenschlager et al. 1997; Zerrahn et al. 1997). This is considered a much higher frequency than would be expected from the generation of receptors through the random rearrangement of gene segments that do not have an inherent predisposition for interactions with MHC, like those involved in the assembly of Ig V regions. Note that although antibodies with TCR-like specificity (i.e. recognizing the MHC-bound peptide as a specificity element) can be selected (Hulsmeyer et al. 2005), it is generally assumed that, among a repertoire of random antibodies, the great majority will fail to bind to a given pMHC ligand in a peptide-specific manner. This assumption is based on the fact that antibodies are likely to treat pMHC ligands in an "opportunistic mode", binding any epitope provided that some energetic prerequisite are met. Therefore, were antibodies V regions ever to be used in lieu of TCR V regions, most of the T-cell clones expressing them would be excluded from TCR αβ selection. Altogether, these considerations suggest that evolution has shaped the germline encoded CDR1–CDR2 loops, so that they have an intrinsic ability to bind MHC α helices, allowing TCR to focus only on MHC and not on any other cell surface proteins. As previously discussed, the postulated interactions between CDR1–CDR2 and MHC are probably weak and can be modulated, in a positive or negative fashion, by the associated CDR3s and the side chains of the bound peptide (Wang and Reinherz 2002). Note that the idea that receptors expressed on T cells, might be intrinsically biased to bind MHC proteins was first proposed more than 30 years ago, at a time when the structure of the TCR was still elusive (Jerne 1971).

After the generation of an MHC allele-neutral TCR repertoire in DP thymocytes, the phase of molecular matching known as positive selection favors those TCRs capable of coping with a composite surface made of self-peptide side chains and of MHC determinants that are both conserved and allele specific. Provided that there is no global repositioning of the TCR V domains on the MHC surface when docking on the self peptide-self MHC ligands responsible for their intrathymic selection and on the foreign peptide-self MHC ligands that are encountered in the periphery, it can be inferred from the crystal structure of TCR bound to foreign peptide-self MHC ligands that during positive selection the TCRs are not particularly "obsessed" with the recognition of the few allele-specific residues that are found on the top of the MHC α helices (Malissen 2003b). In MHCIa molecules, the residues of the α1α2 platform that display the greatest sequence variability point toward the peptide-binding groove resulting in allelic specificity for peptide binding whereas residues on the top of the α helices are more conserved. It has however been argued that these few allele-specific residues constitute a nuisance in that they affect the binding of some Vα or Vβ domains and preclude the selection of the corresponding DP cells (Schumacher and Ploegh 1994). When discussing this kind of steric conflict, Ron Germain (Germain 1990) suggested that it would have been less "wasteful" to conserve those regions of the MHC molecule able to contact the TCR, altering only those that confer peptide-binding specificity. This strategy would maximize the usable fraction of TCRs among those generated by random V(D)J gene segment rearrangements and chain assortment, as no receptor would be excluded from selection because of a failure to interact with the MHC molecules available in the thymus. However, the selective pressure driving MHC polymorphism, likely due to the necessity to diversify the repertoire of sampled peptides, clearly "spilled over" to residues accessible to the TCR, making necessary a phase of epigenetic molecular match.

In contrast to the models that consider the polymorphic residues found on the top of the MHC helices and available to TCR contacts as the nuisance that accounts for most of DP cell loss, we have argued that during TCR αβ selection there are probably fewer constraints linked to matching the few allele-specific residues found on the top of the MHC helices than there are in adapting to the generic features that are imposed on the bound peptides by the architecture of a given MHC peptide-binding groove, and to the diversity of peptide side chains that point toward the TCR (Housset and Malissen 2003). For instance, TCR read-out of peptides that follow a flat and deep course within the MHCI groove, as exemplified by octapeptides bound to H-2K^b and H-2K^k, generates structural constraints distinct from the readout of H-2D^b- and H-2L^d-bound peptides that protrude out of the C-terminal part of the peptide binding-groove (reviewed in Kellenberger et al. 2005). However, despite the malleability of the TCR antigen-binding site and the occurrence of limited conformational adjustments in the pMHC surface, it is likely that the great majority (~70%) of the VαVβ combinations expressed at the surface of the pre-selected DP cells remains "neglected" due to steric conflict with the residues of the self-peptides accessible to TCR contact. Therefore, it is possible that while most germline-encoded CDR1–CDR2 loops are capable of productive interactions with any given MHC allele, the repertoire of somatically generated CDR3 loops has difficulty to match the generic features that the MHC peptide-binding grooves impose on the bound peptides, and coincidently on the peptide residues accessible to TCR contact. The fact that most MHCII-bound peptides adopt a rather deep and flat course in the binding groove might facilitate the interaction of MHCII molecules with the TCR CDR1–CDR2 loops, and make MHCII molecules more "accommodating" than pMHCI molecules to the binding of de novo assembled TCRs.

5.8 Recessive and Dominant Tolerance

The potential to generate random repertoire of αβ TCRs has led to the evolution of complex quality-control mechanisms that eliminate or suppress T-cell clones that strongly react with self-pMHC molecules. It has been suggested that innate-like T lymphocytes (NKT cells and some subsets of γδ T cells) are autoreactive by design (Bendelac et al. 2001). However, it should be stressed that conventional αβ T cells are "trained" on self pMHC molecules and selected to have a low-affinity for such self pMHC complexes. Therefore, conventional αβ T cells are also autoreactive on design and operate on the edge of autoimmunity (Aguado et al. 2002; Lin et al. 1997). For conventional αβ T cells, the immunologic self corresponds to the pMHC molecules that are seen during intrathymic development and on the surface of peripheral, resting dendritic cells (DCs). In the thymus, the pool of MHC-bound self-peptides available for repertoire selection comprises intrathymically expressed ubiquitous antigens and antigens specific to various types of thymic APCs: cortical and medullary thymic epithelial cells (cTECs and mTECs), thymic DCs, macrophages, and thymic B cells. Promiscuous gene expression by mTECs has recently extended the scope of central tolerance to self-constituents expressed by many peripheral tissues, including some that are only encountered during adulthood (Derbinski et al. 2005). This fascinating mechanism is controlled by the *Aire* gene product, and provides a way for developing T cells to anticipate the variety of self-proteins they will encounter while patrolling the body at later time points (Mathis and Benoist 2004). Furthermore, self-antigens can gain access to the thymus either via the blood circulation or by association with immigrating cells (Bonasio et al. 2006). Therefore, in contrast to the cortex that is a compartment genuinely unique to the thymus, there is some functional similarities between the medulla and peripheral secondary lymphoid organs.

Because not all self-antigens are expressed in the thymus, additional tolerance mechanisms exist in the periphery (Goodnow et al. 2005; Klein et al. 2000). In contrast to central tolerance that acts on self-reactive T lymphocytes before they become immunocompetent, peripheral tolerance acts on mature T cells. Peripheral presentation of self-determinants by DCs in the absence of tissue-damage, inflammation, or adjuvants does not prime naive T cells but rather results in abortive immune responses. This last mechanism of peripheral tolerance together with central tolerance are referred to as "recessive tolerance" because deletion of an individual autoreactive T-cell clone according to one of these processes does not affect other self-reactive T-cell clones. In contrast, the recently recognized regulatory T cells (Treg cells) are capable of acting in a dominant, trans-acting way to control self-reactive T cells and terminate conventional immune responses (Coutinho 2005). Treg cells are generated in the thymus upon high-affinity recognition of self-pMHCII ligands, and develop, under the influence of the Foxp3 transcription factor, along a unique differentiative pathway geared to anti-inflammatory and antiproliferative functions (Fontenot and Rudensky 2005). Treg can also be generated in the periphery subsequent to immune stimulation. Therefore, dominant tolerance constitutes an additional mechanism to compensate for incomplete representation of self in the thymus.

5.9 Evolutionary Perspectives

It has been argued that innate cell recognition strategies are more difficult to decipher than those used by B and T cells, where antigen receptors dominate the differentiation, activation, and effector function of these lymphocytes (Lanier 2005). For instance, rather than being regulated by a single receptor, NK cell activation results from the integration of signals emanating from multiple activating and inhibitory receptors that often display overlapping binding specificity. In contrast to the situation observed in T cells and B cells, Syk and ZAP70 are not required for NK cell development, and many NK effector functions are intact in mice that are genetically deficient in both of these kinases. Therefore, unlike B and T lymphocytes, a single signaling pathway does not dominate the differentiation and effector function of NK cells (Chiesa et al. 2006). However, it should be stressed that the BCR and αβ TCR signaling pathways are not autonomous and that their output can be tuned by a number of positive (CD28, CD19) or negative (CD5, CD22, FcγRIIB) regulators. Upon activation, some T cells are also capable of expressing inhibitory or activating receptors identical to the ones used by NK cells, and de novo expression of these inhibitory receptors is capable of aborting potentially productive TCR–pMHC interactions. Likewise, the function of the semi-invariant TCRs found on natural killer T (NKT) cells and on some subsets of γδ T cells is tuned by inhibitory and activating receptors (Bendelac et al. 2001). As documented for the NKG2D activating receptors, it is likely that these inhibitory and activating receptors recognize self molecules that function as generic flags of cell damage or of other nonphysiological processes (Gasser et al. 2005). Therefore, the strategies used to activate both adaptive and innate immune effectors are rather similar and result from the integration of positive and negative signals emanating from a multitude of receptors (Vivier and Malissen 2005).

Natural killer T cells and γδ T cells resemble NK cells in that they are "ready-to-go" cells. In contrast to naïve, conventional αβ T cells that undergo a phase of antigen-driven clonal expansion that is associated with the acquisition of diverse functions (T helper type 1, 2, and 17 in the case of CD4 T cells), the effector functions of NK, NKT, and γδ T cells are triggered without prior proliferation and chromatin remodeling of genes coding for effector molecules (granzyme, perforin, and cytokines) (Lanier 2005). The fact that NK cell receptors are germline-encoded and present on a large proportion of NK cells makes them well suited for such explosive-like defense mechanisms. Likewise, to behave as "ready-to-go" effector cells and thus avoid a lengthy, antigen-driven phase of clonal expansion and differentiation into effector cells, NKT cells and some subsets of γδ T cells require mechanisms of TCR repertoire selection that ensure that a given TCR will be distributed "ab initio" on a relatively large cell subpopulation. The "simplest" solution to this problem is found in the γδ T cells present at body surfaces. The programmed usage of certain V and J gene segments containing short homology repeats near their coding ends in association with an absence of terminal deoxynucleotidyl transferase activity allows them to express TCRs containing a single or a rather limited number of canonical V(D)J junctions. NKT cells also express quasi invariant TCRα chains, however in contrast to the strategy used by γδ T cells, their limited TCR repertoire results from stringent cellular selection mechanisms acting on a population of intrathymic progenitors expressing randomly assembled TCRs. It is thus paradoxical to see that somatic, site-specific DNA recombination can be used to generate TCRs that have almost no diversity and recognize self molecules that function as generic flags of cell infection or other cell dysregulations. Finally, it should be noted that there is probably a link between the existence of somatically assembled and clonally distributed antigen receptors and the appearance of a pool of long-lived naïve circulatory precursors that can undergo antigen-driven clonal expansion and give rise to both effectors and memory T cells (Pancer and Cooper

2006). This unique feature necessitates processes that control independently the numbers of naïve and memory T cells, safeguarding both a diverse repertoire of naïve T cells for the control of newly emerging pathogens as well as a reservoir of memory T cells to rapidly eliminate pathogens that have been already encountered. In contrast, following encounter with their stimulatory ligands, innate-like cells are poised to become effectors without proceeding through a phase of proliferation. However, these contrasted views need probably to be tempered since both hapten-specific NK cells have been observed in a model of contact hypersensitivity, and an NK cell subset bearing an activating receptor specific for cells infected with mouse cytomegalovirus has been found to undergo a phase of "antigen-driven" proliferation akin to clonal expansion of conventional αβ T cells (Yokoyama 2006).

Acknowledgments We thank J. Ewbank and E. Vivier for discussions. Work from the authors was supported by CNRS, INSERM, ANR, ARC, FRM, and the European Communities (MUGEN). C.M. was supported by fellowships from the European Communities (EPI-PEP-VAC) and Fondation pour la Recherche Médicale.

References

Abi-Rached L, Parham P (2005) Natural selection drives recurrent formation of activating killer cell immunoglobulin-like receptor and Ly49 from inhibitory homologues. J Exp Med 201:1319–1332

Aguado E, Richelme S, Nunez-Cruz S, Miazek A, Mura AM, Richelme M, Guo XJ, Sainty D, He HT, Malissen B, Malissen M (2002) Induction of T helper type 2 immunity by a point mutation in the LAT adaptor. Science 296:2036–2040

Allison TJ, Winter CC, Fournie JJ, Bonneville M, Garboczi DN (2001) Structure of a human gammadelta T-cell antigen receptor. Nature 411:820–824

Altan-Bonnet G, Germain RN (2005) Modeling T-cell antigen discrimination based on feedback control of digital ERK responses. PLoS Biol 3:e356

Altfeld M, Allen TM (2006) Hitting HIV where it hurts: an alternative approach to HIV vaccine design. Trends Immunol 27:504–510

Baker BM, Turner RV, Gagnon SJ, Wiley DC, Biddison WE (2001) Identification of a crucial energetic footprint on the alpha1 helix of human histocompatibility leukocyte antigen (HLA)-A2 that provides functional interactions for recognition by tax peptide/HLA-A2-specific T-cell receptors. J Exp Med 193:551–562

Bankovich AJ, Garcia KC (2003) Not just any T-cell receptor will do. Immunity 18:7–11

Batalia MA, Collins EJ (1997) Peptide binding by class I and class II MHC molecules. Biopolymers 43:281–302

Bendelac A, Bonneville M, Kearney JF (2001) Autoreactivity by design: innate B and T lymphocytes. Nat Rev Immunol 1:177–186

Beutler B (2003) Not "molecular patterns" but molecules. Immunity 19:155–156

Bonasio R, Scimone ML, Schaerli P, Grabie N, Lichtman AH, von Andrian UH (2006) Clonal deletion of thymocytes by circulating dendritic cells homing to the thymus. Nat Immunol 7:1092–1100

Bongrand P, Malissen B (1998) Quantitative aspects of T-cell recognition: from within the antigen-presenting cell to within the T cell. Bioessays 20:412–422

Boniface JJ, Reich Z, Lyons DS, Davis MM (1999) Thermodynamics of T-cell receptor binding to peptide-MHC: evidence for a general mechanism of molecular scanning. Proc Natl Acad Sci USA 96:11446–11451

Borg NA, Ely LK, Beddoe T, Macdonald WA, Reid HH, Clements CS, Purcell AW, Kjer-Nielsen L, Miles JJ, Burrows SR, et al (2005) The CDR3 regions of an immunodominant T-cell receptor dictate the "energetic landscape" of peptide-MHC recognition. Nat Immunol 6:171–180

Brdicka T, Kadlecek TA, Roose JP, Pastuszak AW, Weiss A (2005) Intramolecular regulatory switch in ZAP-70: analogy with receptor tyrosine kinases. Mol Cell Biol 25:4924–4933

Burrows SR, Rossjohn J, McCluskey J (2006) Have we cut ourselves too short in mapping CTL epitopes? Trends Immunol 27:11–16

Buslepp J, Wang H, Biddison WE, Appella E. JE C (2003) A correlation between TCR V alpha docking on MHC and CD8 dependence: Implications for T-cell selection. Immunity 19:595–606

Call ME, Pyrdol J, Wiedmann M, Wucherpfennig KW (2002) The organizing principle in the formation of the T-cell receptor-CD3 complex. Cell 111:967–979

Call ME, Schnell JR, Xu C, Lutz RA, Chou JJ, Wucherpfennig KW (2006) The structure of the zetazeta transmembrane dimer reveals features essential for its assembly with the T-cell receptor. Cell 127:355–368

Cebecauer M, Guillaume P, Mark S, Michielin O, Boucheron N, Bezard M, Meyer BH, Segura JM, Vogel H, Luescher IF (2005) CD8+ cytotoxic T lymphocyte activation by soluble major histocompatibility complex-peptide dimers. J Biol Chem 280:23820–23828

Chang HC, Tan K, Ouyang J, Parisini E, Liu JH, Le Y, Wang X, Reinherz EL, Wang JH (2005) Structural and mutational analyses of a CD8alphabeta heterodimer and comparison with the CD8alphaalpha homodimer. Immunity 23:661–671

Chen JL, Stewart-Jones G, Bossi G, Lissin NM, Wooldridge L, Choi EM, Held G, Dunbar PR, Esnouf RM, Sami M, et al (2005) Structural and kinetic basis for heightened immunogenicity of T-cell vaccines. J Exp Med 201:1243–1255

Chiesa S, Mingueneau M, Fuseri N, Malissen B, Raulet DH, Malissen M, Vivier E, Tomasello E (2006) Multiplicity and plasticity of natural killer cell signaling pathways. Blood 107:2364–2372

Choudhuri K, Kearney A, Bakker TR, van der Merwe PA (2005a) Immunology: how do T cells recognize antigen? Curr Biol 15:R382–385

Choudhuri K, Wiseman D, Brown MH, Gould K, van der Merwe PA (2005b) T-cell receptor triggering is critically dependent on the dimensions of its peptide-MHC ligand. Nature 436:578–582

Coutinho A (2005) The Le Douarin phenomenon: a shift in the paradigm of developmental self-tolerance. Int J Dev Biol 49:131–136

Davis SJ, van der Merwe PA (2006) The kinetic-segregation model: TCR triggering and beyond. Nat Immunol 7:803–809

Davis-Harrison RL, Armstrong KM, Baker BM (2005) Two different T-cell receptors use different thermodynamic strategies to recognize the same peptide/MHC ligand. J Mol Biol 346:533–550

Davodeau F, Difilippantonio M, Roldan E, Malissen M, Casanova JL, Couedel C, Morcet JF, Merkenschlager M, Nussenzweig A, Bonneville M, Malissen B (2001) The tight interallelic positional coincidence that distinguishes T-cell receptor Jalpha usage does not result from homologous chromosomal pairing during ValphaJalpha rearrangement. EMBO J 20:4717–4729

Delon J, Gregoire C, Malissen B, Darche S, Lemaitre F, Kourilsky P, Abastado JP, Trautmann A (1998) CD8 expression allows T-cell signaling by monomeric peptide-MHC complexes. Immunity 9:467–473

Derbinski J, Gabler J, Brors B, Tierling S, Jonnakuty S, Hergenhahn M, Peltonen L, Walter J, Kyewski B (2005) Promiscuous gene expression in thymic epithelial cells is regulated at multiple levels. J Exp Med 202:33–45

Diefenbach A, Tomasello E, Lucas M, Jamieson AM, Hsia JK, Vivier E, Raulet DH (2002) Selective associations with signaling proteins determine stimulatory versus costimulatory activity of NKG2D. Nat Immunol 3:1142–1149

Ding YH, Smith KJ, Garboczi DN, Utz U, Biddison WE, Wiley DC (1998) Two human T-cell receptors bind in a similar diagonal mode to the HLA-A2/Tax peptide complex using different TCR amino acids. Immunity 8:403–411

Donermeyer DL, Weber KS, Kranz DM, Allen PM (2006) The study of high-affinity TCRs reveals duality in T-cell recognition of antigen: specificity and degeneracy. J Immunol 177: 6911–6919

Doucey MA, Goffin L, Naeher D, Michielin O, Baumgartner P, Guillaume P, Palmer E, Luescher IF (2003a) CD3 delta establishes a functional link between the T-cell receptor and CD8. J Biol Chem 278:3257–3264

Doucey MA, Legler DF, Faroudi M, Boucheron N, Baumgaertner P, Naeher D, Cebecauer M, Hudrisier D, Ruegg C, Palmer E, et al (2003b) The beta1 and beta3 integrins promote T-cell receptor-mediated cytotoxic T lymphocyte activation. J Biol Chem 278:26983–26991

Ely LK, Beddoe T, Clements CS, Matthews JM, Purcell AW, Kjer-Nielsen L, McCluskey J, Rossjohn J (2006) Disparate thermodynamics governing T-cell receptor-MHC-I interactions implicate extrinsic factors in guiding MHC restriction. Proc Natl Acad Sci USA 103:6641–6646

Feng J, Garrity D, Call ME, Moffett H, Wucherpfennig KW (2005) Convergence on a distinctive assembly mechanism by unrelated families of activating immune receptors. Immunity 22:427–438

Firat H, Cochet M, Rohrlich PS, Garcia-Pons F, Darche S, Danos O, Lemonnier FA, Langlade-Demoyen P (2002) Comparative analysis of the CD8(+) T-cell repertoires of H-2 class I wild-type/HLA-A2.1 and H-2 class I knockout/HLA-A2.1 transgenic mice. Int Immunol 14:925–934

Fontenot JD, Rudensky AY (2005) A well adapted regulatory contrivance: regulatory T-cell development and the forkhead family transcription factor Foxp3. Nat Immunol 6:331–337

Gagnon SJ, Borbulevych OY, Davis-Harrison RL, Baxter TK, Clemens JR, Armstrong KM, Turner RV, Damirjian M, Biddison WE, Baker BM (2005) Unraveling a hotspot for TCR recognition on HLA-A2: evidence against the existence of peptide-independent TCR binding determinants. J Mol Biol 353:556–573

Garcia KC, Adams EJ (2005) How the T-cell receptor sees antigen–a structural view. Cell 122:333–336

Garrett TP, McKern NM, Lou M, Elleman TC, Adams TE, Lovrecz GO, Zhu HJ, Walker F, Frenkel MJ, Hoyne PA, et al (2002) Crystal structure of a truncated epidermal growth factor receptor extracellular domain bound to transforming growth factor alpha. Cell 110:763–773

Gasser S, Orsulic S, Brown EJ, Raulet DH (2005) The DNA damage pathway regulates innate immune system ligands of the NKG2D receptor. Nature 436:1186–1190

Ge Q, Holler PD, Mahajan VS, Nuygen T, Eisen HN, Chen J (2006) Development of CD4+ T cells expressing a nominally MHC class I-restricted T-cell receptor by two different mechanisms. Proc Natl Acad Sci USA 103:1822–1827

Germain RN (1990) Immunology. Making a molecular match. Nature 344:19–22

Germain RN, Stefanova I (1999) The dynamics of T-cell receptor signaling: complex orchestration and the key roles of tempo and cooperation. Annu Rev Immunol 17:467–522

Gil D, Schamel WW, Montoya M, Sanchez-Madrid F, Alarcon B (2002) Recruitment of Nck by CD3 epsilon reveals a ligand-induced conformational change essential for T-cell receptor signaling and synapse formation. Cell 109:901–912

Gil D, Schrum AG, Alarcon B, Palmer E (2005) T-cell receptor engagement by peptide-MHC ligands induces a conformational change in the CD3 complex of thymocytes. J Exp Med 201:517–522

Goldrath AW, Bevan MJ (1999) Selecting and maintaining a diverse T-cell repertoire. Nature 402:255–262

Gonzalez PA, Carreno LJ, Coombs D, Mora JE, Palmieri E, Goldstein B, Nathenson SG, Kalergis AM (2005) T-cell receptor binding kinetics required for T-cell activation depend on the density of cognate ligand on the antigen-presenting cell. Proc Natl Acad Sci USA 102:4824–4829

Goodnow CC, Sprent J, Fazekas de St Groth B, Vinuesa CG (2005) Cellular and genetic mechanisms of self tolerance and autoimmunity. Nature 435:590–597

Goulder PJ, Watkins DI (2004) HIV and SIV CTL escape: implications for vaccine design. Nat Rev Immunol 4:630–640

Guy-Grand D, Rocha B, Mintz P, Malassis-Seris M, Selz F, Malissen B, Vassalli P (1994) Different use of T-cell receptor transducing modules in two populations of gut intraepithelial lymphocytes are related to distinct pathways of T-cell differentiation. J Exp Med 180:673–679

Hahn M, Nicholson MJ, Pyrdol J, Wucherpfennig KW (2005) Unconventional topology of self peptide-major histocompatibility complex binding by a human autoimmune T-cell receptor. Nat Immunol 6:490–496

Hayes SM, Love PE (2002) Distinct structure and signaling potential of the gamma delta TCR complex. Immunity 16:827–838

Hayes SM, Love PE (2006) Stoichiometry of the murine {gamma}{delta} T-cell receptor. J Exp Med 203:47–52

Hennecke J, Wiley DC (2002) Structure of a complex of the human alpha/beta T-cell receptor (TCR) HA1.7, influenza hemagglutinin peptide, and major histocompatibility complex class II molecule, HLA-DR4 (DRA*0101 and DRB1*0401): insight into TCR cross-restriction and alloreactivity. J Exp Med 195:571–581

Hogquist KA, Baldwin TA, Jameson SC (2005) Central tolerance: learning self-control in the thymus. Nat Rev Immunol 5:772–782

Holler PD, Kranz DM (2003) Quantitative analysis of the contribution of TCR/pepMHC affinity and CD8 to T-cell activation. Immunity 18:255–264

Holler PD, Chlewicki LK, Kranz DM (2003) TCRs with high affinity for foreign pMHC show self-reactivity. Nat Immunol 4:55–62

Housset D, Malissen B (2003) What do TCR–pMHC crystal structures teach us about MHC restriction and alloreactivity? Trends Immunol 24:429–437

Hulsmeyer M, Chames P, Hillig RC, Stanfield RL, Held G, Coulie PG, Alings C, Wille G, Saenger W, Uchanska-Ziegler B, et al (2005) A major histocompatibility complex-peptide-restricted antibody and t cell receptor molecules recognize their target by distinct binding modes: crystal structure of human leukocyte antigen (HLA)-A1-MAGE-A1 in complex with FAB-HYB3. J Biol Chem 280:2972–2980

Huseby ES, White J, Crawford F, Vass T, Becker D, Pinilla C, Marrack P, Kappler JW (2005) How the T-cell repertoire becomes peptide and MHC specific. Cell 122:247–260

Huseby ES, Crawford F, White J, Marrack P, Kappler JW (2006) Interface-disrupting amino acids establish specificity between T-cell receptors and complexes of major histocompatibility complex and peptide. Nat Immunol 7:1191–1199

James LC, Tawfik DS (2003) The specificity of cross-reactivity: promiscuous antibody binding involves specific hydrogen bonds rather than nonspecific hydrophobic stickiness. Protein Sci 12:2183–2193

Jerne NK (1971) The somatic generation of immune recognition. Eur J Immunol 1:1–9

Kedl RM, Kappler JW, Marrack P (2003) Epitope dominance, competition and T-cell affinity maturation. Curr Opin Immunol 15:120–127

Kellenberger C, Roussel A, Malissen B (2005) The H-2Kk MHC peptide-binding groove anchors the backbone of an octameric antigenic peptide in an unprecedented mode. J Immunol 175:3819–3825

Kjer-Nielsen L, Clements CS, Purcell AW, Brooks AG, Whisstock JC, Burrows SR, McCluskey J, Rossjohn J (2003) A structural basis for the selection of dominant alphabeta T-cell receptors in antiviral immunity. Immunity 18:53–64

Klein L, Klugmann M, Nave KA, Tuohy VK, Kyewski B (2000) Shaping of the autoreactive T-cell repertoire by a splice variant of self protein expressed in thymic epithelial cells. Nat Med 6:56–61

Krogsgaard M, Davis MM (2005) How T cells "see" antigen. Nat Immunol 6:239–245

Krogsgaard M, Prado N, Adams EJ, He XL, Chow DC, Wilson DB, Garcia KC, Davis MM (2003) Evidence that structural rearrangements and/or flexibility during TCR binding can contribute to T-cell activation. Mol Cell 12:1367–1378

Krogsgaard M, Li QJ, Sumen C, Huppa JB, Huse M, Davis MM (2005) Agonist/endogenous peptide-MHC heterodimers drive T-cell activation and sensitivity. Nature 434:238–243

La Gruta NL, Liu H, Dilioglou S, Rhodes M, Wiest DL, Vignali DA (2004) Architectural changes in the TCR:CD3 complex induced by MHC:peptide ligation. J Immunol 172:3662–3669

Lang HL, Jacobsen H, Ikemizu S, Andersson C, Harlos K, Madsen L, Hjorth P, Sondergaard L, Svejgaard A, Wucherpfennig K, et al (2002) A functional and structural basis for TCR cross-reactivity in multiple sclerosis. Nat Immunol 3:940–943

Lanier LL (2005) NK cell recognition. Annu Rev Immunol 23:225–274

Laugel B, Boulter JM, Lissin N, Vuidepot A, Li Y, Gostick E, Crotty LE, Douek DC, Hemelaar J, Price DA, et al (2005) Design of soluble recombinant T-cell receptors for antigen targeting and T-cell inhibition. J Biol Chem 280:1882–1892

Lee JK, Stewart-Jones G, Dong T, Harlos K, Di Gleria K, Dorrell L, Douek DC, van der Merwe PA, Jones EY, McMichael AJ (2004) T-cell cross-reactivity and conformational changes during TCR engagement. J Exp Med 200:1455–1466

Li Y, Huang Y, Lue J, Quandt JA, Martin R, Mariuzza RA (2005a) Structure of a human autoimmune TCR bound to a myelin basic protein self-peptide and a multiple sclerosis-associated MHC class II molecule. EMBO J 24:2968–2979

Li Y, Moysey R, Molloy PE, Vuidepot AL, Mahon T, Baston E, Dunn S, Liddy N, Jacob J, Jakobsen BK, Boulter JM (2005b) Directed evolution of human T-cell receptors with picomolar affinities by phage display. Nat Biotechnol 23:349–354

Lin SY, Ardouin L, Gillet A, Malissen M, Malissen B (1997) The single positive T cells found in CD3-zeta/eta-/- mice overtly react with self-major histocompatibility complex molecules upon restoration of normal surface density of T-cell receptor-CD3 complex. J Exp Med 185:707–715

Luescher IF, Vivier E, Layer A, Mahiou J, Godeau F, Malissen B, Romero P (1995) CD8 modulation of T-cell antigen receptor-ligand interactions on living cytotoxic T lymphocytes. Nature 373:353–356

Macdonald WA, Purcell AW, Mifsud NA, Ely LK, Williams DS, Chang L, Gorman JJ, Clements CS, Kjer-Nielsen L, Koelle DM, et al (2003) A naturally selected dimorphism within the HLA-B44 supertype alters class I structure, peptide repertoire, and T-cell recognition. J Exp Med 198:679–691

Malissen B (1996) Immunology. Two faces are better than one. Nature 384:518–519

Malissen B (2003a) An evolutionary and structural perspective on T-cell antigen receptor function. Immunol Rev 191:7–27

Malissen B (2003b) Glimpses at TCR trans-species crossreactivity. Immunity 19:463–464

Malissen B, Ardouin L, Lin SY, Gillet A, Malissen M (1999) Function of the CD3 subunits of the pre-TCR and TCR complexes during T-cell development. Adv Immunol 72:103–148

Malissen B, Aguado E, Malissen M (2005) Role of the LAT adaptor in T-cell development and Th2 differentiation. Adv Immunol 87:1–25

Mason D (1998) A very high level of crossreactivity is an essential feature of the T-cell receptor. Immunol Today 19:395–404

Mathis D, Benoist C (2004) Back to central tolerance. Immunity 20:509–516

Maynard J, Petersson K, Wilson DH, Adams EJ, Blondelle SE, Boulanger MJ, Wilson DB, Garcia KC (2005) Structure of an autoimmune T-cell receptor complexed with class II peptide-MHC: insights into MHC bias and antigen specificity. Immunity 22:81–92

Mazza C, Auphan-Anezin N, Gregoire C, Guimezanes A, Kellenberger C, Roussel A, Kearney A, van der Merwe PA, Schmitt-Verhulst AM, Malissen B (2007) How much can a T-cell receptor adapt to structurally distinct antigenic peptides? EMBO J 26:1972–1983

McFarland BJ, Kortemme T, Yu SF, Baker D, Strong RK (2003) Symmetry recognizing asymmetry: analysis of the interactions between the C-type lectin-like immunoreceptor NKG2D and MHC class I-like ligands. Structure 11:411–422

Merkenschlager M, Graf D, Lovatt M, Bommhardt U, Zamoyska R, Fisher AG (1997) How many thymocytes audition for selection? J Exp Med 186:1149–1158

Miley MJ, Messaoudi I, Metzner BM, Wu Y, Nikolich-Zugich J, Fremont DH (2004) Structural basis for the restoration of TCR recognition of an MHC allelic variant by peptide secondary anchor substitution. J Exp Med 200:1445–1454

Mitra AK, Celia H, Ren G, Luz JG, Wilson IA, Teyton L (2004) Supine orientation of a murine MHC class I molecule on the membrane bilayer. Curr Biol 14:718–724

Ogiso H, Ishitani R, Nureki O, Fukai S, Yamanaka M, Kim JH, Saito K, Sakamoto A, Inoue M, Shirouzu M, Yokoyama S (2002) Crystal structure of the complex of human epidermal growth factor and receptor extracellular domains. Cell 110:775–787

Pancer Z, Cooper MD (2006) The evolution of adaptive immunity. Annu Rev Immunol 24:497–518

Pisitkun P, Deane JA, Difilippantonio MJ, Tarasenko T, Satterthwaite AB, Bolland S (2006) Autoreactive B-cell responses to RNA-related antigens due to TLR7 gene duplication. Science 312:1669–1672

Randriamampita C, Boulla G, Revy P, Lemaitre F, Trautmann A (2003) T-cell adhesion lowers the threshold for antigen detection. Eur J Immunol 33:1215–1223

Raz E (2007) Organ-specific regulation of innate immunity. Nat Immunol 8:3–4

Reinherz EL, Tan K, Tang L, Kern P, Liu J, Xiong Y, Hussey RE, Smolyar A, Hare B, Zhang R, et al (1999) The crystal structure of a T-cell receptor in complex with peptide and MHC class II. Science 286:1913–1921

Reiser JB, Darnault C, Guimezanes A, Grégoire C, Mosser T, Schmitt-Verhulst, A.-M., Fontecilla-Camps JC, Malissen B, Housset D, Mazza G (2000) Crystal structure of a T-cell receptor bound to an allogeneic MHC molecule. Nat Immunol 1:291–297

Reiser JB, Grégoire C, Darnault C, Mosser T, Guimezanes A, Schmitt-Verhulst AM, Fontecilla-Camps JC, Mazza G, Malissen B, Housset D (2002) A T-cell receptor CDR3beta loop undergoes conformational changes of unprecedented magnitude upon binding to a peptide/MHC class I complex. Immunity 16:345–354

Reiser JB, Darnault C, Gregoire C, Mosser T, Mazza G, Kearney A, Van Der Merwe PA, Fontecilla-Camps JC, Housset D, Malissen B (2003) CDR3 loop flexibility contributes to the degeneracy of TCR recognition. Nat Immunol 4:241–247

Renard V, Romero P, Vivier E, Malissen B, Luescher IF (1996) CD8 beta increases CD8 coreceptor function and participation in TCR-ligand binding. J Exp Med 184:2439–2444

Rudolph MG, Wilson IA (2002) The specificity of TCR/pMHC interaction. Curr Opin Immunol 14:52–65

Rudolph MG, Stanfield RL, Wilson IA (2006) How TCRs bind MHCs, peptides, and coreceptors. Annu Rev Immunol 24:419–466

Schamel WW, Arechaga I, Risueno RM, van Santen HM, Cabezas P, Risco C, Valpuesta JM, Alarcon B (2005) Coexistence of multivalent and monovalent TCRs explains high sensitivity and wide range of response. J Exp Med 202:493–503

Scharenberg AM, Lin S, Cuenod B, Yamamura H, Kinet JP (1995) Reconstitution of interactions between tyrosine kinases and the high affinity IgE receptor which are controlled by receptor clustering. EMBO J 14:3385–3394

Schumacher TN, Ploegh HL (1994) Are MHC-bound peptides a nuisance for positive selection? Immunity 1:721–723

Selin LK, Cornberg M, Brehm MA, Kim SK, Calcagno C, Ghersi D, Puzone R, Celada F, Welsh RM (2004) CD8 memory T cells: cross-reactivity and heterologous immunity. Semin Immunol 16:335–347

Shih FF, Allen PM (2004) T cells are not as degenerate as you think, once you get to know them. Mol Immunol 40:1041–1046

Spencer DM, Wandless TJ, Schreiber SL, Crabtree GR (1993) Controlling signal transduction with synthetic ligands. Science 262:1019–1024

Sporri R, Reis e Sousa C (2002) Self peptide/MHC class I complexes have a negligible effect on the response of some CD8+ T cells to foreign antigen. Eur J Immunol 32:3161–3170

Springer TA (1990) Adhesion receptors of the immune system. Nature 346:425–434

Stefanova I, Dorfman JR, Germain RN (2002) Self-recognition promotes the foreign antigen sensitivity of naive T lymphocytes. Nature 420:429–434

Stefanova I, Hemmer B, Vergelli M, Martin R, Biddison WE, Germain RN (2003) TCR ligand discrimination is enforced by competing ERK positive and SHP-1 negative feedback pathways. Nat Immunol 4:248–254

Stewart-Jones GB, McMichael AJ, Bell JI, Stuart DI, Jones EY (2003) A structural basis for immunodominant human T-cell receptor recognition. Nat Immunol 4:657–663

Stone JD, Stern LJ (2006) CD8 T cells, like CD4 T cells, are triggered by multivalent engagement of TCRs by MHC-peptide ligands but not by monovalent engagement. J Immunol 176:1498–1505

Strong RK, McFarland BJ (2004) NKG2D and related immunoreceptors. Adv Protein Chem 68:281–312

Sun ZY, Kim ST, Kim IC, Fahmy A, Reinherz EL, Wagner G (2004) Solution structure of the CD3epsilondelta ectodomain and comparison with CD3epsilongamma as a basis for modeling T-cell receptor topology and signaling. Proc Natl Acad Sci USA 101:16867–16872

Szymczak AL, Workman CJ, Gil D, Dilioglou S, Vignali KM, Palmer E, Vignali DA (2005) The CD3epsilon proline-rich sequence, and its interaction with Nck, is not required for T-cell development and function. J Immunol 175:270–275

Taniguchi T, Takaoka A (2001) A weak signal for strong responses: interferon-alpha/beta revisited. Nat Rev Mol Cell Biol 2:378–386

Tynan FE, Burrows SR, Buckle AM, Clements CS, Borg NA, Miles JJ, Beddoe T, Whisstock JC, Wilce MC, Silins SL, et al (2005) T-cell receptor recognition of a "super-bulged" major histocompatibility complex class I-bound peptide. Nat Immunol 6:1114–1122

Vivier E, Malissen B (2005) Innate and adaptive immunity: specificities and signaling hierarchies revisited. Nat Immunol 6:17–21

von Boehmer H (2005) Unique features of the pre-T-cell receptor alpha-chain: not just a surrogate. Nat Rev Immunol 5:571–577

von Boehmer H, Aifantis I, Gounari F, Azogui O, Haughn L, Apostolou I, Jaeckel E, Grassi F, Klein L (2003) Thymic selection revisited: how essential is it? Immunol Rev 191:62–78

Vonakis BM, Haleem-Smith H, Benjamin P, Metzger H (2001) Interaction between the unphosphorylated receptor with high affinity for IgE and Lyn kinase. J Biol Chem 276:1041–1050

Wang JH, Reinherz EL (2002) Structural basis of T-cell recognition of peptides bound to MHC molecules. Mol Immunol 38:1039–1049

Wegener AM, Letourneur F, Hoeveler A, Brocker T, Luton F, Malissen B (1992) The T-cell receptor/CD3 complex is composed of at least two autonomous transduction modules. Cell 68:83–95

Wilson A, Marechal C, MacDonald HR (2001) Biased V beta usage in immature thymocytes is independent of DJ beta proximity and pT alpha pairing. J Immunol 166:51–57

Wu LC, Tuot DS, Lyons DS, Garcia KC, Davis MM (2002) Two-step binding mechanism for T-cell receptor recognition of peptide MHC. Nature 418:552–556

Yachi PP, Ampudia J, Gascoigne NR, Zal T (2005) Nonstimulatory peptides contribute to antigen-induced CD8-T-cell receptor interaction at the immunological synapse. Nat Immunol 6:785–792

Yamasaki S, Ishikawa E, Sakuma M, Ogata K, Sakata-Sogawa K, Hiroshima M, Wiest DL, Tokunaga M, Saito T (2006) Mechanistic basis of pre-T-cell receptor-mediated autonomous signaling critical for thymocyte development. Nat Immunol 7:67–75

Yokoyama WM (2006) Contact hypersensitivity: not just T cells! Nat Immunol 7:437–439

Zerrahn J, Held W, Raulet DH (1997) The MHC reactivity of the T-cell repertoire prior to positive and negative selection. Cell 88:627–636

Zinkernagel RM (2002) Uncertainties—discrepancies in immunology. Immunol Rev 185: 103–125

6
Fc Receptors

Falk Nimmerjahn[1,2] and Jeffrey V. Ravetch[1]

6.1 Introduction

An immune response is characterized by a delicate balance; strong enough to eliminate foreign pathogens but at the same time well controlled and highly specific to prevent destruction of self-tissues. Antibodies are essential to defend the body against invading microorganisms and antibodies bound to their respective antigen in the form of immune complexes (IC) have long been recognized to be potent inflammatory stimuli. The existence of cellular receptors for antibodies was anticipated as early as 1960 (Boyden and Sorkin 1960). Starting with the molecular cloning of cellular receptors for IgG and IgE, subsequently Fc receptors (FcR) for all antibody isotypes (IgM, IgA, IgD, IgE, and IgG) have been identified (Ravetch 2003). Fc receptors are widely expressed throughout the hematopoietic system and are essential regulators of immune cell activation. By recognizing the Fc portion of antibodies in immune complexes, Fc receptors link ancestral pathways of innate immunity to the specificity of the adaptive immune system. Their cellular expression pattern on myeloid effector cells, mast cells and B cells predicted their involvement in different types of inflammatory responses, allergy and B-cell regulation. Indeed, effector mechanisms shown to be triggered by FcR crosslinking include antibody dependent cellular cytotoxicity (ADCC), phagocytosis, release of inflammatory mediators and antigen presentation (Ravetch 2003). Moreover, the potent immunoregulatory functions of ICs (consisting of IgG or IgM antibodies), ranging from a strong enhancement to complete suppression of antibody responses in

[1]Laboratory of Molecular Genetics and Immunology, Rockefeller University, 1230 York Avenue, New York, NY 10021, USA

[2]Laboratory of Experimental Immunology and Immunotherapy, University of Erlangen-Nuernberg, Nikolaus-Fiebiger-Center for Molecular Medicine, Glueckstrasse 6, 91054 Erlangen, Germany

Correspondence to: J.F. Ravetch

addition to their more overt role as effector molecules for the elimination of foreign antigens, can now be ascribed to specific FcRs (Heyman 2000).

At the same time, however, inflammatory processes need to be tightly regulated to prevent destruction of self-tissues. In particular, low-affinity self-reactive antibodies (autoantibodies) found in healthy individuals need to be rendered harmless (Wardemann et al. 2003). Research over the last couple of years has identified multiple checkpoints that function to ensure orderly progression through an immune response (Goodnow et al. 2005; Grimaldi et al. 2005). The basis for these checkpoints is the establishment of discrete thresholds that define narrow windows of response; these thresholds are usually generated by the coexpression of activating and inhibitory molecules on the same immune effector cells. FcRs are a prime example for such regulation as activating FcRs are usually coexpressed with inhibitory counterparts, thereby setting thresholds for antibody mediated immune effector cell activation and release of inflammatory mediators. On B cells the inhibitory Fcγ-receptor IIB (FcγRIIB) regulates activating signals delivered by the B-cell receptor, ensuring that only B cells with a high-affinity B-cell receptor specific for foreign antigens can become activated. Similarly, inhibitory FcR expression on dendritic cells (DC) might prevent spontaneous DC activation and expansion of autoreactive T cells.

The aim of this chapter is to explain that this complexity exists to distinguish between self and nonself, thus avoiding autotoxicity and uncontrolled inflammation. We will focus mainly on the murine IgG and IgE FcRs, for which substantial data concerning their regulation and role in different physiological and pathological conditions in vivo are available. For a more in depth view about recent developments in Fcα-receptor biology the reader is directed to several excellent recent reviews covering that topic (Monteiro and van de Winkel 2003; Otten and van Egmond 2004). We will start with a brief description of basic Fc receptor biology, followed by an overview over activating and inhibitory FcR signaling pathways and examples where thresholds set by these contrasting signals are essential to maintain a balanced immune response and where disturbance of these thresholds leads to an uncontrolled inflammatory response and ultimately to destruction of self-tissues.

6.2 Fc Receptors—Basic Facts

In general Fc receptors can be divided into two classes: the activating and the inhibitory FcRs (Table 1). Whereas the activating receptors cannot signal autonomously and have to associate with additional adaptor molecules to be functional, the inhibitory receptor is a single chain molecule that contains an immunoreceptor tyrosine-based inhibitory motif (ITIM) in its cytosolic tail. A notable exception to this rule is human FcγRIIA that can transmit activating signals by itself. Therefore, a functional FcR consists of a ligand binding α-domain associated with signaling adaptor molecules containing immunoreceptor tyrosine-based activation motifs

Table 1. Human and mouse Fc receptors and their ligands

	FcαR (CD89)	FcεRI	FcεRII (CD23)	FcγRI (CD64)	FcγRIIA (CD32)	FcγRIIB (CD32)	FcγRIIIA (CD16)	FcγRIIIB (CD16)
Protein family	IgG superfamily	IgG superfamily	C-type lectin	IgG superfamily	IgG superfamily	IgG superfamily	IgG superfamily	IgG superfamily
Ligand	$IgA_1 = IgA_2$	IgE	IgE	IgG3,1 > 4 >> 2	IgG3 > 1 >> 2,4*	IgG3 > 1 > 4 > 2	IgG1,3 >> 2,4*	IgG1,3 >> 2,4
Affinity (M^{-1})	10^7	10^{10}	10^6–10^7	10^8–10^9	10^5–10^6	10^5–10^6	10^6	10^6
Signaling (subunit)	Activating (γ-chain)	Activating (γ- and β-chain)	Activating	Activating (γ-chain)	Activating	Inhibitory	Activating (γ- or ζ-chain)	None (gpi-anchor)
Expression	Macrophage Neutrophils Eosinophils Interstitial dendritic cells Kupffer cells	Mast cells Basophils Eosinophils Platelets Dendritic cells	B cells Follicular dendritic cells	Macrophages Neutrophils Eosinophils Dendritic cells	Macrophages Neutrophils Mast cells Eosinophils Platelets Dendritic cells	Macrophages Neutrophils Mast cells Eosinophils Dendritic cells FDC B cells	Macrophages Mast cells Basophils NK cells Dendritic cells	Neutrophils
Mouse receptor	Igα/μ receptor	FcεRI	FcεRII	FcγRI	FcγRIII (extracellular)	FcγRIIB	FcγRIV	Not identified
Ligand	IgA, IgM	IgE	IgE	IgG2a	IgG1, 2a, 2b	IgG1, 2a, 2b	IgG2a, 2b	n.a.
Affinity (M^{-1})	$10^8/10^9$ IgA/IgM	10^{10}	10^6–10^7	10^8	10^5–10^6	10^6	10^7	n.a.
References	Monteiro and van de Winkel, 2003	MacGlashan, 2005	Gould et al., 2003	Hulett and Hogarth, 1994; Ravetch and Kinet, 1991				

*Indicates that allelic variants exist that show different affinities for antibody isotypes (see Dijstelbloem et al., 2001)
n.a., not available

(ITAM). Depending on the cell type the associated signaling adaptor molecules vary. Whereas in the majority of cells, such as monocytes, macrophages, neutrophils, and dendritic cells FcRs are associated with the common gamma chain (γ-chain), in human natural killer (NK) cells FcRs are found in combination with the zeta chain (ζ-chain) (Table 1). Both γ- and ζ-chain are present as dimers linked by disulphide bonds. In the case of the FcεRI and human FcγRIIIA, additional β-chains have been identified in the receptor complex in certain cell types. In addition to the signaling function, these molecules are important for cell surface expression of the respective α-chains. For example, animals that don't express the γ-chain lack cell surface expression of all activating Fcγ-receptors and several other non-FcR-related proteins such as PIR-A and NK cell cytotoxicity receptors (Moretta et al. 2001; Ravetch 2003). As expected, γ-chain knockout animals were demonstrated to have significant defects in antibody-dependent effector cell responses such as phagocytosis of ICs, ADCC and inflammatory responses (Clynes and Ravetch 1995; Park et al. 1998; Sylvestre and Ravetch 1994; Takai et al. 1994; Zhang et al. 2004).

Regarding FcR genetics humans have 8 genes that encode Fcγ-receptors (FcγRIA/IB/IC, FcγRIIA/B/C, and FcγRIIIA/B) located on chromosome 1. In contrast to the duplication and diversification processes that have led to presence of multiple genes in the human genome (Qiu et al. 1990), the majority of other species including the mouse have four different classes of IgG Fc receptors that correspond to their human counterparts: FcγRI (CD64), FcγRII (CD32), FcγRIII (CD16) and FcγRIV. FcγRIV is a recently identified receptor, conserved in all mammalian species with intermediate affinity (10^{-7} M) and restricted subclass specificity (Davis et al. 2002; Mechetina et al. 2003; Nimmerjahn et al. 2005) and is most closely related to human FcγRIIIA (Fig. 1). FcγRI displays high affinity for the antibody constant region (10^{8}–$10^{9}\,M^{-1}$), FcγRII and FcγRIII have a low affinity for the Fc-portion (~$10^{6}\,M^{-1}$) (Hulett and Hogarth 1994; Ravetch and Kinet 1991). The low-affinity Fc receptor genes are clustered in close proximity to each other in syntenic regions on chromosome 1 in humans, chimpanzees and mice. In contrast, the high-affinity FcγRI is located on chromosome 3 in mice and chromosome 1 in humans and chimpanzees. Mirroring this complexity of Fcγ-receptors is the existence of several IgG isotypes that show differential binding to FcγRs (Table 1). In the mouse,

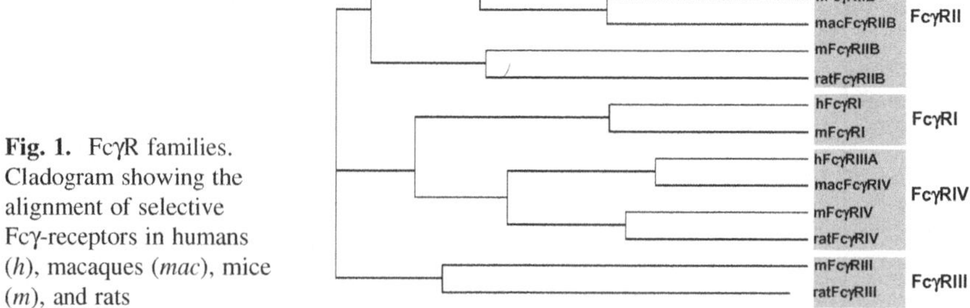

Fig. 1. FcγR families. Cladogram showing the alignment of selective Fcγ-receptors in humans (*h*), macaques (*mac*), mice (*m*), and rats

the high-affinity FcγRI exclusively binds IgG2a, the medium affinity FcγRIV binds IgG2a and IgG2b and the low-affinity receptors FcγRIIB and III bind IgG1, IgG2a and IgG2b. Moreover, it has only recently been appreciated that the IgG Fc receptors show significant differences in their affinity for individual antibody isotypes rendering certain isotypes more strictly regulated than others (Nimmerjahn et al. 2005). This represents a second layer of complexity and is of major importance for understanding Fc receptor dependent antibody mediated effector functions in vivo and for the design of antibody-based therapies. This important point will be discussed in greater detail later.

Although several proteins have been suggested to bind IgA, the best described receptor for IgA to date is CD89 (FcαRI). It is a gamma chain dependent receptor that binds IgA with intermediate affinity and is located on chromosome 19 in humans and chimpanzees; no homologous protein has been found in mice. However, a protein that can bind IgA as well as IgM has been identified in mice and called Fcα/μ-receptor (Shibuya et al. 2000). Like its human homologue, it binds IgA with intermediate affinity and is expressed on the majority of murine B lymphocytes, myeloid and also nonhematopoietic cells like mesangial cells (reviewed in Meteiro and van de Winkel 2003; Otten and van Egmond 2004). In contrast to the prevailing view that IgA antibodies are generally anti-inflammatory due to their capacity to block entry of pathogens at mucosal surfaces, recent evidence suggests that serum IgA can trigger inflammatory reactions as well (Otten and van Egmond 2004).

In contrast, both mice and humans contain two different IgE receptors. FcεRI has a high affinity for IgE ($10^{10} M^{-1}$), is located on chromosome 1 in mice and humans and is expressed on allergic effector cells including basophils and mast cells. Its expression has been detected on other cell types, for example on human antigen presenting cells like dendritic cells and macrophages. In contrast to the tetrameric form of the receptor (α-, β- and γ-chain dimer) on allergic effector cells, the receptor complex on other cells lacks the β-chain (MacGlashan, 2005; Novak et al. 2001). It has been suggested that this additional molecule is important for proper assembly of FcεRI and functions as an amplifier of γ-chain mediated signaling (Dombrowicz et al. 1998; Donnadieu et al. 2003). The low-affinity IgE receptor (FcεRII; CD23) is a structurally unrelated protein that forms trimers and contains a C-type-lectin domain. It is located on chromosome 8 in mice, chromosome 19 in humans and chromosome 20 in chimpanzees. It is important for controlling IgE production and transport and has been implicated in the enhancement as well as the downregulation of IgE responses following immunization with IgE immune complexes (Heyman 2002).

In addition to the classical Fc receptors a new family of FcR-like or homologous proteins has been identified recently. These proteins have significant sequence homology to classical FcRs in their extracellular domains and contain ITAM or ITIM motifs in their cytosolic portions. Several of these molecules, e.g. FcRX and FcRH1–5 are expressed on human and mouse B cells. Despite significant sequence homology to classical Fc receptors, attempts to demonstrate antibody-binding activity for these molecules have been unsuccessful, so FcγRIIB remains the only antibody binding Fc receptor on B cells. Nevertheless, these Fc receptor related molecules show a restricted expression pattern during different stages of B-cell development and might become important new markers for defining B-cell developmental stages or malignancies (reviewed in Davis et al. 2002; Davis et al. 2005).

6.3 Fc Receptor Signaling

In the following sections we will summarize the current models of how activating and inhibitory signals are triggered by crosslinking of cell surface Fc receptors by immune complexes. We will start with the signaling cascades triggered by activating Fc receptors followed by inhibitory signaling pathways and their counter-regulation of activating signals. The outcome of these opposing signaling events (cell activation or inhibition) is determined by several factors including the relative affinity of the antibody isotypes incorporated in immune-complexes, the expression level of activating and inhibitory Fc receptors and the cytokine environment which can influence expression levels of Fc receptors.

6.3.1 Signaling Pathways of Activating Fc Receptors

The first step in triggering signaling pathways by activating Fc receptors is the aggregation or cross linking of these receptors by immune complexes (Fig. 2). The affinity of the majority of activating Fc receptors for monomeric antibodies is not sufficient for stable binding and induction of signaling. High-affinity receptors, like FcγRI and FcεRI, can associate with monomeric IgG or IgE antibodies, but activating signals are only triggered upon replacement of bound IgG by immune complexes (in the case of FcγRI) or upon allergen binding and concomitant crosslinking of cell surface bound IgE (Galli et al. 2005; Gould et al. 2003). Ligands that bind with low affinity cannot trigger sustained receptor aggregation and might even behave as antagonistic ligands (Torigoe et al. 1998).

The overall signaling cascades triggered by the different activating Fc receptors and other activating receptors like the B-cell or T-cell receptor are very similar. As shown for FcαRI aggregation by ICs induces a re-location into cell membrane sub-domains called lipid rafts that are enriched in signaling molecules such as Src-protein kinases (Lang et al. 1999). Tyrosine residues in the ITAM motif of the γ-chain then become phosphorylated by Src kinases creating SH2-domain docking sites for the subsequent recruitment of Syk kinases. Depending on cell type and the receptor in question, different members of the Src-kinase family are involved in phosphorylation of the γ-chain. Whereas Lyn is crucial for the FcεRI pathway in mast cells, Lck is associated with FcγRIIIA in natural killer (NK) cells. In macrophages, both of these kinases and additionally Hck have been suggested to be important for γ-chain phosphorylation after crosslinking of FcγRI or FcγRIIA. This

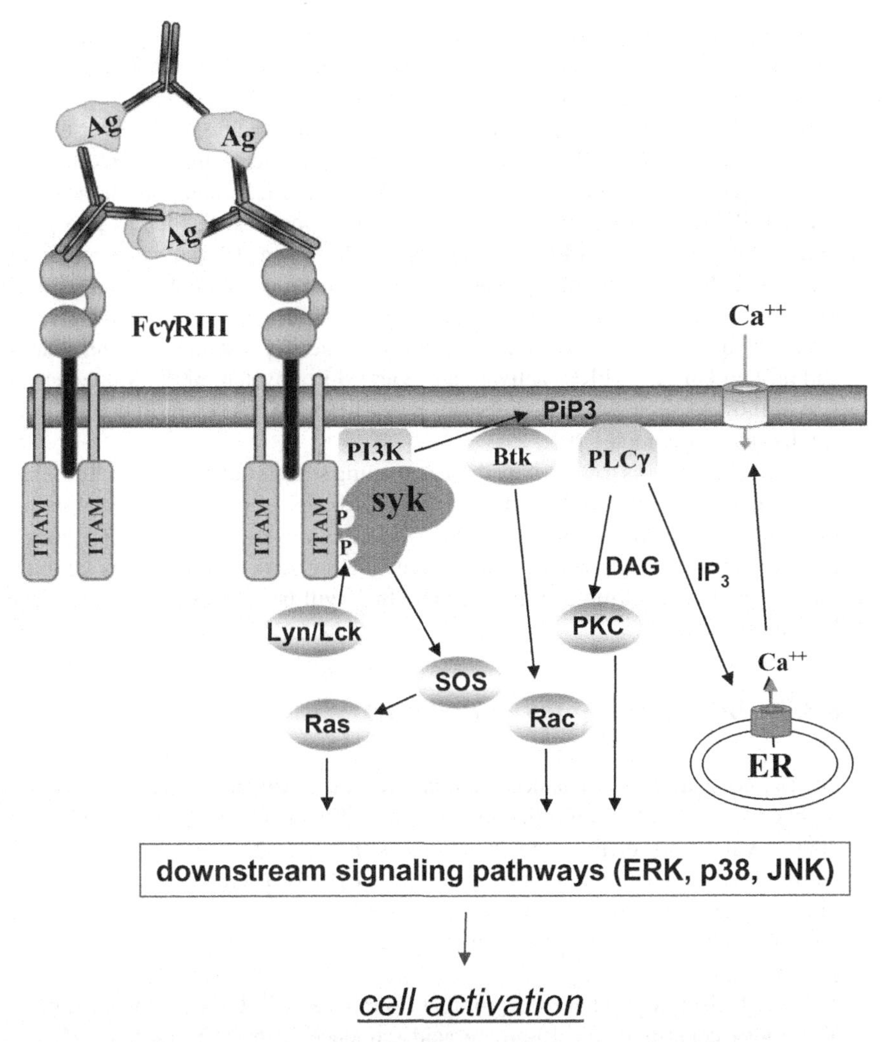

Fig. 2. Signaling pathways triggered by activating Fc receptors. After crosslinking of activating Fc receptors, the receptor associated γ-chains become phosphorylated by Src kinases like Lyn or Lck. This generates Src homology (SH)2-domain docking sites for Syk which in turn activates a number of other signal transduction molecules like phosphatidylinositol 3-kinase (*PI3K*) and SOS (son of sevenless). The generation of phosphatidylinositol 3-phosphate (*PiP$_3$*) recruits Btk and phospholipase-Cγ (*PLCγ*) which lead to activation of downstream kinases and the release of calcium from internal storage sites like the endoplasmic reticulum (*ER*). *DAG*, diacylglycerol; *ITAM*, immunoreceptor tyrosine-based activation motifs

enables members of the Syk-kinase family to bind and to recruit and phosphorylate a number of downstream targets including the linker for activation of T cells (LAT); multi-molecular adaptor complexes consisting of Cbl, Slp-76, Grb-2, Shc, Sos, SHIP; and members of the Btk and Tec kinase family (Gulle et al. 1998; Launay et al. 1998; Park et al. 1999). Important downstream events are triggered by Syk-mediated activation of phosphatidylinositol 3-kinase (PI3K) and phospholipase-Cγ (PLCγ). PI3K generated phosphatidylinositol polyphosphates, such as phosphatidylinositol 3-phosphate (PIP_3), allow pleckstrin homology (PH) domain containing proteins such as Btk and PLCγ to bind to the plasma membrane. Phospholipase Cγ hydrolyzes membrane bound phosphatidylinositol phosphates into diacylglycerol (DAG) and inositol-triphosphate (IP_3) which induces a sustained calcium release and protein kinase C (PKC) activation (reviewed in Ravetch 2003). Moreover, the Ras–Raf–MAPK pathway becomes activated through Sos present in the multimolecular adaptor complex (Fig. 2).

Effector responses triggered by these activating signals most prominently include degranulation of mast cells and neutrophils, release of cytotoxic mediators and inflammatory cytokines, antibody dependent cellular cytotoxicity (ADCC), antigen presentation and phagocytosis. Moreover, activating signals delivered by Fc receptors can induce dendritic cell maturation which will be discussed in greater detail later in this chapter.

6.3.2 Inhibitory Signaling Pathways

For the generation of a balanced immune response activating signals need to be regulated to prevent uncontrolled inflammation (Ravetch 2003; Ravetch and Lanier 2000). A hallmark of Fc receptor biology is that activating receptors are usually coexpressed with their inhibitory counterpart, FcγRIIB. Therefore, FcγRIIB is ubiquitously expressed throughout the hematopoietic system. Exceptions to this rule are NK cells that solely express the activating FcγRIII and T cells that lack Fc receptor expression.

FcγRIIB is a single chain receptor that contains an ITIM in its cytosolic domain. This motif consists of the 13-amino acid sequence AENTITYSLLKHP which is necessary and sufficient for the inhibitory activity of FcγRIIB. Depending on the cell type alternatively spliced forms of FcγRIIB have been described that show a differential capacity to endocytose bound ICs (termed FcγRIIB-1 and FcγRIIB-2). All of these splice forms, however, contain the ITIM motif. On B cells FcγRIIB regulates activating signals transmitted by the B-cell receptor (BCR), whereas on mast cells, neutrophils or macrophages it balances activating signals triggered by Fcε- or Fcγ-receptors.

Upon coaggregation with its activating counterpart, Lyn phosphorylates the ITIM-motif, which leads to the recruitment of SHIP (SH2-domain containing inositol 5′-phosphatase) (Fig. 3). SHIP activation leads to enhanced hydrolysis of phosphatidylinositol intermediates and thereby interferes with the membrane recruitment of Btk and PLCγ, resulting in inhibition of ITAM signaling mediated calcium release and downstream effector functions such as ADCC, cytokine secre-

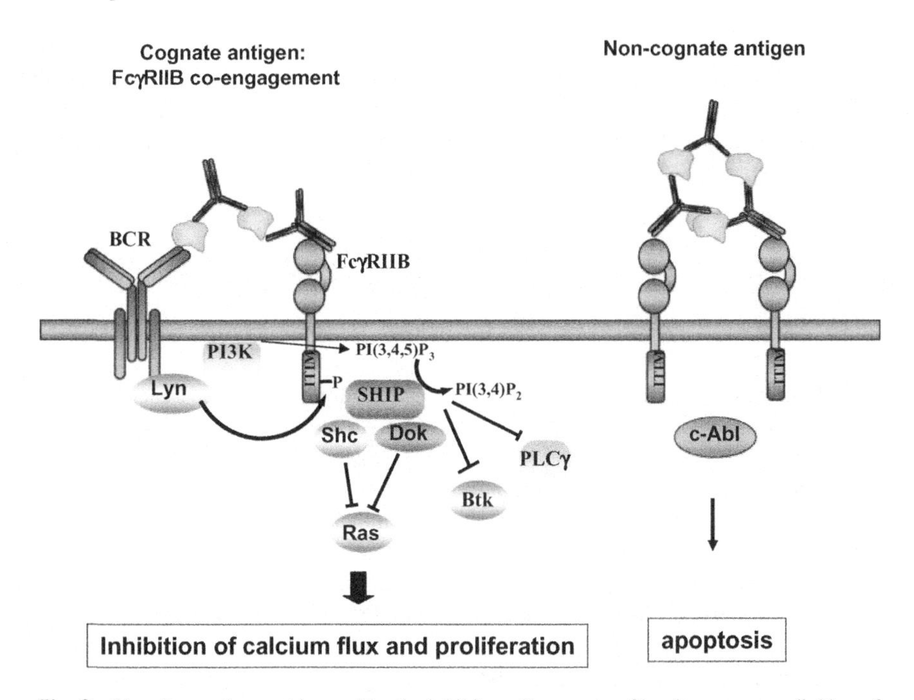

Fig. 3. Signaling pathways triggered by the inhibitory Fc receptor. Simultaneous crosslinking of activating receptors like the B-cell receptor (*BCR*) and the inhibitory Fcγ-receptor IIB (*FcγRIIB*) leads to phosphorylation of the immunoreceptor tyrosine-based inhibitory motif (*ITIM*) in the cytoplasmic tail of FcγRIIB by Lyn. This results in the recruitment of SH2-domain containing inositol 5′-phosphatase (*SHIP*) and the hydrolysis of PIP_3 into PIP_2, which ultimately inhibits recruitment of pleckstrin homology domain containing proteins like Btk and phospholipase-Cγ (*PLCγ*) (*left panel*). Isolated triggering of FcγRIIB leads to B-cell apoptosis via ITIM independent signaling pathways involving the c-Abl kinase family (*right panel*)

tion and release of inflammatory mediators. Moreover, tyrosine phosphorylated SHIP can bind to Shc and Dok, thereby inhibiting activation of the Ras pathway and ultimately cell proliferation. A third ITIM and SHIP independent signaling pathway has been described for crosslinking of FcγRIIB on B cells without concomitant activating signals by the BCR. This pathway leads to B-cell apoptosis via Abl-family kinase dependent pathways (Pearse et al. 1999; Tzeng et al. 2005). This situation may arise during the germinal center reaction when somatic hypermutation generates BCRs that lose specificity for their cognate antigen retained in the form of immune complexes on follicular dendritic cells. Thus, FcγRIIB has been suggested to be important for keeping tolerance.

6.4 Fc Receptor Biology In Vivo

After having described the basic signaling pathways and their interplay to achieve balanced immune responses, we will discuss other intrinsic and extrinsic factors that will determine whether more activating or inhibitory signals are triggered.

These factors include the actual affinity of antibody isotypes for the inhibitory and activating Fc receptors, the relative expression level of these receptors on individual immune effector cells and the local cytokine milieu that can change these relative expression levels. Furthermore, we will describe how this regulation impacts on the afferent and efferent phases of an immune response.

6.4.1 The Role of the Inhibitory Receptor in the Afferent and Efferent Immune Response

FcγRIIB belongs to the family of immune inhibitory receptors. Other prominent members of this family are PIR-B, KIRs, CTLA-4, PD-1, CD5, and CD22. These proteins carry ITIM motifs, are widely expressed on immune effector cells and are important regulators of their activating counterparts (Ravetch and Lanier 2000). The loss of these negative regulators leads to imbalanced immune responses resulting in autoimmunity and overt autoimmune disease (Bolland and Ravetch 2000; O'Keefe et al. 1999; Nishimura et al. 1999; Penninger et al. 1995; Takai et al. 1996; Tivol et al. 1995). Taking into account the cell types where FcγRIIB is expressed the loss or impairment of FcγRIIB mediated negative regulation would be expected to result in aberrant responses in the afferent as well as the efferent phases of an immune response. The generation of a mouse deficient in FcγRIIB over a decade ago has proven invaluable to study these predictions and the majority of the results were generated using this mouse model.

6.4.2 FcγRIIB and Dendritic Cells

Dendritic cells (DC) have long been recognized as central mediators that, depending on their activation state, determine whether an adaptive immune response or tolerance is induced (reviewed in Steinman et al. 2003). Several groups have shown that immune complexes are potent activators of DCs and are important for efficient cross-presentation of endocytosed antigen in the form of ICs on MHC class I molecules to CD8+ cytotoxic T cells (Dhodapkar et al, 2002; Groh et al. 2005; Rafiq et al. 2002; Regnault et al. 1999;). As low levels of ICs are constantly present in the serum, FcγRIIB is crucial to prevent spontaneous activation of DCs. In addition, expression of the inhibitory receptor on DCs present in epithelia has been implicated in establishing tolerance to air-borne and food allergens (Samson et al. 2005).

Although this negative regulation is essential to prevent expansion of autoreactive T cells during the steady state, it also limits immunotherapeutic approaches aimed at generating a strong response, e.g., during tumor therapy with monoclonal antibodies. Consistent with this, DCs derived from FcγRIIB deficient mice showed an enhanced potential to generate antigen-specific T-cell responses in vitro and in vivo (Kalergis et al. 2002). These studies suggest that immunotherapeutic and vaccination approaches can be optimized by overcoming this negative regulatory effect of FcγRIIB on DCs. A recent study using human DCs and an FcγRIIB blocking antibody showed that blocking FcγRIIB was sufficient to induce DC maturation by immune complexes normally present in plasma. Besides upregulation of costimulatory molecules, these DCs were more potent in generating and activating tumor specific T cells (Boruchov et al. 2005; Dhodapkar et al. 2005). This supports the notion that blocking the inhibitory Fc receptor on DCs in vivo might indeed be a strategy for generating stronger and probably longer lasting immune responses.

Before using this strategy in patients it will be important to examine how systemic blocking of FcγRIIB-mediated negative signals impacts on the immune system in the steady state. As indicated before, immature DCs are continuously tolerizing self reactive T cells that escape negative selection in the thymus and FcγRIIB is crucial to control the expansion of autoreactive B cells. To test this hypothesis, the development of novel animal models with targeted deletion of Fc receptors in dendritic cells or other selective cell populations will be essential. Additionally, mice carrying the human Fc receptors instead of the mouse counterparts will become an important animal model for assessing the in vivo activity of blocking antibodies for human Fc receptors.

Whereas the essential role of dendritic cells in regulating T-cell responses is well accepted, it has only recently become clear that DCs are also important for the B-cell response (Bergtold et al. 2005; Kushnir et al 1998; Wykes et al. 1998). In contrast to macrophages, which rapidly degrade phagocytosed material, antigen taken up by DCs is degraded more slowly and therefore present in an intact form for prolonged times (Delamarre et al. 2005). It has been suggested that this allows transport of antigen from the periphery to lymphoid organs where it can be presented to B cells. DC-B-cell interactions have been observed in vivo and it has been suggested that this interaction is important for generation of an IgG response in vitro and in vivo (Wykes et al. 1998). The important role of FcγRIIB in this process is that ICs taken up via FcγRIIB are inefficiently degraded and recycled for cell surface presentation to B cells. In contrast uptake via FcγRIII results in a more rapid degradation of the antigen (Bergtold et al. 2005).

6.4.3 Loss of the Inhibitory Receptor on B Cells—Horror Autotoxicus

B-cell development proceeds through several stages and progression from one stage to another is tightly controlled by numerous checkpoints that ensure that B cells expressing a self-reactive receptor are eliminated (Goodnow et al. 2005; Grimaldi et al. 2005). As indicated before, FcγRIIB regulates activating signals triggered by the B-cell receptor, thus setting a threshold for B-cell activation. The strength of this signal will ultimately determine whether a B-cell proliferates, class switches and matures into an antibody secreting plasma cell. Loss of this negative regulator was therefore predicted to result in uncontrolled B-cell activation. Especially the capacity of FcγRIIB to trigger B-cell apoptosis (Fig. 3) has been proposed to be an important mechanism to maintain self tolerance by deletion of low affinity or self reactive B cells. This was confirmed by the generation of FcγRIIB-deficient mice that spontaneously develop a lupus like disease characterized by the production of autoantibodies and premature death due to severe glomerulonephritis (Bolland and Ravetch 2000; Takai et al. 1996). This autoimmune phenotype is strain dependent—mice on the C57BL/6 but not the Balb/c background develop an autoimmune disease, suggesting that other epistatic modifiers are involved in disease susceptibility and severity (reviewed in Nguyen et al. 2002; Bolland and Ravetch 2002). Supporting this notion it was shown recently that Balb/c mice double deficient in programmed death 1 (PD-1) and FcγRIIB, but not the single knockout mice, developed severe autoimmune hydronephrosis (Okazaki et al. 2005). Moreover, Balb/c-Fcgr2b$^{-/-}$ mice showed enhanced disease phenotypes in a model of pristane-induced lupus (Clynes et al. 2005).

Another line of evidence for the role of FcγRIIB as a gatekeeper of tolerance comes from autoimmune-prone mouse strains such as NZB, NOD, BXSB, and MRL that have been found to express reduced levels of FcγRIIB on activated and germinal-center B cells. It was suggested that this is due to a polymorphism in the FcγRIIB promoter (Jiang et al. 1999, 2000; Pritchard et al. 2000; Xiu et al. 2002). Similarly, a polymorphism in the human FcγRIIB promoter linked to lupus has been identified. This polymorphism leads to decreased transcription and surface expression of FcγRIIB on activated B cells of human lupus patients (Blank et al. 2005). Besides B cells, FcγRIIB is also expressed on other inflammatory immune effector cells such as neutrophils, monocytes and macrophages. As will be discussed below it is very likely, however, that the autoimmunity observed in C57BL/6- FcγRIIB-deficient animals is B-cell autonomous.

One important point in favor of this theory is that animals that only lack FcγRIIB expression on peripheral B cells develop autoimmunity. This was achieved by transferring FcγRIIB deficient bone marrow into irradiated B-cell deficient hosts (RAG or IgH knockouts). In these animals the monocytic compartment still expressed FcγRIIB while it was absent from peripheral B cells (Bolland and Ravetch 2000). Additionally, by doing the reverse experiment it was demonstrated that restoring FcγRIIB expression to wild-type levels in autoimmune prone mouse strains like NZM, BXSB and FcγRIIB knockout animals by retroviral transduction with FcγRIIB, tolerance could be restored (McGaha et al. 2005). Again, B cells represented the majority of cells that showed increased expression of FcγRIIB. A very important result of this study is that restoration of FcγRIIB expression on approximately 40% of peripheral B cells was sufficient to prevent the development of autoantibodies and autoimmune glomerulonephritis (McGaha et al. 2005). This highlights the threshold nature of autoimmunity and suggests that despite the complex nature of autoimmune diseases therapeutic effects are achievable by targeting specific cell populations. Besides polymorphisms that affect FcγRIIB promoter activity, there is evidence that mutations in the transmembrane region of the inhibitory receptor are linked to human lupus in certain racial groups (Chu et al.

2004; Kyogoku et al. 2004; Siriboonrit et al. 2003). Recent evidence suggests that this allelic variant looses its inhibitory capacity due to its inability to associate with lipid rafts (Floto et al. 2005; Kono et al. 2005).

The B-cell stage(s) at which FcγRIIB exerts its function as a gatekeeper of self tolerance has recently been defined. Autoreactive B cells can be generated at several stages during B-cell development (Grimaldi et al. 2005). There is accumulating evidence that FcγRIIB mediates its function during late stages of B-cell maturation thus representing a distal checkpoint. It has been suggested that B cells generating autoreactive or low-affinity BCRs during somatic hypermutation will loose BCR interactions with their cognate antigen retained in the form of immune complexes on follicular dendritic cells (FDC). This results in isolated triggering of FcγRIIB which induces apoptosis (Fig. 3) (Pearse et al. 1999; Ravetch and Bolland 2001). More recently it has been shown that in the absence of FcγRIIB IgG positive plasma cells secreting autoreactive antibody species can accumulate (Fukuyama et al. 2005). FcγRIIB deficiency did not impact on early events in the bone marrow like receptor editing nor did it prevent the development of IgM positive autoreactive B cells. After class switching to IgG, however, FcγRIIB was essential to prevent the expansion of autoreactive B cells and their maturation into plasma cells. Taking the considerably higher pathogenic potential of IgG compared to IgM antibody isotypes into account this relatively late stage of FcγRIIB mediated negative regulation might be sufficient to prevent the initiation of severe autoreactive processes.

These results would support a model in which several central and peripheral checkpoints prevent the emergence of autoreactive B cells and their maturation into plasma cells that could secrete pathogenic antibodies. Central checkpoints including receptor editing, deletion, and anergy of self-reactive BCR species ensure that the majority of B cells with an autoreactive BCR are deleted in the bone marrow (reviewed in Goodnow et al. 2005; Grimaldi et al. 2005; Meffre et al. 2000); this occurs independently of FcγRIIB. It is widely accepted, however, that this process is incomplete and self reactive cells can escape into the periphery, in a background dependent manner. Thus Balb/c are more efficient in editing than C57BL/6 mice, making the later a more permissive strain for the development of autoimmunity. Consistent with this observation, FcγRIIB deficient mice on the Balb/c background did not develop spontaneous autoimmunity; in contrast, this deficiency of an inhibitory receptor on the C57BL/6 background resulted in the emergence of a highly penetrant, fatal lupus-like disease. Moreover, autoreactive B cells can be generated de novo in the periphery during the germinal center reaction (Ray et al. 1996; reviewed in Bona and Stevenson 2004). Therefore, additional checkpoints are of major importance to prevent the accumulation of autoreactive cells in the periphery. Furthermore, the expansion of class switched self reactive antibodies that can trigger a wide variety of inflammatory effector functions needs to be tightly regulated (Dijstelbloem et al. 2001; Ravetch and Bolland, 2001) Here FcγRIIB might serve as the final barrier to prevent these B cells with harmful BCR specificities from maturing into plasma cells that otherwise would induce tissue pathology by secretion of large amounts of self reactive antibodies.

6.4.4 The Role of FcγRIIB in the Efferent Response: Controlling Innate Immune Effector Cell Activation

Besides its autoregulatory role in the afferent response, FcγRIIB is an important modulator of inflammatory effector cells such as mast cells, neutrophils and macrophages during the efferent phase of an immune response (Dijstelbloem et al. 2001; Ravetch and Bolland 2001). On these cell types FcγRIIB is coexpressed with activating Fc receptors of varying affinities and isotype specificities and negatively regulates activating signals delivered by these receptors. Lack of FcγRIIB leads to elevated immune complex mediated inflammation and phagocytosis as demonstrated by an enhanced Arthus reaction, systemic anaphylaxis, anti-GBM glomerulonephritis, immunothrombocytopenia, hemolytic anemia, collagen-induced arthritis, and IgG-mediated clearance of pathogens and tumor cells (Ravetch 2003).

On allergic effector cells such as mast cells and basophils FcγRIIB regulates activating signals triggered by crosslinking FcεRI resulting in enhanced IgE-mediated anaphylaxis and heightened sensitivity to allergic rhinitis (Watanabe et al. 2004; reviewed in Kraft and Novak 2005). Moreover, FcγRIIB deficiency renders otherwise resistant mouse strains susceptible to development of certain forms of collagen induced arthritis (Takai 2002). In some of these models both increased autoantibody production due to FcγRIIB-deficiency on B cells and heightened effector cell responses are likely to contribute to the observed phenotype. As will be discussed below, the magnitude of FcγRIIB modulation is strictly isotype dependent and predictable based on the relative affinities of IgG subtypes for activating and inhibitory receptors.

6.4.5 The Activating Fc Receptors in the Efferent Response

Activating Fc receptors including FcαR, FcεR, and the family of Fcγ-receptors (FcγRI, III, and IV) are expressed on a wide variety of immune effector cells including mast cells, basophils, monocytes, macrophages, neutrophils, and NK cells. The importance of these receptors for effector functions mediated by these cells has been demonstrated by genetic deletion of the common signaling γ-chain used by all of these receptors (Takai et al. 1994). In these animals, immune complex or allergen mediated effector functions, such as antibody dependent cellular cytotoxicity (ADCC), release of inflammatory mediators, cytokine release and phagocytosis of immune complexes are abrogated or heavily impaired. As many of these Fc receptors are coexpressed on the same cell subsequent deletion of the individual ligand binding α-chains was crucial to elucidate the role of the individual Fc receptors.

Expectedly, mice deficient in the high-affinity FcεRI, which is essential for cell surface binding of IgE to mast cells showed a dramatic reduction in IgE-mediated anaphylaxis and allergic reactions. For the family of FcγRs, however, the role of the individual receptors was less clear until recently. Here, deletion of the individual activating Fcγ-receptors I or III resulted in less pronounced phenotypes especially for effector responses involving the IgG2a and IgG2b antibody isotypes. This was surprising as in vitro studies had shown that the high-affinity FcγRI could bind to IgG2a and the low-affinity FcγRIII to IgG1, IgG2a, and IgG2b (reviewed in Hulett and Hogarth 1994; Ravetch and Kinet 1991). Although there is some evidence that FcγRI and III may participate in a limited fashion in IgG2a-mediated effector responses (Barnes et al. 2002; Ioan-Facsinay et al. 2002), the majority of studies concluded that IgG2a and IgG2b triggered effects occur independently of these two receptors, but in a γ-chain dependent manner (Fossati-Jimack et al. 2000; Hazenbos et al. 1996; Meyer et al. 1998; Nimmerjahn et al. 2005; Uchida et al. 2004). In contrast FcγRI, due to its high affinity for IgG2a (K_A: 10^8–$10^9 M^{-1}$), resulting in equal efficiency binding of monomeric IgG2a and immune complexes (ICs), might not be available for newly generated ICs (Fig. 4).

An alternative theory suggested that other effector mechanisms, such as activation of the complement cascade, might mediate the in vivo effects of these isotypes. Indeed, IgG2a and IgG2b can efficiently activate the complement cascade in vitro (Duncan and Winter 1988). However, several studies using mice deficient in a variety of complement proteins such as C3, C4 or CR2 failed to demonstrate a major involvement of the complement cascade (reviewed in Ravetch and Clynes 1998; Uchida et al. 2004). On the other hand, deletion of the γ-chain abrogated

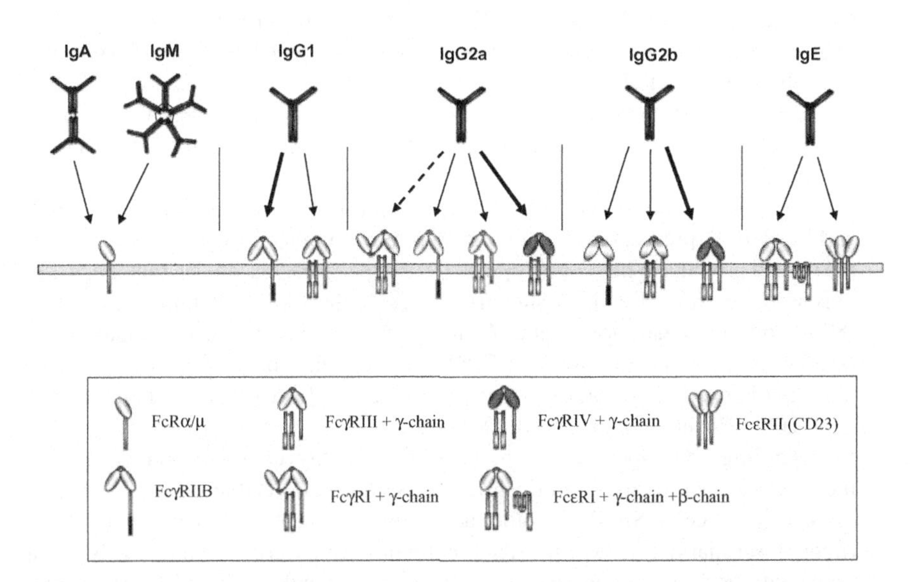

Fig. 4. Antibody isotype binding to Fc receptors. Shown is the interaction of different antibody isotypes (IgA, IgM, IgE, IgG1, IgG2a, and IgG2b) with their respective Fc receptors. *Arrows* indicate interactions and the *thickness of the arrows* indicates preferential interactions. A *broken arrow* indicates that the interaction is hypothetical (see text for details)

IgG2a and IgG2b effector functions strongly arguing for the existence of other γ-chain dependent Fc receptors.

In contrast to the unclear situation for IgG2a and IgG2b, IgG1-mediated effector functions were abrogated in the absence of the low-affinity FcγRIII. These results were confirmed in a variety of models like arthritis, glomerulonephritis, IgG-dependent anaphylaxis, IgG mediated hemolytic anemia and immunothrombocytopenia (ITP) (Bruhns et al. 2003; Fossati-Jimack et al. 2000; Fuji et al. 2003; Hazenbos et al. 1996; Ji et al. 2002; Meyer et al. 1998; Nimmerjahn et al. 2005). However, the most potent antibody isotypes for protection against bacterial or viral infections (Coutelier et al. 1987; Markine-Gorianyoff and Coutelier 2002; Schlageter and Kozel 1990; Taborda et al. 2003), antibody-mediated cytotoxicity or antibody-based therapy (Fossati-Jimack et al. 2000; Kipps et al. 1985; Nimmerjahn et al. 2005 Uchida et al. 2004;) were of the IgG2a and IgG2b isotype. Therefore, a thorough understanding of how these isotypes exert their function is essential.

6.4.6 The Missing Piece in the IgG-Puzzle: Identification of FcγRIV

To identify new Fc receptor homologous or related proteins, several groups started genome database searches on the basis of conserved sequences among the classical Fc receptors. During this search one gene was identified that showed 63% overall amino-acid identity to human FcγRIIIA and an even greater identity in the antibody binding extracellular domain and was called Fc receptor like 3 (Fcrl3), CD16–2, or FcγRIV (Davis et al. 2002; Mechetina et al. 2002; Nimmerjahn et al. 2005). To be consistent with the current nomenclature system for mouse Fcγ-receptors and to avoid confusion with other Fcrl-proteins, we suggest calling this protein FcγRIV.

The FcγRIV gene is located on mouse chromosome 1, tightly linked to FcγRIIB and FcγRIII. A prediction of orthologue proteins in other species shows that there are related proteins in humans (FcγRIIIA), chimpanzees, macaques, rats, dogs, cats, pigs and cows (Fig. 1 and not shown) with the highest level of similarity to the rat (80%) and the human orthologue. As has been described for other γ-chain dependent Fc receptors, cross-linking of FcγRIV by immune complexes induces activating signaling pathways leading to sustained calcium flux (reviewed in Nimmerjahn et al. 2005; Ravetch and Bolland 2001).

Regarding cell type restriction, FcγRIV is highly expressed on neutrophils, monocytes, macrophages and dendritic cells and undetectable on mast cells, NK cells, T and B cells. Similar to other activating Fc receptors, inflammatory stimuli (lipopolysaccharide; LPS) and Th-1 cytokines (interferon gamma; IFN-γ) can upregulate FcγRIV; in contrast Th-2 cytokines interleukin (IL)-4, IL-10, or transforming growth factor beta (TGF-β) downregulate FcγRIV cell surface expression (Nimmerjahn et al. 2005). After induction of DC maturation FcγRIV together with other activating Fc receptors is downregulated. An important difference between mouse FcγRIV and the human FcγRIIIA is that FcγRIV is not expressed on NK cells. Human neutrophils do not express FcγRIIIA but rather FcγRIIA as their dominant activating FcR.

In vitro analysis revealed that FcγRIV bound IgG2a and IgG2b with intermediate affinity (K_A: 2–3 × $10^7 M^{-1}$), but not IgG1 or IgG3 antibody isotypes. In contrast to the high-affinity FcγRI, this affinity was not sufficient to enable stable binding to monomeric IgG, leaving it accessible for immune complex binding. Remarkably its higher affinity for IgG2a and IgG2b compared to the inhibitory FcγRIIB predicted that it would be less sensitive to FcγRIIB-mediated negative regulation (Table 1). More importantly, even if coexpressed with FcγRIII, IgG2a and IgG2b ICs would preferentially engage FcγRIV due to its 20–40 times higher affinity and due to the strong FcγRIIB imposed negative regulation of FcγRIII. Consistent with this notion, blocking FcγRIV function in vivo greatly impairs the pathogenic effects of IgG2a and IgG2b antibodies in passive models of antibody mediated platelet depletion or tumor cell destruction (Hamaguchi et al. in press; Nimmerjahn and Ravetch 2005; Nimmerjahn et al. 2005) and in an active model of glomerulonephritis induced by administration of nephrotoxic serum (Kaneko et al. 2006). Neither deficiency in members of the complement cascade nor in FcγRI or III had a significant effect on IgG2a/2b mediated responses.

6.4.7 Isotype Specific Fcγ-Receptor Engagement and Differential Regulation by FcγRIIB

Taken together, these studies suggest that even if several activating Fc receptors with the same isotype specificity are present on the same cell only those Fc receptors will be engaged that show the optimal affinity for the respective isotype (Fig. 4). Therefore, IgG1 immune complexes will only trigger FcγRIII as it is the only activating Fc receptor that can bind IgG1 (Hazenbos et al. 1996; Meyer et al. 1998; Nimmerjahn et al. 2005; Takai 1994); IgG2a and IgG2b, despite their ability to bind FcγRI and FcγRIII, respectively, will mainly engage FcγRIV as FcγRI will be occupied by monomeric IgG2a. Interestingly, human FcγRIIIA also has a higher affinity for IgG compared to human FcγRIIA, indicating that the above rules might also apply for the human system. It should be noted, however, that allelic Fc receptor variants that show differential affinities for the specific antibody isotypes exist in humans, which creates a more complex scenario (Dijstelbloem et al. 2001).

If antibody activity in vivo is mostly dependent on the interaction with activating and inhibitory Fc receptors, antibody activity might be predictable if one knows the affinity differences (A/I-ratio) of antibody isotypes for the individual activating and the inhibitory FcR. This question was addressed recently and it was shown that individual antibody isotypes have significantly different A/I ratios (Table 2). IgG1 for example had an A/I ratio of 0.1, indicating a strong influence of the inhibitory receptor on antibody activity. In contrast, IgG2a and IgG2b had A/I ratios of 7 and 70 making them less sensitive to FcγRIIB-mediated negative regulation

Table 2. Activating Fc receptors are differentially regulated by FcγRIIB

	Ratio of activating to inhibitory receptor(A/I)	
	FcγRIII/FcγRIIB	FcγRIV/FcγRIIB
IgG1	0.1[a]	n.b.
IgG2a	1.6	70
IgG2b	0.5	7
IgG3	n.b.	n.b.

[a] A/I ratios were calculated based on the affinities of the indicated Fc receptors as described (Nimmerjahn et al. 2005). n.b. indicates that the isotype is not binding to the indicated Fc receptor

(Nimmerjahn et al. 2005). Testing these predictions in in vivo model systems of antibody-mediated platelet depletion or tumor cell destruction with antibody switch variants showed that IgG2a and IgG2b antibodies were the most efficient isotypes and that deletion of the inhibitory receptor impacted most strongly on IgG1 activity (Clynes et al. 2000; Nimmerjahn and Ravetch 2005). This hierarchy of antibody isotype activity has been observed in other model systems where isotype switch variants were used showing that IgG2a and IgG2b variants were more potent than IgG1or IgG3 (Fossati-Jimack et al. 2000; Kipps et al. 1985).

6.4.8 The Effect of Cytokines on FcR Expression

Many studies have addressed the question how cytokines can impact on Fc receptor expression. It was shown that depending on cell type and Fc receptor these effects can be different. Frequently cytokines regulate expression of the associated signaling adaptors (β- and γ-chains), which leads to a concomitant change in α-chain expression, as shown for TGF-β, IL-4, and IL-10 for example (Gillespie et al. 2004; Tridandapani et al. 2003). Moreover, inflammatory cytokines/stimuli, such as tumor necrosis factor (TNF)-α or LPS, tend to upregulate activating receptors, such as Fcα-, Fcε- and Fcγ-receptors, whereas TGF-β, L-4, and IL-10 seem to have the opposite effect (Nimmerjahn et al. 2005; Otten and van Egmond 2004; Tridandapani et al. 2003). Importantly, these effects can be cell type specific. Interleukin-4 for example upregulates the inhibitory FcγRIIB on myeloid cells. On activated B cells, however, this cytokine downregulates receptor expression (Rudge et al. 2002).

Regarding the different IgG isotypes it is interesting to consider how cytokines will influence the basic A/I ratios discussed before. It seems likely that a change in basal expression levels of activating versus inhibitory receptors will impact differentially on the various IgG isotypes. Thus, IgG2a antibodies are relatively insensitive to these effects, while IgG1 is quite sensitive to modest changes in A/I ratios. For example, in active models of antibody mediated inflammation the steady state ratios will be changed in favor of the activating Fc receptors. Inflammatory mediators, such as IFN-γ and C5a, can upregulate activating Fcγ-receptors and at the same time reduce FcγRIIB expression levels (Guyre et al. 1983; Shushakova et al. 2002). Under these circumstances autoreactive IgG1 antibodies are capable of triggering severe damage. In contrast Th-2 cytokines like IL-4, IL-10, or TGF-β upregulate the inhibitory Fcγ-receptor and decrease expression of the activating Fcγ-receptors (Nimmerjahn et al. 2005; Okayama et al. 2000; Pricop et al. 2001; Radeke et al. 2002; Tridandapani et al. 2003). Under these conditions isotypes that have a low or moderate A/I ratio (IgG1 and IgG2b) would be expected to loose more activity than those with a high ratio, such as IgG2a. High-dose intravenous gamma globulin (IVIG) provides a compelling example of the validity of this approach. The anti-inflammatory activity of this preparation has recently been shown to be linked to its ability to upregulate FcγRIIB expression on effector macrophages in models of ITP, rheumatoid arthritis (RA) and glomerulonephritis (Bruhns et al. 2003; Kaneko et al. 2006; Samuelsson et al. 2001). For IgG1, this modulation alone is sufficient to convert a pathogenic antibody to a nonpathogenic isotype, consistent with the low A/I ratio for this subclass in which modest changes in FcγRIIB expression will raise the threshold required for effective IgG1 crosslinking of FcγRIII. In contrast, in an active model of nephrotoxic nephritis, IgG2b antibodies were demonstrated to mediate the inflammatory response. IVIG protects in this disease model by upregulating FcγRIIB, as observed previously, and by downregulating FcγRIV, thus altering the A/I ratio to a level where inhibition dominates and inflammation is blocked. FcγRIIB modulation alone is insufficient to provide protection for this subclass, requiring the additional contribution of FcγRIV downregulation to lower the threshold to prevent IgG2b activation of FcγRIV (Kaneko et al. 2006).

Interestingly there is evidence that this paradigm holds true for human IgG antibodies as well. Human clinical trials with antibodies directed against tumor cell surface proteins, such as CD20, revealed that patients with the high-affinity FcγRIIIA allele resulting in a higher A/I ratio responded significantly better to antibody therapy with increased survival and time to relapse (Cartron et al. 2002; Weng and Levy 2003; Weng et al. 2004). The challenge for future antibody based immunotherapeutic approaches will be to select, or preferably design antibody isotypes that can trigger effector functions such as ADCC reactions even under unfavorable A/I ratios. A critical factor in such Fc engineering is knowledge of the FcR expression profile of the relevant effector cells mediating the biologically relevant response and the impact of both disease and therapy on the modulation of these receptors. Moreover, knowledge of the dominant antibody isotypes involved in an autoimmune disease might be predictive for the involvement of a specific activating Fc receptor, thereby allowing the identification of therapeutic targets. In line with this a recent study using a nephrotoxic nephritis model has shown that the predominant pathogenic antibody isotype under these experimental conditions was IgG2b (Kaneko et al. 2006). Despite the possibility that this subclass could bind FcγRIII and FcγRIV, only FcγRIV was found to be responsible for triggering the pathogenic effects; blocking of RIV abrogated disease. Similarly, since murine lupus models display a dominance of IgG2a and IgG2b subclasses, it is tempting to speculate that FcγRIV is the responsible activating Fc receptor in these disease models, as well. The regulation of FcR expression by cytokines is coupled to the regulation of isotypes by these same cytokines. Thus, Th1 cytokines such as IFN-γ induce class switching to IgG2a whereas Th2-type cytokines (IL-4) induce class switching to IgG1; TGF-β in contrast will induce switching to IgG2b (Coffman et al. 1989; Finkelman et al. 1990). As these cytokines also influence Fc receptor expression the pathogenicity of an autoimmune response will be determined by both cytokine mediated regulation of class-switching and the changes of expression levels of the responsible activating versus inhibitory Fc receptors.

6.5 Summary

Every day the human immune system is confronted with billions of pathogenic and nonpathogenic microorganisms. The decision when to react and with what strength is crucial for survival of the host. Therefore, the immune system needs to distinguish self from nonself and especially from dangerous nonself. Research in the field of Fc receptor biology over the last decade has established the important role of these receptors for making such decisions. There are several levels of regulation that will determine whether Fc receptor triggering on immune cells will result in cell activation or inhibition. The most important is the relative affinity of individual antibody isotypes for activating and inhibitory receptors (their A/I ratio), the relative expression level of activating and inhibitory receptors and the cytokine environment which can influence this expression level. Factors that impair this threshold established by the delicate balance of activating and inhibitory signals often result in uncontrolled immune cell activation ultimately leading to destruction of self tissues and autoimmune disease. Thus, Fc receptors together with other proteins such as Toll-like receptors, NK cell receptors or costimulatory molecules provide the basis for distinguishing between self and nonself and for the generation of a well balanced immune response that destroys pathogens without concomitant damage to the host

Acknowledgments As this is a very broad field, and due to limited space, we could cite only select articles and reviews. We would like to apologize for not directly citing all of the important work on which the review is based. This work was supported by grants form the Cancer Research Institute and the Bayerisches Genomforschungsnetzwerk (BayGene) to F.N. and from the NIH to J.V.R.

References

Barnes N, Gavin AL, Tan PS, Mottram P, Koentgen F, Hogarth PM (2002) FcgammaRI-deficient mice show multiple alterations to inflammatory and immune responses. Immunity 16:379–389

Bergtold A, Desai DD, Gavhane A, Clynes R (2005) Cell surface recycling of internalized antigen permits dendritic cell priming of B cells. Immunity 23:503–514

Blank MC, Stefanescu RN, Masuda E, Marti F, King PD, Redecha PB, Wurzburger RJ, Peterson MG, Tanaka S, Pricop L (2005) Decreased transcription of the human FCGR2B gene mediated by the -343 G/C promoter polymorphism and association with systemic lupus erythematosus. Hum Genet 117:220–227

Bolland S, Ravetch JV (2000) Spontaneous autoimmune disease in Fc(gamma)RIIB-deficient mice results from strain-specific epistasis. Immunity 13:277–285

Bolland S, Yim YS, Tus K, Wakeland EK, Ravetch JV (2002) Genetic modifiers of systemic lupus erythematosus in FcgammaRIIB(–/–) mice. J Exp Med 195:1167–1174

Bona CA, Stevenson FK (2004) B cells producing pathogenic autoantibodies. In: Honjo T, Alt FW, Neuberger MS (eds) Molecular biology of B cells. Elsevier, Boston, pp 381–402

Boruchov AM, Heller G, Veri MC, Bonvini E, Ravetch JV, Young JW (2005) Activating and inhibitory IgG Fc receptors on human DCs mediate opposing functions. J Clin Invest 115:2914–2923

Boyden SV, Sorkin E (1960) The adsorption of antigen by spleen cells previously treated with antiserum in vitro. Immunology 3:272–283

Bruhns P, Samuelsson A, Pollard JW, Ravetch JV (2003) Colony-stimulating factor-1-dependent macrophages are responsible for IVIG protection in antibody-induced autoimmune disease. Immunity 18:573–581

Cartron G, Dacheux L, Salles G, Solal-Celigny P, Bardos P, Colombat P, Watier H (2002) Therapeutic activity of humanized anti-CD20 monoclonal antibody and polymorphism in IgG Fc receptor FcgammaRIIIa gene. Blood 99:754–758

Chu ZT, Tsuchiya N, Kyogoku C, Ohashi J, Qian YP, Xu SB, Mao CZ, Chu JY, Tokunaga K (2004) Association of Fcgamma receptor IIb polymorphism with susceptibility to systemic lupus erythematosus in Chinese: a common susceptibility gene in the Asian populations. Tissue Antigens 63:21–27

Clynes R, Calvani N, Croker BP, Richards HB (2005) Modulation of the immune response in pristane-induced lupus by expression of activation and inhibitory Fc receptors. Clin Exp Immunol 141:230–237

Clynes R, Ravetch JV (1995) Cytotoxic antibodies trigger inflammation through Fc receptors. Immunity 3:21–26

Clynes RA, Towers TL, Presta LG, Ravetch JV (2000) Inhibitory Fc receptors modulate in vivo cytotoxicity against tumor targets. Nat Med 6:443–446

Coffman RL, Savelkoul HF, Lebman DA (1989) Cytokine regulation of immunoglobulin isotype switching and expression. Semin Immunol 1:55–63

Coutelier JP, van der Logt JT, Heessen FW, Warnier G, Van Snick J (1987) IgG2a restriction of murine antibodies elicited by viral infections. J Exp Med 165:64–69

Davis RS, Dennis G, Jr., Odom MR, Gibson AW, Kimberly RP, Burrows PD, Cooper MD (2002) Fc receptor homologs: newest members of a remarkably diverse Fc receptor gene family. Immunol Rev 190:123–136

Davis RS, Ehrhardt GR, Leu CM, Hirano M, Cooper MD (2005) An extended family of Fc receptor relatives. Eur J Immunol 35:674–680

Delamarre L, Pack M, Chang H, Mellman I, Trombetta ES (2005) Differential lysosomal proteolysis in antigen-presenting cells determines antigen fate. Science 307:1630–1634

Dhodapkar KM, Krasovsky J, Williamson B, Dhodapkar MV (2002) Antitumor monoclonal antibodies enhance cross-presentation of Cellular antigens and the generation of myeloma-specific killer T cells by dendritic cells. J Exp Med 195:125–133

Dhodapkar KM, Kaufman JL, Ehlers M, Banerjee DK, Bonvini E, Koenig S, Steinman RM, Ravetch JV, Dhodapkar MV (2005) Selective blockade of inhibitory Fcgamma receptor enables human dendritic cell maturation with IL-12p70 production and immunity to antibody-coated tumor cells. Proc Natl Acad Sci USA 102:2910–2915

Dijstelbloem HM, van de Winkel JG, Kallenberg CG (2001) Inflammation in autoimmunity: receptors for IgG revisited. Trends Immunol 22:510–516

Dombrowicz D, Lin S, Flamand V, Brini AT, Koller BH, Kinet JP (1998) Allergy-associated FcRbeta is a molecular amplifier of IgE- and IgG-mediated in vivo responses. Immunity 8:517–529

Donnadieu E, Jouvin MH, Rana S, Moffatt MF, Mockford EH, Cookson WO, Kinet JP (2003) Competing functions encoded in the allergy-associated F(c)epsilonRIbeta gene. Immunity 18:665–674

Duncan AR, Winter G (1988) The binding site for C1q on IgG. Nature 332:738–740

Finkelman FD, Holmes J, Katona IM, Urban JF, Jr., Beckmann MP, Park LS, Schooley KA, Coffman RL, Mosmann TR, Paul WE (1990) Lymphokine control of in vivo immunoglobulin isotype selection. Annu Rev Immunol 8:303–333

Floto RA, Clatworthy MR, Heilbronn KR, Rosner DR, MacAry PA, Rankin A, Lehner PJ, Ouwehand WH, Allen JM, Watkins NA, Smith KG (2005) Loss of function of a lupus-associated FcgammaRIIb polymorphism through exclusion from lipid rafts. Nat Med 11:1056–1058

Fossati-Jimack L, Ioan-Facsinay A, Reininger L, Chicheportiche Y, Watanabe N, Saito T, Hofhuis FM, Gessner JE, Schiller C, Schmidt RE, et al (2000) Markedly different pathogenicity of four immunoglobulin G isotype-switch variants of an antierythrocyte autoantibody is based on their capacity to interact in vivo with the low-affinity Fcgamma receptor III. J Exp Med 191:1293–1302

Fujii T, Hamano Y, Ueda S, Akikusa B, Yamasaki S, Ogawa M, Saisho H, Verbeek JS, Taki S, Saito T (2003) Predominant role of FcgammaRIII in the induction of accelerated nephrotoxic glomerulonephritis. Kidney Int 64:1406–1416

Fukuyama H, Nimmerjahn F, Ravetch JV (2005) The inhibitory Fcgamma receptor modulates autoimmunity by limiting the accumulation of immunoglobulin G+ anti-DNA plasma cells. Nat Immunol 6:99–106

Galli SJ, Kalesnikoff J, Grimbaldeston MA, Piliponsky AM, Williams CM, Tsai M (2005) Mast cells as "tunable" effector and immunoregulatory cells: recent advances. Annu Rev Immunol 23:749–786

Gillespie SR, DeMartino RR, Zhu J, Chong HJ, Ramirez C, Shelburne CP, Bouton LA, Bailey DP, Gharse A, Mirmonsef P, et al (2004) IL-10 inhibits Fc epsilon RI expression in mouse mast cells. J Immunol 172:3181–3188

Goodnow CC, Sprent J, de St Groth BF, Vinuesa CG (2005) Cellular and genetic mechanisms of self tolerance and autoimmunity. Nature 435:590–597

Gould HJ, Sutton BJ, Beavil AJ, Beavil RL, McCloskey N, Coker HA, Fear D, Smurthwaite L (2003) The biology of IGE and the basis of allergic disease. Annu Rev Immunol 21:579–628

Grimaldi CM, Hicks R, Diamond B (2005) B-cell selection and susceptibility to autoimmunity. J Immunol 174:1775–1781

Groh V, Li YQ, Cioca D, Hunder NN, Wang W, Riddell SR, Yee C, Spies T (2005) Efficient cross-priming of tumor antigen-specific T cells by dendritic cells sensitized with diverse anti-MICA opsonized tumor cells. Proc Natl Acad Sci USA 102:6461–6466

Gulle H, Samstag A, Eibl MM, Wolf HM (1998) Physical and functional association of Fc alpha R with protein tyrosine kinase Lyn. Blood 91:383–391

Guyre PM, Morganelli PM, Miller R (1983) Recombinant immune interferon increases immunoglobulin G Fc receptors on cultured human mononuclear phagocytes. J Clin Invest 72: 393–397

Hamaguchi Y, Xiu Y, Komura K, Nimmerjahn F, Tedder TF (2006) Antibody isotype-specific engagement of Fc gamma receptors regulates B lymphocyte depletion during CD20 immunotherapy. J Exp Med 203:743–753

Hazenbos WL, Gessner JE, Hofhuis FM, Kuipers H, Meyer D, Heijnen IA, Schmidt RE, Sandor M, Capel PJ, Daeron M, et al (1996) Impaired IgG-dependent anaphylaxis and Arthus reaction in Fc gamma RIII (CD16) deficient mice. Immunity 5:181–188

Heyman B (2000) Regulation of antibody responses via antibodies, complement, and Fc receptors. Annu Rev Immunol 18:709–737

Heyman B (2002) IgE-mediated enhancement of antibody responses: the beneficial function of IgE? Allergy 57:577–585

Hulett MD, Hogarth PM (1994) Molecular basis of Fc receptor function. Adv Immunol 57:1–127

Ioan-Facsinay A, de Kimpe SJ, Hellwig SM, van Lent PL, Hofhuis FM, van Ojik HH, Sedlik C, da Silveira SA, Gerber J, de Jong YF, et al (2002) FcgammaRI (CD64) contributes substantially to severity of arthritis, hypersensitivity responses, and protection from bacterial infection. Immunity 16:391–402

Ji H, Ohmura K, Mahmood U, Lee DM, Hofhuis FM, Boackle SA, Takahashi K, Holers VM, Walport M, Gerard C, et al (2002) Arthritis critically dependent on innate immune system players. Immunity 16:157–168

Jiang Y, Hirose S, Sanokawa-Akakura R, Abe M, Mi X, Li N, Miura Y, Shirai J, Zhang D, Hamano Y, Shirai T (1999) Genetically determined aberrant down-regulation of FcgammaRIIB1 in germinal center B cells associated with hyper-IgG and IgG autoantibodies in murine systemic lupus erythematosus. Int Immunol 11:1685–1691

Jiang Y, Hirose S, Abe M, Sanokawa-Akakura R, Ohtsuji M, Mi X, Li N, Xiu Y, Zhang D, Shirai J, et al (2000) Polymorphisms in IgG Fc receptor IIB regulatory regions associated with autoimmune susceptibility. Immunogenetics 51:429–435

Kalergis AM, Ravetch JV (2002) Inducing tumor immunity through the selective engagement of activating Fcgamma receptors on dendritic cells. J Exp Med 195:1653–1659

Kaneko Y, Nimmerjahn F, Madaio M, Ravetch JV (2006) Pathology and protection in nephrotoxic nephritis is determined by selective engagement of specific FcRs. J Exp Med 203:789–797

Kipps TJ, Parham P, Punt J, Herzenberg LA (1985) Importance of immunoglobulin isotype in human antibody-dependent, cell-mediated cytotoxicity directed by murine monoclonal antibodies. J Exp Med 161:1–17

Kono H, Kyogoku C, Suzuki T, Tsuchiya N, Honda H, Yamamoto K, Tokunaga K, Honda Z (2005) FcgammaRIIB Ile232Thr transmembrane polymorphism associated with human systemic lupus erythematosus decreases affinity to lipid rafts and attenuates inhibitory effects on B-cell receptor signaling. Hum Mol Genet 14:2881–2892

Kraft S, Novak N (2005) Fc receptors as determinants of allergic reactions. Trends Immunol 27:88–95

Kushnir N, Liu L, MacPherson GG (1998) Dendritic cells and resting B cells form clusters in vitro and in vivo: T-cell independence, partial LFA-1 dependence, and regulation by cross-linking surface molecules. J Immunol 160:1774–1781

Kyogoku C, Tsuchiya N, Wu H, Tsao BP, Tokunaga K (2004) Association of Fcgamma receptor IIA, but not IIB and IIIA, polymorphisms with systemic lupus erythematosus: A family-based association study in Caucasians. Arthritis Rheum 50:671–673

Lang ML, Shen L, Wade WF (1999) Gamma-chain dependent recruitment of tyrosine kinases to membrane rafts by the human IgA receptor Fc alpha R. J Immunol 163:5391–5398

Launay P, Lehuen A, Kawakami T, Blank U, Monteiro RC (1998) IgA Fc receptor (CD89) activation enables coupling to syk and Btk tyrosine kinase pathways: differential signaling after IFN-gamma or phorbol ester stimulation. J Leukoc Biol 63:636–642

MacGlashan D Jr (2005) IgE and Fc{epsilon}RI regulation. Ann N Y Acad Sci 1050:73–88

Markine-Goriaynoff D, Coutelier JP (2002) Increased efficacy of the immunoglobulin G2a subclass in antibody-mediated protection against lactate dehydrogenase-elevating virus-induced polioencephalomyelitis revealed with switch mutants. J Virol 76:432–435

McGaha TL, Sorrentino B, Ravetch JV (2005) Restoration of tolerance in lupus by targeted inhibitory receptor expression. Science 307:590–593

Mechetina LV, Najakshin AM, Alabyev BY, Chikaev NA, Taranin AV (2002) Identification of CD16–2, a novel mouse receptor homologous to CD16/Fc gamma RIII. Immunogenetics 54:463–468

Meffre E, Casellas R, Nussenzweig MC (2000) Antibody regulation of B-cell development. Nat Immunol 1:379–385

Meyer D, Schiller C, Westermann J, Izui S, Hazenbos WL, Verbeek JS, Schmidt RE, Gessner JE (1998) FcgammaRIII (CD16)-deficient mice show IgG isotype-dependent protection to experimental autoimmune hemolytic anemia. Blood 92:3997–4002

Monteiro RC, Van De Winkel JG (2003) IgA Fc receptors. Annu Rev Immunol 21:177–204

Moretta A, Bottino C, Vitale M, Pende D, Cantoni C, Mingari MC, Biassoni R, Moretta L (2001) Activating receptors and coreceptors involved in human natural killer cell-mediated cytolysis. Annu Rev Immunol 19:197–223

Nguyen C, Limaye N, Wakeland EK (2002) Susceptibility genes in the pathogenesis of murine lupus. Arthritis Res 4:S255–263

Nimmerjahn F, Ravetch JV (2005) Divergent immunoglobulin g subclass activity through selective Fc receptor binding. Science 310:1510–1512

Nimmerjahn F, Bruhns P, Horiuchi K, Ravetch JV (2005) FcgammaRIV: a novel FcR with distinct IgG subclass specificity. Immunity 23:41–51

Nishimura H, Nose M, Hiai H, Minato N, Honjo T (1999) Development of lupus-like autoimmune diseases by disruption of the PD-1 gene encoding an ITIM motif-carrying immunoreceptor. Immunity 11:141–151

Novak N, Kraft S, Bieber T (2001) IgE receptors. Curr Opin Immunol 13:721–726

Okayama Y, Kirshenbaum AS, Metcalfe DD (2000) Expression of a functional high-affinity IgG receptor, Fc gamma RI, on human mast cells: Up-regulation by IFN-gamma. J Immunol 164:4332–4339

Okazaki T, Otaka Y, Wang J, Hiai H, Takai T, Ravetch JV, Honjo T (2005) Hydronephrosis associated with antiurothelial and antinuclear autoantibodies in BALB/c-Fcgr2b–/–Pdcd1–/– mice. J Exp Med 202:1643–1648

O'Keefe TL, Williams GT, Batista FD, Neuberger MS (1999) Deficiency in CD22, a B cell-specific inhibitory receptor, is sufficient to predispose to development of high affinity autoantibodies. J Exp Med 189:1307–1313

Otten MA, van Egmond M (2004) The Fc receptor for IgA (FcalphaRI, CD89). Immunol Lett 92:23–31

Park SY, Ueda S, Ohno H, Hamano Y, Tanaka M, Shiratori T, Yamazaki T, Arase H, Arase N, Karasawa A, et al (1998) Resistance of Fc receptor- deficient mice to fatal glomerulonephritis. J Clin Invest 102:1229–1238

Park RK, Izadi KD, Deo YM, Durden DL (1999) Role of Src in the modulation of multiple adaptor proteins in FcalphaRI oxidant signaling. Blood 94:2112–2120

Pearse RN, Kawabe T, Bolland S, Guinamard R, Kurosaki T, Ravetch JV (1999) SHIP recruitment attenuates Fc gamma RIIB-induced B-cell apoptosis. Immunity 10:753–760

Penninger JM, Timms E, Shahinian A, Jezo-Bremond A, Nishina H, Ionescu J, Hedrick SM, Mak TW (1995) Alloreactive gamma delta thymocytes utilize distinct costimulatory signals from peripheral T cells. J Immunol 155:3847–3855

Pricop L, Redecha P, Teillaud JL, Frey J, Fridman WH, Sautes-Fridman C, Salmon JE (2001) Differential modulation of stimulatory and inhibitory Fc gamma receptors on human monocytes by Th1 and Th2 cytokines. J Immunol 166:531–537

Pritchard NR, Cutler AJ, Uribe S, Chadban SJ, Morley BJ, Smith KG (2000) Autoimmune-prone mice share a promoter haplotype associated with reduced expression and function of the Fc receptor FcgammaRII. Curr Biol 10:227–230

Qiu WQ, de Bruin D, Brownstein BH, Pearse R, Ravetch JV (1990) Organization of the human and mouse low-affinity Fc gamma R genes: duplication and recombination. Science 248:732–735

Radeke HH, Janssen-Graalfs I, Sowa EN, Chouchakova N, Skokowa J, Loscher F, Schmidt RE, Heeringa P, Gessner JE (2002) Opposite regulation of type II and III receptors for immunoglobulin G in mouse glomerular mesangial cells and in the induction of anti-glomerular basement membrane (GBM) nephritis. J Biol Chem 277:27535–27544

Rafiq K, Bergtold A, Clynes R (2002) Immune complex-mediated antigen presentation induces tumor immunity. J Clin Invest 110:71–79

Ravetch JV (2003) Fc receptors. In: Paul WE (ed) Fundamental immunology. Lippincott-Raven, Philadelphia, pp 685–700

Ravetch JV, Bolland S (2001) IgG Fc receptors. Annu Rev Immunol 19:275–290

Ravetch JV, Clynes RA (1998) Divergent roles for Fc receptors and complement in vivo. Annu Rev Immunol 16:421–432

Ravetch JV, Kinet JP (1991) Fc receptors. Annu Rev Immunol 9:457–492

Ravetch JV, Lanier LL (2000) Immune inhibitory receptors. Science 290:84–89

Ray SK, Putterman C, Diamond B (1996) Pathogenic autoantibodies are routinely generated during the response to foreign antigen: a paradigm for autoimmune disease. Proc Natl Acad Sci USA 93:2019–2024

Regnault A, Lankar D, Lacabanne V, Rodriguez A, Thery C, Rescigno M, Saito T, Verbeek S, Bonnerot C, Ricciardi-Castagnoli P, Amigorena S (1999) Fcgamma receptor-mediated induction of dendritic cell maturation and major histocompatibility complex class I-restricted antigen presentation after immune complex internalization. J Exp Med 189:371–380

Rudge EU, Cutler AJ, Pritchard NR, Smith KG (2002) Interleukin 4 reduces expression of inhibitory receptors on B cells and abolishes CD22 and Fc gamma RII-mediated B-cell suppression. J Exp Med 195:1079–1085

Samsom JN, van Berkel LA, van Helvoort JM, Unger WW, Jansen W, Thepen T, Mebius RE, Verbeek SS, Kraal G (2005) Fc gamma RIIB regulates nasal and oral tolerance: a role for dendritic cells. J Immunol 174:5279–5287

Samuelsson A, Towers TL, Ravetch JV (2001) Anti-inflammatory activity of IVIG mediated through the inhibitory Fc receptor. Science 291:484–486

Schlageter AM, Kozel TR (1990) Opsonization of Cryptococcus neoformans by a family of isotype-switch variant antibodies specific for the capsular polysaccharide. Infect Immun 58:1914–1918

Shibuya A, Sakamoto N, Shimizu Y, Shibuya K, Osawa M, Hiroyama T, Eyre HJ, Sutherland GR, Endo Y, Fujita T, et al (2000) Fc alpha/mu receptor mediates endocytosis of IgM-coated microbes. Nat Immunol 1:441–446

Shushakova N, Skokowa J, Schulman J, Baumann U, Zwirner J, Schmidt RE, Gessner JE (2002) C5a anaphylatoxin is a major regulator of activating versus inhibitory FcgammaRs in immune complex-induced lung disease. J Clin Invest 110:1823–1830

Siriboonrit U, Tsuchiya N, Sirikong M, Kyogoku C, Bejrachandra S, Suthipinittharm P, Luangtrakool K, Srinak D, Thongpradit R, Fujiwara K, et al (2003) Association of Fcgamma receptor IIb and IIIb polymorphisms with susceptibility to systemic lupus erythematosus in Thais. Tissue Antigens 61:374–383

Steinman RM, Hawiger D, Liu K, Bonifaz L, Bonnyay D, Mahnke K, Iyoda T, Ravetch J, Dhodapkar M, Inaba K, Nussenzweig M (2003) Dendritic cell function in vivo during the steady state: a role in peripheral tolerance. Ann N Y Acad Sci 987:15–25

Sylvestre DL, Ravetch JV (1994) Fc receptors initiate the Arthus reaction: redefining the inflammatory cascade. Science 265:1095–1098

Taborda CP, Rivera J, Zaragoza O, Casadevall A (2003) More is not necessarily better: prozone-like effects in passive immunization with IgG. J Immunol 170:3621–3630

Takai T (2002) Roles of Fc receptors in autoimmunity. Nat Rev Immunol 2:580–592

Takai T, Li M, Sylvestre D, Clynes R, Ravetch JV (1994) FcR gamma chain deletion results in pleiotrophic effector cell defects. Cell 76:519–529

Takai T, Ono M, Hikida M, Ohmori H, Ravetch JV (1996) Augmented humoral and anaphylactic responses in Fc gamma RII-deficient mice. Nature 379:346–349

Tivol EA, Borriello F, Schweitzer AN, Lynch WP, Bluestone JA, Sharpe AH, Penninger JM, Timms E, Shahinian A, Jezo-Bremond A, et al (1995) Loss of CTLA-4 leads to massive lymphoproliferation and fatal multiorgan tissue destruction, revealing a critical negative regulatory role of CTLA-4. Immunity 3:541–547

Torigoe C, Inman JK, Metzger H (1998) An unusual mechanism for ligand antagonism. Science 281:568–572

Tridandapani S, Wardrop R, Baran CP, Wang Y, Opalek JM, Caligiuri MA, Marsh CB (2003) TGF-beta 1 supresses myeloid Fc gamma receptor function by regulating the expression and function of the common gamma-subunit. J Immunol 170:4572–4577

Tzeng SJ, Bolland S, Inabe K, Kurosaki T, Pierce SK (2005) The B-cell inhibitory Fc receptor triggers apoptosis by a novel c-Abl-family kinase dependent pathway. J Biol Chem 280:35247–35254

Uchida J, Hamaguchi Y, Oliver JA, Ravetch JV, Poe JC, Haas KM, Tedder TF (2004) The innate mononuclear phagocyte network depletes B lymphocytes through Fc receptor-dependent mechanisms during anti-CD20 antibody immunotherapy. J Exp Med 199:1659–1669

Wardemann H, Yurasov S, Schaefer A, Young JW, Meffre E, Nussenzweig MC (2003) Predominant autoantibody production by early human B-cell precursors. Science 301:1374–1377

Watanabe T, Okano M, Hattori H, Yoshino T, Ohno N, Ohta N, Sugata Y, Orita Y, Takai T, Nishizaki K (2004) Roles of FcgammaRIIB in nasal eosinophilia and IgE production in murine allergic rhinitis. Am J Respir Crit Care Med 169:105–112

Weng WK, Levy R (2003) Two immunoglobulin G fragment C receptor polymorphisms independently predict response to rituximab in patients with follicular lymphoma. J Clin Oncol 21:3940–3947

Weng WK, Czerwinski D, Timmerman J, Hsu FJ, Levy R (2004) Clinical outcome of lymphoma patients after idiotype vaccination is correlated with humoral immune response and immunoglobulin G Fc receptor genotype. J Clin Oncol 22:4717–4724

Wykes M, Pombo A, Jenkins C, MacPherson GG (1998) Dendritic cells interact directly with naive B lymphocytes to transfer antigen and initiate class switching in a primary T-dependent response. J Immunol 161:1313–1319

Xiu Y, Nakamura K, Abe M, Li N, Wen XS, Jiang Y, Zhang D, Tsurui H, Matsuoka S, Hamano Y, et al (2002) Transcriptional regulation of Fcgr2b gene by polymorphic promoter region and its contribution to humoral immune responses. J Immunol 169:4340–4346

Zhang M, Zhang Z, Garmestani K, Goldman CK, Ravetch JV, Brechbiel MW, Carrasquillo JA, Waldmann TA (2004) Activating Fc receptors are required for antitumor efficacy of the antibodies directed toward CD25 in a murine model of adult T-cell leukemia. Cancer Res 64:5825–5829

7
Self and Nonself Recognition by Coreceptors on B Lymphocytes: Regulation of B Lymphocytes by CD19, CD21, CD22, and CD72

Kozo Watanabe and Takeshi Tsubata

7.1 Introduction

B-cell antigen receptor (BCR) signaling plays an essential role in regulation of the development and function of B lymphocytes (see Chapter 4). These signals are modulated by coreceptors expressed on the surface of B cells, including CD19, CD21 (complement receptor 2, CR2), CD22, CD72 (Lyb-2), and low-affinity receptor for IgG (FcγRII) (see Chapter 6 on Fc receptors). When CD19/CD21 complex is colligated with BCR, this complex positively regulates BCR signaling through the recruitment of signaling molecules, such as a phosphatidylinositol 3-kinase (PI3K), to their phosphorylated tyrosines. On the other hand, CD22 and CD72 negatively regulate BCR signaling through the recruitment of the SH2-domain-containing protein tyrosine phosphatase-1 (SHP-1) to their phosphorylated immunoreceptor tyrosine-based inhibitory motifs (ITIMs). From the characterization of each of these coreceptors deficient mice, now it is clear that the regulations by these coreceptors physiologically play a significant role in the development, activation, proliferation and differentiation of B cells. Importantly, these coreceptors are able to detect and respond to extracellular environment and modulate BCR signaling by their ligand bindings. These functions contribute to the recognition of self- and nonself-antigens (Ags), in some cases, in cooperation with the innate immune system, such as the recognition of activated complement fragments by CD21 (see Chapter 1 on innate immunity). In this chapter, we focus on the physiological function of the CD19, CD21, CD22, and CD72 and their molecular mechanisms involving signal transduction by these coreceptors, which will help us to understand how B cells recognize self and nonself.

Laboratory of Immunology, School of Biomedical Science, and Department of Immunology, Medical Research Institute, Tokyo Medical and Dental University, 1-5-45 Yushima, Bunkyo-ku, Tokyo 113-8510, Japan

Correspondence to: K. Watanabe

7.2 Regulation of B Cells by CD19/CD21 Complex

7.2.1 Structure and Expression of CD19 and CD21

CD19 is an approximately 95 kDa membrane protein and its extracellular region contains two or three immunoglobulin (Ig) like domains (Stamenkovic and Seed 1988; Tedder and Isaacs 1989) and its cytoplasmic region is extensively conserved among human, mouse and pig, including nine conserved tyrosine residues (Tedder and Isaacs 1989; Zhou et al. 1991), which is consistent with its critical role in CD19 function (as described in section 7.2.5). Expression of human CD19 is restricted to B lineage cells and follicular dendritic cells (FDCs) (Tedder et al. 1997) and its expression is regulated by B-cell-specific transcription factor BSAP (Kozmik et al. 1992). CD19 interacts with CD21 and other proteins (Fig. 1).

CD21, an approximately 150kDa protein, possesses 15 repeating structures termed short consensus repeat sequences (SCRs) in its extracellular domain and a 34-amino-acid short cytoplasmic domain. In mice both CD21 and CD35 (complement receptor 1, CR1) are encoded by Cr2 gene and mouse CD21 and CD35 are produced by alternative splicing. However, in human, distinct genes encode CD21 and CD35 (Kurtz et al. 1990; Molina et al. 1990). Murine CD35 shares 15 SCRs with CD21 and possess 6 additional SCRs. Expression of human CD21 is not restricted to B lineage cells. FDCs (Reynes et al. 1985), a subset of T lymphocytes (Fischer et al. 1991) and epithelial cells (Birkenbach et al. 1992), also express CD21. Thus, CD21 expression is different from CD19 and is regulated by promoter

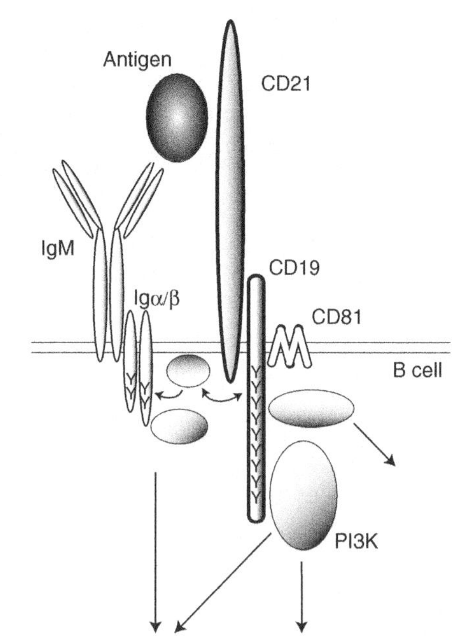

Fig. 1. CD19/CD21 complex-mediated positive regulation of B-cell antigen receptor (*BCR*) signaling. Complement receptor CD21 interacts with activated C3 fragments that are bound to antigens. This coligation of BCR and CD19/CD21 complex by antigens induces activation of protein tyrosine kinases and also induces tyrosine phosphorylation of CD19. Tyrosine phosphorylation of CD19 induces recruitment of SH2-domain containing proteins, including phosphatidylinositol 3-kinase (*PI3K*), and activation of these proteins. The protein kinases and PI3K activated by CD19 positively regulates BCR signaling. PI3K also promotes cell survival. CD19/CD21 complex contains CD81, which is required for efficient cell surface expression of CD19

and intronic sequences of CD21 in mice and humans (Hu et al. 1997; Makar et al. 1998). Pro-, pre- and immature B cells express CD19 but not CD21 (Tedder et al. 1984). In the plasma, soluble form of CD21 exists (Lowe et al. 1989), which is produced not by alternative splicing (Illges et al. 1997) but by proteolysis of its extracellular portion from peripheral B cells (Masilamani et al. 2003). Besides the complex formation with CD19, as described in section 7.2.3, CD21 acts as a complement receptor by recognizing cleavage fragments of the third complement component (C3), that is iC3b, C3dg and C3d.

7.2.2 CD19/CD21/CD81 Complex

To date, several components of the CD19/CD21 complex are reported. Interactions of some of them depend on detergents used for solubilization of cells and may depend on the developmental stage of B cells. CD19 is a part of the complex including CD21, CD81 (the target of an anti-proliferative antibody, TAPA-1), and Leu-13. CD19 directly interacts with CD21 and CD81 through its extracellular domain (Bradbury et al. 1992; Matsumoto et al. 1991, 1993). Moreover in B cell lines, treatments of monoclonal antibodies (mAbs) against these molecules enhance the increase in intracellular Ca^{2+} concentration following suboptimal dose of IgM crosslinking (Bradbury et al. 1992). In this complex, CD21 recognizes cleaved complement fragments, CD81 regulates localization of them and CD19 mediates intracellular signaling.

CD81, a member of the tetraspan family, is a ubiquitously expressed 26 kDa cell surface protein with four transmembrane domains and a short intracellular tail lacking signaling motifs. CD81 is expressed at all stages of B-cell development (Levy et al. 1998). On B cells, CD81 interacts with CD19, CD21, Leu-13, MHC class II, tetraspan family proteins including CD37, CD53, R2/C33, CD82, and some of the integrin family proteins (Angelisova et al. 1994; Horvath et al. 1998; Takahashi et al. 1990). In CD81 deficient mice, the development of B cells is normal but the expression of CD19 is reduced (Maecker and Levy 1997; Miyazaki et al. 1997; Tsitsikov et al. 1997). The reduction of CD19 does not depend on transcriptional regulation of CD19 gene because mRNA levels of CD19 are not affected and the reduction of CD19 rather depends on trafficking to surface membrane or instability on surface membrane (Miyazaki et al. 1997; Shoham et al. 2003). Recently, it has been demonstrated that coligation of BCR and CD19/CD21/CD81 complex induces rapid CD81 palmitoylation and this results in stable partitioning of colligated complex into membrane lipid rafts (Cherukuri et al. 2004a,b). This function of CD81 can contribute to the CD19/CD21 complex-mediated enhancement of BCR signaling. CD81 also regulates cellular adhesion of B cells (Behr and Schriever 1995; Mittelbrunn et al. 2002).

The earliest B-cell precursors express CD19 but not CD21 and expression of CD19 is molar excess of that of CD21 at all developmental stages of B cells. CD19 interacts with equimolar CD21. This implicates that most of CD21 forms complex with CD19, but not necessarily all CD19 form complex with CD21 (Fearon and Carroll 2000) and that CD19 can act independently of CD21 as described in 7.2.6 in this chapter.

7.2.3 Complement Receptor CD21

CD21 acts as a complement receptor and it recognizes cleaved C3 fragments iC3b, C3d, and C3dg. In human, CD21 also acts as the Epstein–Barr virus receptor (Fingeroth et al. 1984) and has been reported to interact with the low affinity IgE receptor, CD23 (FcεRII) (Aubry et al. 1992). Many works show the biological significance of the interaction between CD21 and C3 fragments. In addition to an important role in innate immune responses, C3 also plays an important role in T-dependent (TD) Ag responses (Bottger et al. 1986). Characterization of C3 deficient mice demonstrated that impairment of TD Ag responses was not at T-cell but at B-cell level, and due to impaired retention of Ags by FDCs (Fischer et al. 1996). Mice deficient in Cr2 gene which encodes both CD21 and CD35 as alternatively spliced gene products also exhibit the reduced responses to TD Ags, which is similar to the phenotype of C3 or C4 deficient animals (Ahearn et al. 1996; Croix et al. 1996; Molina et al. 1996). Moreover, B cells from C3 deficient mice show reduced antibody (Ab) responses and FDCs from C3 deficient mice are unable to promote Ab responses of B cells. These results can be explained by the notion that the C3 fragments that bound to Ags on FDCs interact with CD21 on B cells (Qin et al. 1998). Expression of CD21/CD35 on FDCs also plays an important role in immune complexes (ICs) binding to FDCs, IgG production and normal TD responses (Fang et al. 1998). CD35 on marginal zone (MZ) B cells also contributes to an initial phase of TD response by promoting efficient transport of IgM-containing ICs from MZ B cells to FDCs because the initial IgM-ICs binding to MZ B cells is dependent on complement and its receptor, CD21/CD35 (Ferguson et al. 2004).

It is noteworthy that murine systemic lupus erythematosus (SLE) susceptibility locus contains mutated *Cr2* and this mutation located in the C3d binding domain causes the reduction of ligand-binding ability (Boackle et al. 2001). Moreover, CD21/CD35 deficiency concomitant with Fas (Apo-1/CD95) defective *lpr* mutation in C57BL/6 background is sufficient to increase the serum concentrations of IgG1, IgG2b, antinuclear and anti-double stranded DNA (dsDNA) Abs (Wu et al. 2002). Fas, a member of the death receptor family, plays a role in induction of apoptosis and its defect is associated with SLE susceptibility.

The generation of C3 fragments that interact with CD21 is as follows. By the activation of the classical complement pathway, alternative pathway, or mannose-binding lectin pathway, C3 is cleaved to C3a and C3b. C3b is then converted to iC3b and subsequently converted to C3dg. These cleavages are promoted by CD35 (Rickert 2005). Cleavage of C3dg generates C3d. Because of the high reactivity of C3b, it promptly binds various molecules and cells with covalent bonds.

Activation of C3b bound to host cells is regulated by many factors that exist on surface of host cells and protect host cells from complement systems. On human B cells, part of CD21 interacts with CD35 (Delibrias et al. 1992) but CD35 is not a component of complex including CD19/CD21 (Tuveson et al. 1991). In contrast, on mouse B cells, CD19 interacts with both CD21 and CD35. This is probably dependent on structure of mouse CD35 that contains six additional SCRs to the amino-terminal end of the mouse CD21. The conversion of C3b to iC3b or C3dg decreases the affinity for CD35 and increases the affinity for CD21 (Kalli et al. 1991; Pramoonjago et al. 1993). So it is possible that the interaction between CD35 and CD21 may plays a role in facilitating the transfer of C3b bound Ags that interact with CD35 to CD21.

7.2.4 Coligation of CD19/CD21 Complex with BCR

As we saw in the previous section, CD21 has ability to interact with the cleaved complement fragments. Importantly, this function of CD21 enables the coligation of CD19/CD21/CD81 complexes with BCR when C3d bound Ags are recognized by BCR. In fact, it has been demonstrated that crosslinking of CD19/CD21 with BCR on B cells using bi-specific Abs, which is able to interact with both CD19 or CD21 and BCR, enhances the elevation of intracellular Ca^{2+} concentrations and proliferation, and decreased the threshold for BCR-mediated stimulation by 100 fold (Carter et al. 1991; Carter and Fearon 1992). Similarly, C3d bound Ags become significantly immunogenic and enhance B-cell responses to these Ags (Dempsey et al. 1996). The BCR stimulation with anti-Ig Abs alone also induces tyrosine phosphorylation of CD19 (Chalupny et al. 1993) and recruitment of CD19 to BCR (Pesando et al. 1989). CD19 interacts directly with BCR through juxtamembrane region of its cytoplasmic domain (Carter et al. 1997). However, the effect of the BCR stimulation on B-cell response in the absence of CD19/CD21 complex coligation is quite small, compared to that of the stimulation causing direct CD19/CD21 complex coligation with BCR (Fearon and Carroll 2000).

7.2.5 CD19-Mediated Signal Transduction

We have described CD19/CD21/CD81 complexes, where CD81 contributes controlling the localization of them, CD21 detects extracellular environment by recognizing cleaved complement fragments, and CD19 enables the coligation of CD19/CD21 complex with BCR. Now we focus on the role of CD19 in this complex especially in signal transduction. CD19 possesses relatively large intracellular region and this region contains nine conserved tyrosine residues, through which CD19 interacts with various proteins and regulates the intracellular signaling pathways.

B-cell receptor antigen stimulation and/or CD19 complex ligation induces phosphorylation of these tyrosines and leads recruitment of Src homology 2 (SH2) domain-containing proteins. Y330 (in the human CD19 sequence) of CD19 binds Grb2, which forms a complex with Sos and activates Ras (Brooks et al. 2000), Y391 binds Vav, which is Rho-family guanine nucleotide exchange factor (O'Rourke et al. 1998), Y391 and Y421 bind both Vav and phospholipase C-gamma 2 (PLCγ2) (Brooks et al. 2000), and Y403 and Y443 bind Fyn and Lyn, which are members of the Src-family protein tyrosine kinases (Chalupny et al. 1995). Y482 and Y513 bind PI3K (Chalupny et al. 1995; Tuveson et al. 1993). Importantly, the analysis of CD19 deficient mice expressing CD19 mutant transgene demonstrated the essential role of Y482 and Y513 for normal B-cell differentiation and function. In mice that have mutated Y482 and Y513 CD19 (Y482F/Y513F), normal B-cell differentiation and function are blocked and this is similar to the phenotype of CD19 deficient mice. In contrast, in mice that have mutated Y330 and Y360 CD19 or mutated Y391 and Y421 CD19, the differentiation of B cells, Ab responses and germinal center reaction is relatively normal and is similar to the phenotype of CD19 hetero deficient mice (Wang et al. 2002).

Several studies have also demonstrated the importance of Y482 and Y513 and have revealed the mechanisms of how these tyrosines regulate intracellular signaling. First, Y482 and Y513 preferentially interact with PI3K (Chalupny et al. 1995; Tuveson et al. 1993). Moreover, when the BCR stimulation that is independent of CD19/CD21 coligation induces PI3K activation, CD19 plays a critical role in PI3K activation and subsequent Bruton's tyrosine kinase (Btk) activation and these events are dependent on Y482 and Y513 (Buhl et al. 1997; Buhl and Cambier 1999). Btk is a critical kinase in BCR signaling, as indicated by X-linked immunodeficiency phenotype of mice that express mutant forms of the Btk and X-linked agammaglobulinemia phenotypes of men who also express the mutated Btk. PI3K is also known as a critical activator of Akt/PKB that provides survival signal and inhibits apoptosis. Consistently, CD19 is required for the enhanced and prolonged activation of Akt following the BCR stimulation independent of CD19 coligation (Otero et al. 2001) and plays an important role in the BCR-mediated survival signaling (Barrington et al. 2005; Otero et al. 2003).

The results that PI3K, Vav1, and Vav2 play a critical role in the increase of intracellular Ca^{2+} concentration, and Vav1 and Vav2 play a critical role in the activation of PI3K following crosslinking of BCR and CD19, also support the significance of PI3K (Vigorito et al. 2004). Although strong BCR stimulation independent of CD19 coligation induces PI3K activation, coligation of BCR and CD19 induces the more efficient activation of PI3K. This study also indicates that the activation of PI3K following strong BCR stimulation independent of CD19 coligation does not require Vav (Vigorito et al. 2004).

Following BCR ligation, Lyn also phosphorylates Y513, subsequently Y482 of CD19, and then phosphorylated Y513 and Y482 provides a binding site for Lyn (Fujimoto et al. 2000). CD19 functions as a specialized adapter protein for the Src-family protein kinases and these kinases phosphorylates CD19 (Fujimoto et al. 1999, 2000; Hasegawa et al. 2001). However, there is room for the consideration on the connection of Lyn and CD19. Whereas one report indicated that the activity of Lyn is diminished following BCR ligation in CD19 deficient mice (Fujimoto et al. 1999), another report indicated that the phosphorylation of CD19 after BCR ligation was not affected in Lyn deficient mice, and Lyn kinase activity is not affected in CD19 deficient B cells, either (Xu et al. 2002).

7.2.6 Regulation of B Cells by CD19 and CD21

As the result of the cooperation of CD19 and CD21, CD19/CD21 complex acts as a complement receptor and regulates intracellular signaling following coligation of this complex with BCR. This notion is supported by many studies although a part of the CD19 function is considered to be independent of CD21. For example, the characterizations of CD19 and CD21/CD35 deficient mice have revealed that CD19 and CD21/CD35 play a crucial role in TD Ag responses by lowering the threshold of BCR activation by several fold (Ahearn et al. 1996; Chen et al. 2000; Croix et al. 1996; Engel et al. 1995a,b; Molina et al. 1996; Rickert et al. 1995; Sato et al. 1995). In contrast to TD Ag responses, T-cell independent (TI) type II (TI-II) Ag responses are not significantly affected in CD19 deficient mice (Rickert et al. 1995; Sato et al. 1995). In CD21/CD35 deficient mice, however, TI responses to low dose Ags are significantly reduced including IgG3 Ab responses, which is important in response to TI-II Ags (Haas et al. 2002).

CD19 plays a critical role in the development of B-1 B cells. In CD19-deficient mice, B-1 cells are significantly reduced, early B-cell precursors are not significantly different and the number of peripheral B cells is reduced compared with those of wild-type mice (Engel et al. 1995a,b; Rickert et al. 1995). As for CD21, one line of CD21/CD35-deficient mice exhibit reduction of peritoneal B-1 cells (Ahearn et al. 1996) and demonstrated the importance of CD21/CD35 in maintenance of the B-1 cell repertoire to some, but not all, specificities (Reid et al. 2002), whereas another line of CD21/CD35 deficient mice exhibit normal number of peritoneal B-1 cells (Molina et al. 1996).

Proliferation of B cells in germinal centers (GCs) also requires CD19 and CD21. GCs play an important role in generation of memory B cells and Ab-forming cells (AFCs). The characterizations of CD21/CD35 deficient mice have demonstrated that even high-affinity antigen-activated B cells require CD21/CD35 for proliferation and survival in Germinal centers (GCs) (Fischer et al. 1998) and that CD21/CD35 is required for maintenance of serum Abs and long-lived AFCs (Chen et al. 2000) and the induction of Blimp-1, XBP-1, and Bcl-2 (Gatto et al. 2005). Blimp-1 and XBP-1 drive plasma cell differentiation, and Bcl-2 is important for inhibition of apoptosis. In addition, the analysis of CD19 deficient mice that express mutated Y482 and Y513 CD19 also indicates that normal function of CD19 is required for expansion and positive selection of GC B cells (Callea et al. 1997). Clonal selection of B cells in GCs requires a death receptor Fas that induces apoptosis (Takahashi et al. 2001), and deficiency of CD21/CD35 results in reduction of c-FLIP, which inhibits death receptor-mediated apoptosis, and induction of Fas-mediated apoptosis (Barrington et al. 2005).

CD19 contributes to B-cell generation in the absence of CD21 at the early Ag-independent stages of B-cell development, which is consistent with its early expression in B-cell development. Bone marrow reconstitution experiments demonstrated competitive disadvantage of CD19 deficient early B cells in mixed bone marrow chimeras and in CD19 deficient mice: pre-B-cell proliferation is reduced after sublethal irradiation and pre-BCR signaling is impaired (Otero and Rickert 2003). The experiments using mAb to CD19 also indicate the involvement of CD19 to pre-BCR signaling (Krop et al. 1996).

7.3 Regulation of B Cells by CD22

7.3.1 Structure and Expression of CD22

CD22 is an approximately 140 kDa type I membrane protein and its expression is B-cell specific. CD22-deficient mice show the inhibitory function of CD22 in BCR signaling (O'Keefe et al. 1996; Nitschke et al. 1997; Otipoby et al. 1996; Sato et al. 1996). CD22 is a member of the Siglecs, sialic acid-binding Ig-like lectins. CD22 contains seven Ig-like domains in extracellular region and six tyrosines in cytoplasmic region (Wilson et al. 1991) and three of these tyrosines resides in ITIM sequence (Fig. 2). Although the form termed CD22α containing five Ig-like domains

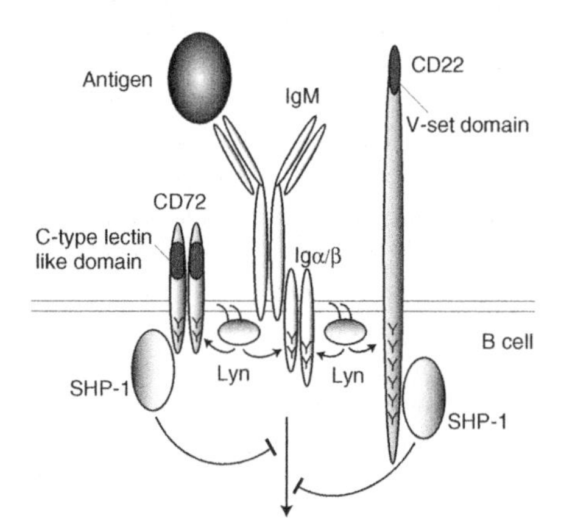

Fig. 2. B-cell antigen receptor (BCR) signaling inhibition by CD22 and CD72. Following antigen ligation, protein kinases, including Lyn, are activated. Lyn phosphorylates tyrosines in CD22 and CD72. This phosphorylation leads SH2-domain-containing protein tyrosine phosphatase-1 (*SHP-1*) recruitment to CD22 and CD72. SHP-1 negatively regulates BCR signaling by dephosphorylating intracellular signaling molecules. CD22 can interact with sialic acid containing proteins including IgM by recognizing sialic acids. V-set domain in CD22, which is conserved in the sialic acid-binding Ig-like lectins (siglecs), plays an essential role in interaction with sialic acids. CD72 contains a C-type lectin like domain. However, its lectin activity is not yet clear

exists (Stamenkovic and Seed 1990), the form termed CD22β containing seven Ig-like domains is dominant. Murine CD22 is expressed at high levels on mature recirculating B cells and in the peripheral B-cell subsets including follicular, marginal zone, B1, and class-switched B cells, at low levels on immature IgM^{hi} B cells, but is absent on pro-B cells, pre-B cells, newly emerging IgM^{+} B cells, and plasma cells (Erickson et al. 1996). Another work demonstrated the low level murine CD22 surface expression from Pre-B cells (Nitschke et al. 1997). The cell surface expression pattern of human CD22 is similar to that of mouse CD22, although human CD22 is detected in cytoplasm of pro-B and pre-B cells (Dorken et al. 1986).

7.3.2 Interaction of CD22 with Sialic Acid

It has been demonstrated that CD22 specifically interacts with Neu5Acα2–6Galβ1–4GlcNAc or α2,6-linked sialic acid (α2–6 sialic acid) (Powell and Varki 1994). Sialic acids include *N*-acetylneuraminic acid (Neu5Ac), *N*-glycolylneuraminic acid (Neu5Gc), 3-deoxy-D-glycero-D-galacto-nonulo-pyranosonic acid (KDN), and their alkyl, acyl, and deoxy derivatives. Using CD22-Ig fusion protein (CD22RG) and truncating the side chain of sialic acids by mild periodate oxidation, CD22 was demonstrated to be a sialic acid binding-lectin (Sgroi et al. 1993). Then, it was demonstrated that oligosaccharides that are able to bind CD22 contain α2–6 sialic acids (Powell et al. 1993). Furthermore, the oligosaccharide binding specificity of CD22 has been demonstrated by the study using CD22RG conjugated column and a number of naturally and enzymatically sialylated oligosaccharides and sialoglycoproteins. CD22 recognizes Neu5Acα2–6Galβ1–4GlcNAc as a minimal structure. The recognition by CD22 is not affected by the presence or absence of α2–3 sialic acids (Powell and Varki 1994), whereas 9-*O*-acetylation of sialic acids, a natural modification of sialic acid, was demonstrated to prevent CD22 recognition (Sjoberg et al. 1994). Moreover, the subsequent study using equilibrium dialysis and enzyme-linked immunosorbent assay (ELISA) demonstrated that dimeric CD22RG has two sialic acid binding sites and the apparent binding affinity for α2–6 sialyl-lactose is low (32 μM) (Powell et al. 1995). Another study using mAbs against CD22 demonstrated that a single region of CD22 mediates adhesion of B cells to T cells, monocytes and erythrocytes, and a treatment with neuraminidase, which catalyzes hydrolysis of *N*-acetylneuraminic acid residues from glycoproteins and oligosaccharides, inhibits these adhesions (Engel et al. 1993). Subsequently, the membrane-distal parts of CD22, the first and/or the second Ig-like domains of CD22, each corresponds to V-set domain and C2-set domain, respectively, were identified as ligand binding domains (Engel et al. 1995a,b; Law et al. 1995), and residues including arginine in V-set domain were demonstrated to be essential for CD22 lectin activity (van der Merwe et al. 1996). The V-set domain, including a key arginine residue, and the C2 set domain are conserved in the siglec family. In contrast to human CD22, mouse CD22 strongly prefers a Neu5Gc form of the ligand. In human Neu5Gc is absent and only Neu5Ac forms of the ligands are expressed because of a mutation in CMP-sialic acid hydroxylase (Brinkman-Van der Linden et al. 2000).

7.3.3 Sialic Acid-Mediated Glycoprotein Interaction of CD22

Although CD22 specifically interacts with α2–6 sialic acids, it must be noted that this structure of oligosaccharides is common to many glycoprotein. So far, several proteins including CD45, soluble IgM, haptoglobin, Ly-6 have been reported as CD22 ligands (Hanasaki et al. 1995a,b; Pflugh et al. 2002; Stamenkovic et al. 1991). Importantly, CD22 interacts with surface IgM (Leprince et al. 1993; Peaker and Neuberger 1993; Zhang and Varki 2004). In addition, CD22 interacts with oligosaccharides of other CD22 or itself, which was suggested by the experiments that shared α2,6-sialyltransferase-mediated sialylation of CD22RG abrogates CD22RG-ligand interaction (Braesch-Andersen and Stamenkovic 1994). The recent study also suggests that CD22 interacts with oligosaccharides of neighboring CD22 and forms homomultimeric complexes by photoaffinity crosslinking of oligosaccharide ligands to CD22 (Han et al. 2005). On the ligand binding preference of CD22, it is suggested that interaction between CD22 and CD45 is not dependent on higher affinity but on multiple ligand existence on CD45 because the binding affinity of CD22 to native CD45 and a synthetic α2–6 sialoglycoconjugate is similar (Bakker et al. 2002). On the other hand, it has been demonstrated that the interaction of CD22 with surface IgM and/or CD45 and the internalization of CD22 are independent of sialic acid (Zhang and Varki 2004). It has been suggested that *cis* ligands, ligands on the same cellular surface, regulate the ligand activity of CD22. The *cis* ligands on B cells can occupy the binding site of CD22 and this occupancy results in masking of CD22. Ligand interaction of CD22 is affected by induction of α2–6 sialic acid on B cells and target cells (Hanasaki et al. 1995a,b). *Cis* interaction restricts CD22 lectin activity in human resting B cells; however in vitro activation of these cells induces partial desialylation and unmasking of CD22 (Razi and Varki 1998, 1999). Other works have demonstrated that masked CD22 by *cis* ligands on resting B cells redistributes to the site of contact with other B or T cells and this redistribution is dependent on *trans* ligands and is independent of CD45 (Collins et al. 2004) although CD45 was reported as both a *cis* and *trans* ligand of CD22.

7.3.4 Ligand Binding of CD22 and Its Effects

CD22 interacts with sialic acid containing proteins and this function of CD22 plays an important role in the regulation of B cells. Although early works using anti-CD22 mAbs reported that simultaneous treatment with anti-CD22 mAb and anti-Ig

Ab or anti-CD22 mAb treatment prior to anti-Ig Ab stimulation enhances B-cell proliferation, the elevation in intracellular Ca^{2+} concentration and decreases the threshold for BCR stimulation (Doody et al. 1995; Pezzutto et al. 1987, 1988). Later studies demonstrated the significance of ligand binding activity of CD22 in negative regulation of BCR signaling by the experiments using synthetic sialic acids (Kelm et al. 2002) or cell lines expressing mutant CD22 that was not able to interact with sialic acids (Jin et al. 2002). On the other hand, in vivo study using CD22-deficient mice expressing mutant CD22 demonstrated that the ligand binding ability is not required for negative regulation of BCR signaling, although it should be considered that IgM and CD22 expression is decreased while MHC class II expression on B cells are increased due to ligand binding inability of mutant CD22 (Poe et al. 2004).

Importantly, B-cell activation is depressed when antigen-presenting cells express α2–6 sialic acid abundantly, and this depression is dependent on CD22. This may help the recognition of self and nonself because sialylation is a feature of higher eukaryotes (Lanoue et al. 2002).

7.3.5 CD22-Mediated Signal Transduction

CD22 contains three ITIMs and two immunoreceptor tyrosine-based activation motif (ITAM)-like regions. Through these regions CD22 has been demonstrated to interact with various proteins and regulate the intracellular signaling. Following BCR stimulation, CD22 cytoplasmic tyrosines are rapidly phosphorylated (Schulte et al. 1992) and recruit SHP-1 (Campbell and Klinman 1995; Doody et al. 1995; Lankester et al. 1995), which negatively regulates BCR signaling. In this signaling pathway, BCR-associated tyrosine kinase Lyn is required for the tyrosine phosphorylation and the inhibitory function of CD22 (Cornall et al. 1998; Smith et al. 1998). For the SHP-1 recruitment, carboxyl-terminal two tyrosines in ITIMs of CD22 are required (Otipoby et al. 2001). Furthermore, CD22 is reported to interact with plasma membrane calcium-ATPase (PMCA) and to augment calcium efflux after BCR crosslinking by potentiating PMCA. This interaction requires cytoplasmic tyrosine residues of CD22, and CD22-mediated calcium efflux requires SHP-1 (Chen et al. 2004). In addition to SHP-1, CD22 also interacts with Syk, PLCγ, PI3K, Grb2, Shc and SH2-domain-containing inositol 5-phosphatase-1 (SHIP) (Law et al. 1996; Poe et al. 2000; Wienands et al. 1995; Yohannan et al. 1999). Although the overall function of CD22 in BCR signaling is inhibitory as demonstrated by CD22-deficient mice, B cells in CD22-deficient mice also show a shorter life span, reduced proliferation, and enhanced apoptosis following BCR stimulation (Nitschke et al. 1997; O'Keefe et al. 1996; Otipoby et al. 1996; Sato et al. 1996). Thus, it is possible that the interactions of some of these molecules with CD22 may also function in positive regulation of BCR signaling and contribute to promoting cell survival and enhancing proliferation.

7.3.6 Regulation of B Cells by CD22

The characterization of CD22-deficient mice demonstrated that CD22 plays an important role in negative regulation of BCR signaling in mature B cells whereas the development of B cells in CD22-deficient mice is relatively normal. In CD22-deficient mice, the increase in intracellular Ca^{2+} concentration following BCR stimulation and tyrosine phosphorylation is enhanced (Nitschke et al. 1997; O'Keefe et al. 1996; Otipoby et al. 1996; Sato et al. 1996). On the other hand, CD22-deficient mice show reduced response to TI-II Ags and CD22-deficient B cells show the shorter life span, reduced proliferation and enhanced apoptosis following BCR stimulation. CD22-deficient mice also display decreased number of recirculating B cells (Nitschke et al. 1997; Otipoby et al. 1996) and reduction of MZ B cells (Samardzic et al. 2002; Sato et al. 1996). Since MZ B cells perform TI-II responses, it explains the reduced TI-II responses in CD22-deficient mice, and the maintenance of MZ B cells requires CD22 ligand-binding activity (Poe et al. 2004). Furthermore, hyper-responsiveness of B cells in CD22-deficient mice is correlated with the development of serum IgG against double-stranded DNA (O'Keefe et al. 1999). Importantly, negative regulation of BCR signaling by CD22 is dependent on the BCR isotype. Although CD22 negatively regulates signaling through IgM or IgD, CD22 is not phosphorylated on ligation of membrane bound IgG by Ags and does not inhibit signaling through IgG because of an IgG cytoplasmic tail. In contrast to IgM and IgD, which contain only three amino acids in their cytoplasmic domains, IgG contains 28 amino acid residues in its cytoplasmic tail. This BCR-isotype specific regulation is not the case of inhibitory coreceptor CD72. CD72 negatively regulates signaling through IgG as well as IgM and IgD (Wakabayashi et al. 2002). In fact, IgG-positive B cells show enhanced clonal expansion compared to IgM-positive B cells (Martin and Goodnow 2002). Thus, in addition to other factors, BCR isotype-specific regulation of CD22 may be involved in efficient switching from IgM to IgG production at the cellular level and the efficient response of IgG-positive memory B cells.

In mice, the deficiency of ST6Gal sialyltransferase, which appears to be solely responsible for producing the α2–6 sialic acids terminus on various *N*-glycans, causes a severe and widespread immunodeficiency, including reduced serum IgM levels and attenuated Ab production to TI and TD Ags, unlike CD22-deficient mice. These results indicate that ST6Gal sialyltransferase have additional function other than the CD22 ligand production and regulation of CD22 (Hennet et al. 1998).

7.4 Regulation of B Cells by CD72

7.4.1 Ligand Binding of CD72

Inhibitory coreceptor CD72 (Lyb-2) is an approximately 45 kDa type II membrane protein and is a member of C-type lectin superfamily (Fig. 2). CD72 contains

C-type lectin like domain in the extracellular region and an ITIM and an ITIM-like sequence in the cytoplasmic region. However, lectin activity of CD72 is not yet clear. CD72 forms a homodimer by disulfide linking. CD72 is expressed on all B cells but is downregulated on terminally differentiated plasma cells (Nakayama et al. 1989) and is also expressed on dendritic cells. Although CD72 was suggested to be a ligand of CD5 (Van de Velde et al. 1991), other works have demonstrated that CD5-Ig fusion protein containing an extracellular domain of CD5 does not bind CD72 (Biancone et al. 1996; Bikah et al. 1998). As a CD72 ligand, CD100/Sema4D, a member of the semaphorin family, was identified (Kumanogoh et al. 2000). In addition to nervous systems, CD100 is also abundantly expressed on T cells (Furuyama et al. 1996; Hall et al. 1996). CD100 can be released from the surface of T cells by proteolysis (Elhabazi et al. 2001). Interaction of CD72 with membrane anchored CD100 or soluble CD100 appears to inhibit negative signaling by CD72 because this interaction causes the reduction of both CD72 phosphorylation and recruitment of SHP-1 to CD72 (Kumanogoh et al. 2000; Shi et al. 2000). This resembles the treatment with anti-CD72 Ab, which promotes B-cell activation and survival (Nomura et al. 1996; Subbarao and Mosier 1984) and inhibits both CD72 phosphorylation and recruitment of SH2-domain-containing protein tyrosine phosphatase-1 (SHP-1) following BCR stimulation (Hokazono et al. 2003; Nitschke and Tsubata 2004)

7.4.2 Signal Transduction by CD72

Early works using anti-CD72 mAb have suggested the positive role of CD72 in B-cell activation. However, the cytoplasmic region of CD72 contains an ITIM and an ITIM-like sequence. In fact, BCR stimulation enhances both the tyrosine phosphorylation of CD72 and recruitment of tyrosine phosphatase SHP-1 to tyrosine phosphorylated ITIM in CD72 (Adachi et al. 1998; Wu et al. 1998). CD72 inhibits extracellular signal-related kinase (ERK) activation and the increase in intracellular Ca^{2+} concentration depending on its ITIM following BCR stimulation (Adachi et al. 2000) probably by dephosphorylating Igα/β and its down stream signaling molecules Syk and SLP-65 (Adachi et al. 2001). In consistent with above results, B cells from CD72-deficient mice are hyper-reactive to various stimuli and show enhanced Ca^{2+} mobilization following BCR stimulation (Pan et al. 1999). CD72 also interacts with Grb2 through its ITIM-like sequence and this region does not interact with SHP-1, although the functional significance of Grb2 binding is not yet clear (Wu et al. 1998). In vitro, Tyrosine kinase Lyn phosphorylates CD72 (Adachi et al. 1998). The phenotype of CD72-deficient mice is relatively similar to that of CD22-deficient mice and Lyn deficient mice (Chan et al. 1998; Hibbs et al. 1995; Nishizumi et al. 1995). In case of CD22, CD22 can interact with IgM by recognizing sialic acids that exist on IgM. This interaction of CD22 with BCR seems to be important because CD22 is phosphorylated by BCR-associated kinase Lyn and this phosphorylation enables the recruitment of tyrosine phosphatase

SHP-1 which negatively regulates BCR signaling upon BCR stimulation. However, as for CD72, it is not yet known how CD72 associates with BCR although the signal transduction mechanism for inhibitory function of CD72 is similar to that of CD22.

7.4.3 Regulation of B Cells by CD72

The characterization of CD72-deficient mice demonstrated that the development of B cells in CD72-deficient mice is largely normal and CD72 negatively regulates BCR signaling (Pan et al. 1999). In the bone marrow of CD72-deficient mice, the number of mature recirculating $IgM^{+}IgD^{+}$ B cells is reduced, whereas the number of $IgM^{-}IgD^{-}$ pre-B cells increased. This may reflect the role of CD72 in the transition from pre-B to immature B cells and from immature to mature B cells. In the periphery of CD72-deficient mice, there is a decrease in mature and long-lived follicular B cells and an increase in B-1 cells (Pan et al. 1999; Parnes and Pan 2000).

7.5 Self and Nonself Recognition by Coreceptors on B cells

Coreceptors on the surface of B cells modulate BCR signaling positively or negatively, and control the outcome of an immune response by setting thresholds for B-cell activation. Importantly, this modulation by coreceptors is regulated by ligand bindings to coreceptors and ligands for coreceptors include *cis* ligands on B cells and *trans* ligands on targets. For example, CD19/CD21 complex interacts with both IgM and C3 fragments and positively regulates BCR signaling in the presence of C3 fragments conjugated Ags. This function facilitates B-cell activation to nonself-Ags. CD22 interacts with both IgM and other proteins by sialic acids recognition and negatively regulates BCR signaling. CD22 has been demonstrated to negatively regulate BCR signaling when Ag-presenting cells express sialic acids abundantly. This CD22 function may contributes inhibition of B-cell activation by self-Ags. CD72 also regulates BCR signaling negatively. CD72 interacts with CD100 and this interaction appears to prevent the negative function of CD72. These *cis*- and *trans*-ligand-mediated regulation of coreceptors have an advantage in selective regulation of immune response of B cells because of the function of recognizing extracellular environment, which enables the recognition of self- and nonself-Ag, in some cases, by cooperating with other mechanisms. However, details of the mechanisms of ligand bindings to coreceptors are not yet clear and many things on the mechanism of ligand bindings, especially on inhibitory coreceptors, remain to be elucidated.

References

Adachi T, Flaswinkel H, Yakura H, Reth M, Tsubata T (1998) The B-cell surface protein CD72 recruits the tyrosine phosphatase SHP-1 upon tyrosine phosphorylation. J Immunol 160:4662–4665

Adachi T, Wakabayashi C, Nakayama T, Yakura H, Tsubata T (2000) CD72 negatively regulates signaling through the antigen receptor of B cells. J Immunol 164:1223–1229

Adachi T, Wienands J, Wakabayashi C, Yakura H, Reth M, Tsubata T (2001) SHP-1 requires inhibitory co-receptors to down-modulate B-cell antigen receptor-mediated phosphorylation of cellular substrates. J Biol Chem 276:26648–26655

Ahearn JM, Fischer MB, Croix D, Goerg S, Ma M, Xia J, Zhou X, Howard RG, Rothstein TL, Carroll MC (1996) Disruption of the Cr2 locus results in a reduction in B-1a cells and in an impaired B-cell response to T-dependent antigen. Immunity 4:251–262

Angelisova P, Hilgert I, Horejsi V (1994) Association of four antigens of the tetraspans family (CD37, CD53, TAPA-1, and R2/C33) with MHC class II glycoproteins. Immunogenetics 39:249–256

Aubry JP, Pochon S, Graber P, Jansen KU, Bonnefoy JY (1992) CD21 is a ligand for CD23 and regulates IgE production. Nature 358:505–507

Bakker TR, Piperi C, Davies EA, Merwe PA (2002) Comparison of CD22 binding to native CD45 and synthetic oligosaccharide. Eur J Immunol 32:1924–1932

Barrington RA, Zhang M, Zhong X, Jonsson H, Holodick N, Cherukuri A, Pierce SK, Rothstein TL, Carroll MC (2005) CD21/CD19 coreceptor signaling promotes B-cell survival during primary immune responses. J Immunol 175:2859–2867

Behr S, Schriever F (1995) Engaging CD19 or target of an antiproliferative antibody 1 on human B lymphocytes induces binding of B cells to the interfollicular stroma of human tonsils via integrin alpha 4/beta 1 and fibronectin. J Exp Med 182:1191–1199

Biancone L, Bowen MA, Lim A, Aruffo A, Andres G, Stamenkovic I (1996) Identification of a novel inducible cell-surface ligand of CD5 on activated lymphocytes. J Exp Med 184: 811–819

Bikah G, Lynd FM, Aruffo AA, Ledbetter JA, Bondada S (1998) A role for CD5 in cognate interactions between T cells and B cells, and identification of a novel ligand for CD5. Int Immunol 10:1185–1196

Birkenbach M, Tong X, Bradbury LE, Tedder TF, Kieff E (1992) Characterization of an Epstein-Barr virus receptor on human epithelial cells. J Exp Med 176:1405–1414

Boackle SA, Holers VM, Chen X, Szakonyi G, Karp DR, Wakeland EK, Morel L (2001) Cr2, a candidate gene in the murine Sle1c lupus susceptibility locus, encodes a dysfunctional protein. Immunity 15:775–785

Bottger EC, Metzger S, Bitter-Suermann D, Stevenson G, Kleindienst S, Burger R (1986) Impaired humoral immune response in complement C3-deficient guinea pigs: absence of secondary antibody response. Eur J Immunol 16:1231–1235

Bradbury LE, Kansas GS, Levy S, Evans RL, Tedder TF (1992) The CD19/CD21 signal transducing complex of human B lymphocytes includes the target of antiproliferative antibody-1 and Leu-13 molecules. J Immunol 149:2841–2850

Braesch-Andersen S, Stamenkovic I (1994) Sialylation of the B lymphocyte molecule CD22 by alpha 2,6-sialyltransferase is implicated in the regulation of CD22-mediated adhesion. J Biol Chem 269:11783–11786

Brinkman-Van der Linden EC, Sjoberg ER, Juneja LR, Crocker PR, Varki N, Varki A (2000) Loss of N-glycolylneuraminic acid in human evolution. Implications for sialic acid recognition by siglecs. J Biol Chem 275:8633–8640

Brooks SR, Li X, Volanakis EJ, Carter RH (2000) Systematic analysis of the role of CD19 cytoplasmic tyrosines in enhancement of activation in Daudi human B cells: clustering of phospholipase C, Vav and of Grb2 and Sos with different CD19 tyrosines. J Immunol 164:3123–3131

Buhl AM, Cambier JC (1999) Phosphorylation of CD19 Y484 and Y515, and linked activation of phosphatidylinositol 3-kinase, are required for B-cell antigen receptor-mediated activation of Bruton's tyrosine kinase. J Immunol 162:4438–4446

Buhl AM, Pleiman CM, Rickert RC, Cambier JC (1997) Qualitative regulation of B-cell antigen receptor signaling by CD19: selective requirement for PI3-kinase activation, inositol-1,4,5-trisphosphate production and Ca2+ mobilization. J Exp Med 186:1897–1910

Callea V, Comis M, Iaria G, Sculli G, Morabito F, Lombardo VT (1997) Clinical significance of HLA-DR+, CD19+, CD10+ immature B-cell phenotype and CD34+ cell detection in bone marrow lymphocytes from children affected with immune thrombocytopenic purpura. Haematologica 82:471–473

Campbell MA, Klinman NR (1995) Phosphotyrosine-dependent association between CD22 and protein tyrosine phosphatase 1C. Eur J Immunol 25:1573–159

Carter RH, Fearon DT (1992) CD19: lowering the threshold for antigen receptor stimulation of B lymphocytes. Science 256:105–107

Carter RH, Tuveson DA, Park DJ, Rhee SG, Fearon DT (1991) The CD19 complex of B lymphocytes. Activation of phospholipase C by a protein tyrosine kinase-dependent pathway that can be enhanced by the membrane IgM complex. J Immunol 147:3663–3671

Carter RH, Doody GM, Bolen JB, Fearon DT (1997) Membrane IgM-induced tyrosine phosphorylation of CD19 requires a CD19 domain that mediates association with components of the B-cell antigen receptor complex. J Immunol 158:3062–3069

Chalupny NJ, Kanner SB, Schieven GL, Wee SF, Gilliland LK, Aruffo A, Ledbetter JA (1993) Tyrosine phosphorylation of CD19 in pre-B and mature B cells. EMBO J 12:2691–2696

Chalupny NJ, Aruffo A, Esselstyn JM, Chan PY, Bajorath J, Blake J, Gilliland LK, Ledbetter JA, Tepper MA (1995) Specific binding of Fyn and phosphatidylinositol 3-kinase to the B-cell surface glycoprotein CD19 through their src homology 2 domains. Eur J Immunol 25:2978–2984

Chan VW, Lowell CA, DeFranco AL (1998) Defective negative regulation of antigen receptor signaling in Lyn-deficient B lymphocytes. Curr Biol 8:545–553

Chen Z, Koralov SB, Gendelman M, Carroll MC, Kelsoe G (2000) Humoral immune responses in Cr2-/- mice: enhanced affinity maturation but impaired antibody persistence. J Immunol 164:4522–4532

Chen J, McLean PA, Neel BG, Okunade G, Shull GE, Wortis HH (2004) CD22 attenuates calcium signaling by potentiating plasma membrane calcium-ATPase activity. Nat Immunol 5:651–657

Cherukuri A, Carter RH, Brooks S, Bornmann W, Finn R, Dowd CS, Pierce SK (2004a) B-cell signaling is regulated by induced palmitoylation of CD81. J Biol Chem 279:31973–31982

Cherukuri A, Shoham T, Sohn HW, Levy S, Brooks S, Carter R, Pierce SK (2004b) The tetraspanin CD81 is necessary for partitioning of coligated CD19/CD21-B-cell antigen receptor complexes into signaling-active lipid rafts. J Immunol 172:370–380

Collins BE, Blixt O, DeSieno AR, Bovin N, Marth JD, Paulson JC (2004) Masking of CD22 by cis ligands does not prevent redistribution of CD22 to sites of cell contact. Proc Natl Acad Sci USA 101:6104–6109

Cornall RJ, Cyster JG, Hibbs ML, Dunn AR, Otipoby KL, Clark EA, Goodnow CC (1998) Polygenic autoimmune traits: Lyn, CD22, and SHP-1 are limiting elements of a biochemical pathway regulating BCR signaling and selection. Immunity 8:497–508

Croix DA, Ahearn JM, Rosengard AM, Han S, Kelsoe G, Ma M, Carroll MC (1996) Antibody response to a T-dependent antigen requires B-cell expression of complement receptors. J Exp Med 183:1857–1864

Delibrias CC, Fischer E, Bismuth G, Kazatchkine MD (1992) Expression, molecular association, and functions of C3 complement receptors CR1 (CD35) and CR2 (CD21) on the human T-cell line HPB-ALL. J Immunol 149:768–774

Dempsey PW, Allison ME, Akkaraju S, Goodnow CC, Fearon DT (1996) C3d of complement as a molecular adjuvant: bridging innate and acquired immunity. Science 271:348–350

Doody GM, Justement LB, Delibrias CC, Matthews RJ, Lin J, Thomas ML, Fearon DT (1995) A role in B-cell activation for CD22 and the protein tyrosine phosphatase SHP. Science 269:242–244

Dorken B, Moldenhauer G, Pezzutto A, Schwartz R, Feller A, Kiesel S, Nadler LM (1986) HD39 (B3), a B lineage-restricted antigen whose cell surface expression is limited to resting and activated human B lymphocytes. J Immunol 136:4470–4479

Elhabazi A, Delaire S, Bensussan A, Boumsell L, Bismuth G (2001) Biological activity of soluble CD100. I. The extracellular region of CD100 is released from the surface of T lymphocytes by regulated proteolysis. J Immunol 166:4341–4347

Engel P, Nojima Y, Rothstein D, Zhou LJ, Wilson GL, Kehrl JH, Tedder TF (1993) The same epitope on CD22 of B lymphocytes mediates the adhesion of erythrocytes, T, B lymphocytes, neutrophils, and monocytes. J Immunol 150:4719–4732

Engel P, Wagner N, Miller AS, Tedder TF (1995a) Identification of the ligand-binding domains of CD22, a member of the immunoglobulin superfamily that uniquely binds a sialic acid-dependent ligand. J Exp Med 181:1581–1586

Engel P, Zhou LJ, Ord DC, Sato S, Koller B, Tedder TF (1995b) Abnormal B lymphocyte development, activation, and differentiation in mice that lack or overexpress the CD19 signal transduction molecule. Immunity 3:39–50

Erickson LD, Tygrett LT, Bhatia SK, Grabstein KH, Waldschmidt TJ (1996) Differential expression of CD22 (Lyb8) on murine B cells. Int Immunol 8:1121–1129

Fang Y, Xu C, Fu YX, Holers VM, Molina H (1998) Expression of complement receptors 1 and 2 on follicular dendritic cells is necessary for the generation of a strong antigen-specific IgG response. J Immunol 160:5273–5279

Fearon DT, Carroll MC (2000) Regulation of B lymphocyte responses to foreign and self-antigens by the CD19/CD21 complex. Annu Rev Immunol 18:393–422

Ferguson AR, Youd ME, Corley RB (2004) Marginal zone B cells transport and deposit IgM-containing immune complexes onto follicular dendritic cells. Int Immunol 16:1411–1422

Fingeroth JD, Weis JJ, Tedder TF, Strominger JL, Biro PA, Fearon DT (1984) Epstein-Barr virus receptor of human B lymphocytes is the C3d receptor CR2. Proc Natl Acad Sci USA 81:4510–4514

Fischer E, Delibrias C, Kazatchkine MD (1991) Expression of CR2 (the C3dg/EBV receptor, CD21) on normal human peripheral blood T lymphocytes. J Immunol 146:865–869

Fischer MB, Ma M, Goerg S, Zhou X, Xia J, Finco O, Han S, Kelsoe G, Howard RG, Rothstein TL, Kremmer E, Rosen FS, Carroll MC (1996) Regulation of the B-cell response to T-dependent antigens by classical pathway complement. J Immunol 157:549–556

Fischer MB, Goerg S, Shen L, Prodeus AP, Goodnow CC, Kelsoe G, Carroll MC (1998) Dependence of germinal center B cells on expression of CD21/CD35 for survival. Science 280:582–585

Fujimoto M, Poe JC, Jansen PJ, Sato S, Tedder TF (1999) CD19 amplifies B lymphocyte signal transduction by regulating Src-family protein tyrosine kinase activation. J Immunol 162:7088–7094

Fujimoto M, Fujimoto Y, Poe JC, Jansen PJ, Lowell CA, DeFranco AL, Tedder TF (2000) CD19 regulates Src family protein tyrosine kinase activation in B lymphocytes through processive amplification. Immunity 13:47–57

Furuyama T, Inagaki S, Kosugi A, Noda S, Saitoh S, Ogata M, Iwahashi Y, Miyazaki N, Hamaoka T, Tohyama M (1996) Identification of a novel transmembrane semaphorin expressed on lymphocytes. J Biol Chem 271:33376–33381

Gatto D, Pfister T, Jegerlehner A, Martin SW, Kopf M, Bachmann MF (2005) Complement receptors regulate differentiation of bone marrow plasma cell precursors expressing transcription factors Blimp-1 and XBP-1. J Exp Med 201:993–1005

Haas KM, Hasegawa M, Steeber DA, Poe JC, Zabel MD, Bock CB, Karp DR, Briles DE, Weis JH, Tedder TF (2002) Complement receptors CD21/35 link innate and protective immunity during Streptococcus pneumoniae infection by regulating IgG3 antibody responses. Immunity 17:713–123

Hall KT, Boumsell L, Schultze JL, Boussiotis VA, Dorfman DM, Cardoso AA, Bensussan A, Nadler LM, Freeman GJ (1996) Human CD100, a novel leukocyte semaphorin that promotes B-cell aggregation and differentiation. Proc Natl Acad Sci USA 93:11780–11785
Han S, Collins BE, Bengtson P, Paulson JC (2005) Homomultimeric complexes of CD22 in B cells revealed by protein-glycan cross-linking. Nat Chem Biol 1:93–97
Hanasaki K, Powell LD, Varki A (1995) Binding of human plasma sialoglycoproteins by the B cell-specific lectin CD22. Selective recognition of immunoglobulin M and haptoglobin. J Biol Chem 270:7543–7550
Hanasaki K, Varki A, Powell LD (1995) CD22-mediated cell adhesion to cytokine-activated human endothelial cells. Positive and negative regulation by alpha 2–6-sialylation of cellular glycoproteins. J Biol Chem 270:7533–7542
Hasegawa M, Fujimoto M, Poe JC, Steeber DA, Lowell CA, Tedder TF (2001) A CD19-dependent signaling pathway regulates autoimmunity in Lyn-deficient mice. J Immunol 167:2469–2478
Hennet T, Chui D, Paulson JC, Marth JD (1998) Immune regulation by the ST6Gal sialyltransferase. Proc Natl Acad Sci USA 95:4504–4509
Hibbs ML, Tarlinton DM, Armes J, Grail D, Hodgson G, Maglitto R, Stacker SA, Dunn AR (1995) Multiple defects in the immune system of Lyn-deficient mice, culminating in autoimmune disease. Cell 83:301–311
Hokazono Y, Adachi T, Wabl M, Tada N, Amagasa T, Tsubata T (2003) Inhibitory coreceptors activated by antigens but not by anti-Ig heavy chain antibodies install requirement of costimulation through CD40 for survival and proliferation of B cells. J Immunol 171:1835–1843
Horvath G, Serru V, Clay D, Billard M, Boucheix C, Rubinstein E (1998) CD19 is linked to the integrin-associated tetraspans CD9, CD81, and CD82. J Biol Chem 273:30537–30543
Hu H, Martin BK, Weis JJ, Weis JH (1997) Expression of the murine CD21 gene is regulated by promoter and intronic sequences. J Immunol 158:4758–4768
Illges H, Braun M, Peter HH, Melchers I (1997) Analysis of the human CD21 transcription unit reveals differential splicing of exon 11 in mature transcripts and excludes alternative splicing as the mechanism causing solubilization of CD21. Mol Immunol 34:683–693
Jin L, McLean PA, Neel BG, Wortis HH (2002) Sialic acid binding domains of CD22 are required for negative regulation of B-cell receptor signaling. J Exp Med 195:1199–1205
Kalli KR, Ahearn JM, Fearon DT (1991) Interaction of iC3b with recombinant isotypic and chimeric forms of CR2. J Immunol 147:590–594
Kelm S, Gerlach J, Brossmer R, Danzer CP, Nitschke L (2002) The ligand-binding domain of CD22 is needed for inhibition of the B-cell receptor signal, as demonstrated by a novel human CD22-specific inhibitor compound. J Exp Med 195:1207–1213
Kozmik Z, Wang S, Dorfler P, Adams B, Busslinger M (1992) The promoter of the CD19 gene is a target for the B-cell-specific transcription factor BSAP. Mol Cell Biol 12:2662–2672
Krop I, Shaffer AL, Fearon DT, Schlissel MS (1996) The signaling activity of murine CD19 is regulated during cell development. J Immunol 157:48–56
Kumanogoh A, Watanabe C, Lee I, Wang X, Shi W, Araki H, Hirata H, Iwahori K, Uchida J, Yasui T, Matsumoto M, Yoshida K, Yakura H, Pan C, Parnes JR, Kikutani H (2000) Identification of CD72 as a lymphocyte receptor for the class IV semaphorin CD100: a novel mechanism for regulating B-cell signaling. Immunity 13:621–631
Kurtz CB, O'Toole E, Christensen SM, Weis JH (1990) The murine complement receptor gene family. IV. Alternative splicing of Cr2 gene transcripts predicts two distinct gene products that share homologous domains with both human CR2 and CR1. J Immunol 144: 3581–3591
Lankester AC, van Schijndel GM and van Lier RA (1995) Hematopoietic cell phosphatase is recruited to CD22 following B-cell antigen receptor ligation. J Biol Chem 270: 20305–20308
Lanoue A, Batista FD, Stewart M, Neuberger MS (2002) Interaction of CD22 with alpha2,6-linked sialoglycoconjugates: innate recognition of self to dampen B-cell autoreactivity? Eur J Immunol 32:348–355

Law CL, Aruffo A, Chandran KA, Doty RT, Clark EA (1995) Ig domains 1 and 2 of murine CD22 constitute the ligand-binding domain and bind multiple sialylated ligands expressed on B, T cells. J Immunol 155:3368–3376

Law CL, Sidorenko SP, Chandran KA, Zhao Z, Shen SH, Fischer EH, Clark EA (1996) CD22 associates with protein tyrosine phosphatase 1C, Syk, and phospholipase C-gamma(1) upon B-cell activation. J Exp Med 183:547–560

Leprince C, Draves KE, Geahlen RL, Ledbetter JA, Clark EA (1993) CD22 associates with the human surface IgM-B-cell antigen receptor complex. Proc Natl Acad Sci USA 90:3236–3240

Levy S, Todd SC, Maecker HT (1998) CD81 (TAPA-1): a molecule involved in signal transduction and cell adhesion in the immune system. Annu Rev Immunol 16:89–109

Lowe J, Brown B, Hardie D, Richardson P, Ling N (1989) Soluble forms of CD21 and CD23 antigens in the serum in B-cell chronic lymphocytic leukaemia. Immunol Lett 20:103–109

Maecker HT, Levy S (1997) Normal lymphocyte development but delayed humoral immune response in CD81-null mice. J Exp Med 185:1505–1510

Makar KW, Pham CT, Dehoff MH, O'Connor SM, Jacobi SM, Holers VM (1998) An intronic silencer regulates B lymphocyte cell- and stage-specific expression of the human complement receptor type 2 (CR2, CD21) gene. J Immunol 160:1268–1278

Martin SW, Goodnow CC (2002) Burst-enhancing role of the IgG membrane tail as a molecular determinant of memory. Nat Immunol 3:182–188

Masilamani M, Kassahn D, Mikkat S, Glocker MO, Illges H (2003) B-cell activation leads to shedding of complement receptor type II (CR2/CD21). Eur J Immunol 33:2391–2397

Matsumoto AK, Kopicky-Burd J, Carter RH, Tuveson DA, Tedder TF, Fearon DT (1991) Intersection of the complement and immune systems: a signal transduction complex of the B lymphocyte-containing complement receptor type 2 and CD19. J Exp Med 173:55–64

Matsumoto AK, Martin DR, Carter RH, Klickstein LB, Ahearn JM, Fearon DT (1993) Functional dissection of the CD21/CD19/TAPA-1/Leu-13 complex of B lymphocytes. J Exp Med 178:1407–1417

Mittelbrunn M, Yanez-Mo M, Sancho D, Ursa A, Sanchez-Madrid F (2002) Cutting edge: dynamic redistribution of tetraspanin CD81 at the central zone of the immune synapse in both T lymphocytes and APC. J Immunol 169:6691–6695

Miyazaki T, Muller U, Campbell KS (1997) Normal development but differentially altered proliferative responses of lymphocytes in mice lacking CD81. EMBO J 16:4217–4225

Molina H, Kinoshita T, Inoue K, Carel JC, Holers VM (1990) A molecular and immunochemical characterization of mouse CR2. Evidence for a single gene model of mouse complement receptors 1 and 2. J Immunol 145:2974–2983

Molina H, Holers VM, Li B, Fung Y, Mariathasan S, Goellner J, Strauss-Schoenberger J, Karr RW, Chaplin DD (1996) Markedly impaired humoral immune response in mice deficient in complement receptors 1 and 2. Proc Natl Acad Sci USA 93:3357–3361

Nakayama E, von Hoegen I, Parnes JR (1989) Sequence of the Lyb-2 B-cell differentiation antigen defines a gene superfamily of receptors with inverted membrane orientation. Proc Natl Acad Sci USA 86:1352–1356

Nishizumi H, Taniuchi I, Yamanashi Y, Kitamura D, Ilic D, Mori S, Watanabe T, Yamamoto T (1995) Impaired proliferation of peripheral B cells and indication of autoimmune disease in lyn-deficient mice. Immunity 3:549–560

Nitschke L, Tsubata T (2004) Molecular interactions regulate BCR signal inhibition by CD22 and CD72. Trends Immunol 25:543–550

Nitschke L, Carsetti R, Ocker B, Kohler G, Lamers MC (1997) CD22 is a negative regulator of B-cell receptor signalling. Curr Biol 7:133–143

Nomura T, Han H, Howard MC, Yagita H, Yakura H, Honjo T, Tsubata T (1996) Antigen receptor-mediated B-cell death is blocked by signaling via CD72 or treatment with dextran sulfate and is defective in autoimmunity-prone mice. Int Immunol 8:867–875

O'Keefe TL, Williams GT, Batista FD, Neuberger MS (1999) Deficiency in CD22, a B cell-specific inhibitory receptor, is sufficient to predispose to development of high affinity autoantibodies. J Exp Med 189:1307–1313
O'Keefe TL, Williams GT, Davies SL, Neuberger MS (1996) Hyperresponsive B cells in CD22-deficient mice. Science 274:798–801
O'Rourke LM, Tooze R, Turner M, Sandoval DM, Carter RH, Tybulewicz VL, Fearon DT (1998) CD19 as a membrane-anchored adaptor protein of B lymphocytes: costimulation of lipid and protein kinases by recruitment of Vav. Immunity 8:635–645
Otero DC, Rickert RC (2003) CD19 function in early and late B-cell development. II. CD19 facilitates the pro-B/pre-B transition. J Immunol 171:5921–5930
Otero DC, Omori SA, Rickert RC (2001) Cd19-dependent activation of Akt kinase in B-lymphocytes. J Biol Chem 276:1474–1478
Otero DC, Anzelon AN, Rickert RC (2003) CD19 function in early and late B-cell development: I. Maintenance of follicular and marginal zone B cells requires CD19-dependent survival signals. J Immunol 170:73–83
Otipoby KL, Andersson KB, Draves KE, Klaus SJ, Farr AG, Kerner JD, Perlmutter RM, Law CL, Clark EA (1996) CD22 regulates thymus-independent responses and the lifespan of B cells. Nature 384:634–637
Otipoby KL, Draves KE, Clark EA (2001) CD22 regulates B-cell receptor-mediated signals via two domains that independently recruit Grb2 and SHP-1. J Biol Chem 276:44315–44322
Pan C, Baumgarth N, Parnes JR (1999) CD72-deficient mice reveal nonredundant roles of CD72 in B-cell development and activation. Immunity 11:495–506
Parnes JR, Pan C (2000) CD72, a negative regulator of B-cell responsiveness. Immunol Rev 176:75–85
Peaker CJ, Neuberger MS (1993) Association of CD22 with the B-cell antigen receptor. Eur J Immunol 23:1358–1363
Pesando JM, Bouchard LS, McMaster BE (1989) CD19 is functionally and physically associated with surface immunoglobulin. J Exp Med 170:2159–2164
Pezzutto A, Dorken B, Moldenhauer G, Clark EA (1987) Amplification of human B-cell activation by a monoclonal antibody to the B cell-specific antigen CD22, Bp 130/140. J Immunol 138: 98–103
Pezzutto A, Rabinovitch PS, Dorken B, Moldenhauer G, Clark EA (1988) Role of the CD22 human B-cell antigen in B-cell triggering by anti-immunoglobulin. J Immunol 140:1791–1795
Pflugh DL, Maher SE, Bothwell AL (2002) Ly-6 superfamily members Ly-6A/E, Ly-6C, and Ly-6I recognize two potential ligands expressed by B lymphocytes. J Immunol 169:5130–5136
Poe JC, Fujimoto M, Jansen PJ, Miller AS, Tedder TF (2000) CD22 forms a quaternary complex with SHIP, Grb2, and Shc. A pathway for regulation of B lymphocyte antigen receptor-induced calcium flux. J Biol Chem 275:17420–17427
Poe JC, Fujimoto Y, Hasegawa M, Haas KM, Miller AS, Sanford IG, Bock CB, Fujimoto M, Tedder TF (2004) CD22 regulates B lymphocyte function in vivo through both ligand-dependent and ligand-independent mechanisms. Nat Immunol 5:1078–1087
Powell LD, Varki A (1994) The oligosaccharide binding specificities of CD22 beta, a sialic acid-specific lectin of B cells. J Biol Chem 269:10628–10636
Powell LD, Sgroi D, Sjoberg ER, Stamenkovic I, Varki A (1993) Natural ligands of the B-cell adhesion molecule CD22 beta carry N-linked oligosaccharides with alpha-2,6-linked sialic acids that are required for recognition. J Biol Chem 268:7019–7027
Powell LD, Jain RK, Matta KL, Sabesan S, Varki A (1995) Characterization of sialyloligosaccharide binding by recombinant soluble and native cell-associated CD22. Evidence for a minimal structural recognition motif and the potential importance of multisite binding. J Biol Chem 270:7523–7532
Pramoonjago P, Takeda J, Kim YU, Inoue K, Kinoshita T (1993) Ligand specificities of mouse complement receptor types 1 (CR1) and 2 (CR2) purified from spleen cells. Int Immunol 5:337–343

Qin D, Wu J, Carroll MC, Burton GF, Szakal AK, Tew JG (1998) Evidence for an important interaction between a complement-derived CD21 ligand on follicular dendritic cells and CD21 on B cells in the initiation of IgG responses. J Immunol 161:4549–4554
Razi N, Varki A (1998) Masking and unmasking of the sialic acid-binding lectin activity of CD22 (Siglec-2) on B lymphocytes. Proc Natl Acad Sci USA 95:7469–7474
Razi N, Varki A (1999) Cryptic sialic acid binding lectins on human blood leukocytes can be unmasked by sialidase treatment or cellular activation. Glycobiology 9:1225–1234
Reid RR, Woodcock S, Shimabukuro-Vornhagen A, Austen WG, Jr., Kobzik L, Zhang M, Hechtman HB, Moore FD, Jr. and Carroll MC (2002) Functional activity of natural antibody is altered in Cr2-deficient mice. J Immunol 169:5433–5440
Reynes M, Aubert JP, Cohen JH, Audouin J, Tricottet V, Diebold J, Kazatchkine MD (1985) Human follicular dendritic cells express CR1, CR2, and CR3 complement receptor antigens. J Immunol 135:2687–2694
Rickert RC (2005) Regulation of B lymphocyte activation by complement C3 and the B-cell coreceptor complex. Curr Opin Immunol 17:237–243
Rickert RC, Rajewsky K, Roes J (1995) Impairment of T-cell-dependent B-cell responses and B-1 cell development in CD19-deficient mice. Nature 376:352–355
Samardzic T, Marinkovic D, Danzer CP, Gerlach J, Nitschke L, Wirth T (2002) Reduction of marginal zone B cells in CD22-deficient mice. Eur J Immunol 32:561–567
Sato S, Steeber DA, Tedder TF (1995) The CD19 signal transduction molecule is a response regulator of B-lymphocyte differentiation. Proc Natl Acad Sci USA 92:11558–11562
Sato S, Miller AS, Inaoki M, Bock CB, Jansen PJ, Tang ML, Tedder TF (1996) CD22 is both a positive and negative regulator of B lymphocyte antigen receptor signal transduction: altered signaling in CD22-deficient mice. Immunity 5:551–562
Schulte RJ, Campbell MA, Fischer WH, Sefton BM (1992) Tyrosine phosphorylation of CD22 during B-cell activation. Science 258:1001–1004
Sgroi D, Varki A, Braesch-Andersen S, Stamenkovic I (1993) CD22, a B cell-specific immunoglobulin superfamily member, is a sialic acid-binding lectin. J Biol Chem 268:7011–7018
Shi W, Kumanogoh A, Watanabe C, Uchida J, Wang X, Yasui T, Yukawa K, Ikawa M, Okabe M, Parnes JR, Yoshida K, Kikutani H (2000) The class IV semaphorin CD100 plays nonredundant roles in the immune system: defective B, T-cell activation in CD100-deficient mice. Immunity 13:633–642
Shoham T, Rajapaksa R, Boucheix C, Rubinstein E, Poe JC, Tedder TF, Levy S (2003) The tetraspanin CD81 regulates the expression of CD19 during B-cell development in a postendoplasmic reticulum compartment. J Immunol 171:4062–4072
Sjoberg ER, Powell LD, Klein A, Varki A (1994) Natural ligands of the B-cell adhesion molecule CD22 beta can be masked by 9-O-acetylation of sialic acids. J Cell Biol 126:549–562
Smith KG, Tarlinton DM, Doody GM, Hibbs ML, Fearon DT (1998) Inhibition of the B-cell by CD22: a requirement for Lyn. J Exp Med 187:807–811
Stamenkovic I, Seed B (1988) CD19, the earliest differentiation antigen of the B-cell lineage, bears three extracellular immunoglobulin-like domains and an Epstein-Barr virus-related cytoplasmic tail. J Exp Med 168:1205–1210
Stamenkovic I, Seed B (1990) The B-cell antigen CD22 mediates monocyte and erythrocyte adhesion. Nature 345:74–77
Stamenkovic I, Sgroi D, Aruffo A, Sy MS, Anderson T (1991) The B lymphocyte adhesion molecule CD22 interacts with leukocyte common antigen CD45RO on T cells and alpha 2–6 sialyltransferase, CD75, on B cells. Cell 66:1133–1144
Subbarao B, Mosier DE (1984) Activation of B lymphocytes by monovalent anti-Lyb-2 antibodies. J Exp Med 159:1796–1801
Takahashi S, Doss C, Levy S, Levy R (1990) TAPA-1, the target of an antiproliferative antibody, is associated on the cell surface with the Leu-13 antigen. J Immunol 145:2207–2213
Takahashi Y, Ohta H, Takemori T (2001) Fas is required for clonal selection in germinal centers and the subsequent establishment of the memory B-cell repertoire. Immunity 14:181–192

Tedder TF, Isaacs CM (1989) Isolation of cDNAs encoding the CD19 antigen of human and mouse B lymphocytes. A new member of the immunoglobulin superfamily. J Immunol 143: 712–717

Tedder TF, Clement LT, Cooper MD (1984) Expression of C3d receptors during human B-cell differentiation: immunofluorescence analysis with the HB-5 monoclonal antibody. J Immunol 133:678–683

Tedder TF, Inaoki M, Sato S (1997) The CD19-CD21 complex regulates signal transduction thresholds governing humoral immunity and autoimmunity. Immunity 6:107–118

Tsitsikov EN, Gutierrez-Ramos JC, Geha RS (1997) Impaired CD19 expression and signaling, enhanced antibody response to type II T independent antigen and reduction of B-1 cells in CD81-deficient mice. Proc Natl Acad Sci USA 94:10844–10849

Tuveson DA, Ahearn JM, Matsumoto AK, Fearon DT (1991) Molecular interactions of complement receptors on B lymphocytes: a CR1/CR2 complex distinct from the CR2/CD19 complex. J Exp Med 173:1083–1089

Tuveson DA, Carter RH, Soltoff SP, Fearon DT (1993) CD19 of B cells as a surrogate kinase insert region to bind phosphatidylinositol 3-kinase. Science 260:986–989

Van de Velde H, von Hoegen I, Luo W, Parnes JR, Thielemans K (1991) The B-cell surface protein CD72/Lyb-2 is the ligand for CD5. Nature 351:662–665

van der Merwe PA, Crocker PR, Vinson M, Barclay AN, Schauer R, Kelm S (1996) Localization of the putative sialic acid-binding site on the immunoglobulin superfamily cell-surface molecule CD22. J Biol Chem 271:9273–9280

Vigorito E, Bardi G, Glassford J, Lam EW, Clayton E, Turner M (2004) Vav-dependent and vav-independent phosphatidylinositol 3-kinase activation in murine B cells determined by the nature of the stimulus. J Immunol 173:3209–3214

Wakabayashi C, Adachi T, Wienands J, Tsubata T (2002) A distinct signaling pathway used by the IgG-containing B-cell antigen receptor. Science 298:2392–2395

Wang Y, Brooks SR, Li X, Anzelon AN, Rickert RC, Carter RH (2002) The physiologic role of CD19 cytoplasmic tyrosines. Immunity 17:501–514

Wienands J, Freuler F, Baumann G (1995) Tyrosine-phosphorylated forms of Ig beta, CD22, TCR zeta and HOSS are major ligands for tandem SH2 domains of Syk. Int Immunol 7: 1701–1708

Wilson GL, Fox CH, Fauci AS, Kehrl JH (1991) cDNA cloning of the B-cell membrane protein CD22: a mediator of B-B-cell interactions. J Exp Med 173:137–146

Wu Y, Nadler MJ, Brennan LA, Gish GD, Timms JF, Fusaki N, Jongstra-Bilen J, Tada N, Pawson T, Wither J, Neel BG, Hozumi N (1998) The B-cell transmembrane protein CD72 binds to and is an in vivo substrate of the protein tyrosine phosphatase SHP-1. Curr Biol 8:1009–1017

Wu X, Jiang N, Deppong C, Singh J, Dolecki G, Mao D, Morel L, Molina HD (2002) A role for the Cr2 gene in modifying autoantibody production in systemic lupus erythematosus. J Immunol 169:1587–1592

Xu Y, Beavitt SJ, Harder KW, Hibbs ML, Tarlinton DM (2002) The activation and subsequent regulatory roles of Lyn and CD19 after B-cell receptor ligation are independent. J Immunol 169:6910–6918

Yohannan J, Wienands J, Coggeshall KM, Justement LB (1999) Analysis of tyrosine phosphorylation-dependent interactions between stimulatory effector proteins and the B-cell co-receptor CD22. J Biol Chem 274:18769–18776

Zhang M, Varki A (2004) Cell surface sialic acids do not affect primary CD22 interactions with CD45 and surface IgM nor the rate of constitutive CD22 endocytosis. Glycobiology 14:939–949

Zhou LJ, Ord DC, Hughes AL, Tedder TF (1991) Structure and domain organization of the CD19 antigen of human, mouse, and guinea pig B lymphocytes. Conservation of the extensive cytoplasmic domain. J Immunol 147:1424–1432

8 Co-Receptors in the Positive and Negative Regulation of T-Cell Immunity

Helga Schneider and Christopher E. Rudd

8.1 Introduction

Engagement of the antigen-receptor complex (TcRζ/CD3) on T cells is insufficient for the development of an optimal response against foreign antigen. Instead, T cells require at least two signals for optimal T-cell expansion (Bretscher 1999). Peptides presented to antigen-specific T cells in the context of major histocompatibility complex (MHC) molecules deliver signal 1, whereas a co-stimulatory signal through a distinct T-cell surface molecule triggers signal 2. Signaling through signal 1, in the absence of co-stimulation, leads to aborted activation and often a state of anergy or clonal unresponsiveness (Schwartz 1990). T-cell activation can occur in the absence of signal 2 under conditions where the TcR signal is very strong (i.e., high avidity peptide). With the advancement of the field, it is now apparent that there exists an array of different co-receptors on T cells (Fig. 1). These include CD28, inducible T-cell co-stimulator (ICOS) and cytotoxic T-cell antigen (CTLA-4), programmed death-1 (PD-1), B and T lymphocyte attenuator (BTLA) and T-cell immunoglobulin and mucin-domain-containing molecule- 3 (Tim-3). As will be outlined in this review, each of these receptors provides unique and overlapping signals that modulate different aspects of T-cell immunity.

CD28 is one of the best-characterized co-receptors on T cells (June et al. 1994). It comprises a single immunoglobulin domain, and is expressed on resting and activated T cells. During antigen-presentation, the TcRζ/CD3 complexes bind to peptides as presented by MHC antigens. The CD4 and CD8 receptors also engage MHC antigens leading to the activation of the associated src kinase p56lck and the initiation of the tyrosine phosphorylation activation cascade (Rudd 1990). At the same time, CD28 binds to ligands B7-1 (CD80) and B7-2 (CD86) on antigen-

---

Cell Signalling Section, Division of Immunology, Department of Pathology, University of Cambridge, Tennis Court Road, Cambridge CB2 1QP, UK

Correspondence to: H. Schneider

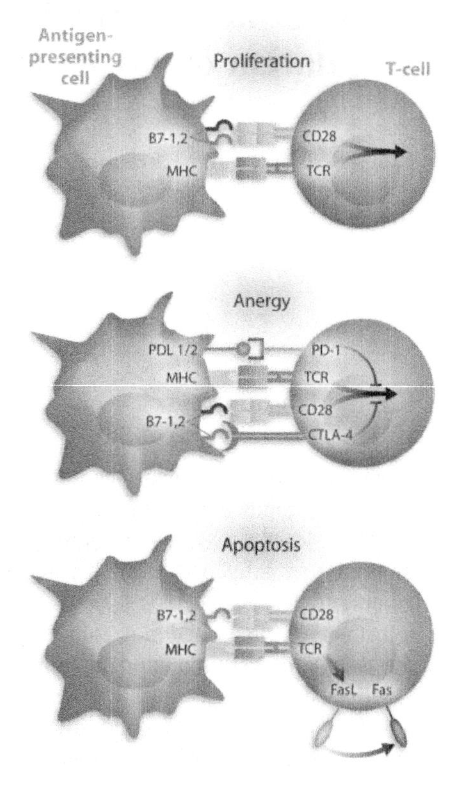

Fig. 1. Co-receptors determine multiple aspects of T-cell function. *Upper panel*, co-ligation of CD28 with TcR/CD3 lowers the threshold of TcR signaling needed to induce T-cell activation. This leads to increased cytokine production and proliferation. *Middle panel*, ligation of the TcR in the absence of CD28 can lead to an induction of anergy or non-responsiveness. By contrast, expression of CTLA-4 is needed for the induction of anergy, while programmed death-1 (*PD-1*) ligation by programmed death-ligand 1/ ligand 2(*PD-L1/PD-L2*) can inhibit T-cell proliferation and reduces cytokine production. This can lead to peripheral tolerance. *Lower panel*, TcR stimulation without engagement of CD28 leads to apoptosis via the Fas/FasL pathway. (See Color Plates)

presenting cells (APCs). B7-2 is constitutively expressed at low levels and rapidly up-regulated, whereas B7-1 is inducibly expressed later than B7-2. The ligation of the TcRζ/CD3 complex and CD28 provides the optimal signals for interleukin-2 (IL-2) production and proliferation (Fig. 1, upper panel). By contrast, engagement of TcRζ/CD3 in the absence of CD28 can cause non-responsiveness or anergy (Fig. 1, middle panel). The related family members ICOS and CTLA-4 are expressed only on activated cells (Fig. 2). In this way, they are thought to play specialized roles in the function of activated and effector T cells. While CTLA-4 also binds to CD80 and CD86, ICOS lacks the so-called MYPPPY sequence needed for CD80/86 binding, and instead binds to ICOSL (B7h GL50, B7RP-1, LICOS) (Ling et al. 2000; Swallow et al. 1999; Wang et al. 2000; Yoshinaga et al. 2000).

While CD28 provides signals from the general activation of T cells, ICOS is preferentially expressed on T-helper 2 (Th2) cells and provides positive signals for various aspects of Th2 cell function (Coyle et al. 2000; Dong et al. 2001; McAdam et al. 2000, 2001; Tafuri et al. 2001; Tesciuba et al. 2001). Within the T-cell compartment exist T helper 1 (Th1) and Th2 subsets that express different cytokines and regulate distinct functions. Th1 cells produce interferon (IFN)-γ, interleukin (IL)-2 and tumor necrosis factor (TNF)-α, and are involved in cell-mediated immu-

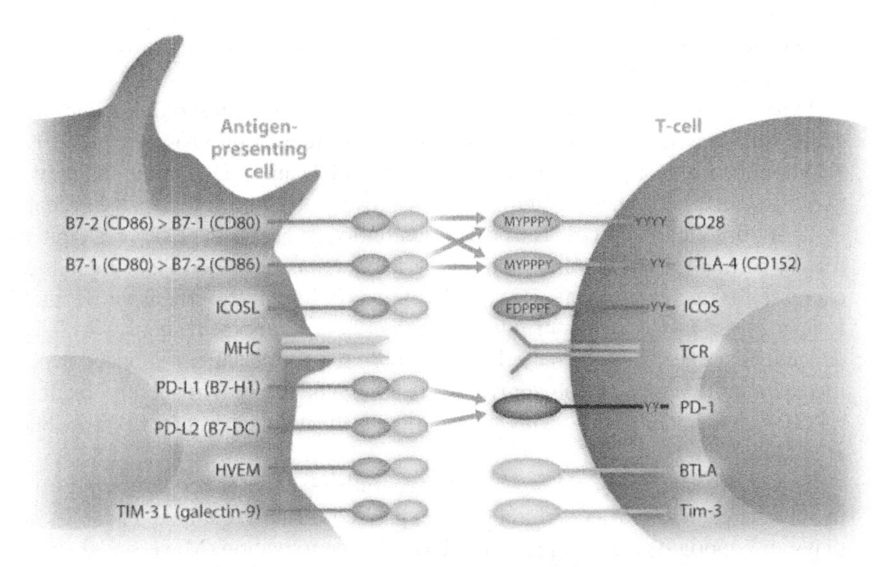

Fig. 2. Receptor–ligand pairs during the conjugate formation between T cells and antigen-presenting cells. T cells express an array of co-receptors that include CD28, inducible T-cell co-stimulator (*ICOS*), and cytotoxic T-cell antigen (*CTLA-4*) as well as co-receptors programmed death-1 (*PD-1*), B- and T-lymphocyte attenuator (*BTLA*), and T-cell immunoglobulin and mucin-domain-containing molecule-3 (*Tim-3*). CD28 and CTLA-4 bind to ligands B7-1 (CD80) and B7-2 (CD86) on antigen-presenting cells. Both possess MYPPPY sequence needed for B7-1/B7-2 binding. By contrast, ICOS has a FDPPPF motif and binds to ICOSL but not to B7-1 and B7-2. PD-1 is a receptor for both PD-L1 and PD-L2. The herpes virus entry mediator (*HVEM*), a member of the tumor-necrosis factor receptors (TNFR) superfamily, is the unique ligand for BTLA. Tim 3 binds galectin-9, a member of the galectin family that is expressed on lymphocytes and other cell types

nity and delayed-type hypersensitivity reactions. Autoimmune and chronic inflammatory diseases such as multiple sclerosis, type II diabetes and rheumatoid arthritis have been described as Th1-dominant diseases. By contrast, Th2 cells are characterized by the production of interleukins IL-4, IL-5, IL-10, and IL-13, and regulate humoral responses. Th2 responses dominate in defense against parasitic infections, such as helminth infections, affect the activation of mast cells and eosinophils, and are associated with atopy and allergy. ICOS expression and regulation of Th2 cells implicates this co-receptor in the regulation of these functions (Coyle et al. 2000).

On the other hand, positive co-signals are counterbalanced by signals that dampen the immune responses. These are provided by CTLA-4, PD-1 and the recently described Ig superfamily members BTLA and Tim-3 (Fig. 2). Surface CTLA-4 can only be detected in activated T cells, and CTLA-4-deficient mice show a profound post-thymic autoimmune phenotype with death by 3 weeks due to massive tissue infiltration and organ destruction (Tivol et al. 1995; Waterhouse et al. 1995). Recently, it has been suggested that CTLA-4 is constitutively expressed on CD4+CD25+ regulatory T cells (Tregs) where it plays a role in regulating their function (Read et al. 2000; Takahashi et al. 2000). However, the relationship is not obligatory since CTLA-4-deficient mice possess Tregs with normal function (Takahashi et al. 2000).

PD-1 can be induced not only on T cells, but also on B cells and myeloid cells, suggesting that it has broader roles on immune regulation (Agata et al. 1996; Nishimura and Honjo 2001). The ligands PD-L1 and PDL-2 are inductively expressed on the lymphoid system and constitutively on parenchymal cells (Freeman et al. 2000; Latchman et al. 2001). The inhibitory co-signal may suppress inflammatory responses and autoimmunity in various organs on which PD-1 ligands are expressed (Greenwald et al. 2005). BTLA exerts inhibitory effects on B and T lymphocytes. BTLA-deficient mice show increased humoral responses to T-cell dependent antigens, and increased susceptibility to peptide antigen-induced experimental autoimmune encephalomyelitis (EAE). The herpesvirus entry mediator (HVEM), a member of the TNFR superfamily, is the unique ligand for BTLA (Sedy et al. 2005).

The Ig superfamily member Tim-3 has recently been described as a novel transmembrane protein that is preferentially expressed on differentiated Th1 cells and plays an important role in suppressing Th1 effector activation (Meyers et al. 2005). Galectin-9, a member of the galectin family which is expressed on lymphocytes and other cell types, has been identified as a ligand for Tim-3 (Zhu et al. 2005). Blockade of the Tim-3 pathway by treatment with anti-Tim-3 antibody or Tim-3Ig fusion proteins led to increased Th1 cell proliferation and cytokine responses, in addition to loss of tolerance (Sabatos et al. 2003).

Overall, co-receptors not only provide critical positive second signals that promote and sustain T-cell responses, but they can also down-regulate T-cell responses. These negative signals function to limit and terminate T-cell responses, and they appear to be especially important for regulating T-cell tolerance and autoimmunity. Any subtle imbalance can be exaggerated and lead to an inappropriate immune response, which may cause serious autoimmune diseases. The balance of stimulatory and inhibitory signals is crucial to maximize protective immune responses while maintaining immunological tolerance and preventing autoimmunity.

8.2 Positive Co-stimulatory Molecules

8.2.1 CD28

8.2.1.1 Expression and Function

Human CD28 is expressed constitutively on most CD4 positive T cells and approximately 50% of CD8 positive T cells. Murine CD28 is expressed on all T cells, and CD28 is not down-regulated with age or chronic disease (Riley and June 2005). The importance of CD28 has been underscored by the phenotype of the CD28-

deficient mouse. T cells from these mice are impaired in their ability to proliferate in vitro in response to anti-TcR/CD3 stimulation, allo-stimulation and stimulation with specific antigen. In vivo, CD28-deficient mice exhibit impaired T helper B-cell response and fail to form germinal centers after immunization with protein antigens (Ferguson et al. 1996). These cells have a reduced capacity to produce IL-2 and to proliferate on subsequent stimulations. However, longer-term engagement of the TcRζ/CD3 complex with repeated antigen stimulation (i.e., high virulence infection, repeated injection of peptide) can bypass the requirement for CD28 (Kündig et al. 1996).

CD28 can regulate an array of events including cytokine production, proliferation, anergy, apoptosis, glucose metabolism, and Th1 vs. Th2 cell differentiation (Acuto and Michel 2003; Rudd and Schneider 2003). Enhanced cytokine production occurs via the potentiation of TcR signaling, the enhancement of NFAT activity as well as the stabilization of messenger RNA (Acuto et al. 2003; Riley and June 2005). The ability to augment glucose metabolism is related to the up-regulation of Glut 4 receptors on the surface of T cells (Frauwirth et al. 2002). Some reports indicate that CD28 ligation might skew the differentiation of T cells into IL-4-producing Th2 cells, although the exact molecular mechanism by which this occurs is not known (Schulze-Koops et al. 1998).

8.2.1.2 Signaling Pathways

CD28, ICOS, and CTLA-4 possess small cytoplasmic domains of 41, 35, and 36 residues, respectively (Fig. 3). Common to each is the presence of a YxxM consensus motif: a YMNM motif for CD28, a YMFM motif for ICOS and a YVKM motif for CTLA-4. Phosphorylation of the tyrosine creates conditions for the binding of one of two SH2 domains of the p85 subunit of the phosphatidylinositol 3-kinase (PI3K) (August and Dupont 1994; Coyle et al. 2000; Pages et al. 1994; Prasad et al. 1994; Truitt et al. 1994; Schneider et al. 1995b). Specificity is determined by residues adjacent to the tyrosine, specifically a methionine (M) in the plus 3 position. CD28 also carries an asparagine (N) in the plus 2 position (i.e., pYxNM) that is not found in ICOS or CTLA-4. This is a signature residue for SH2 domain binding of the adaptor Grb-2 (Kim et al. 1998; Okkenhaug and Rottapel 1998; Schneider et al. 1995a). Grb-2 is an adaptor protein that is comprised of one SH2 domain flanked by two SH3 domains that bind to the exchange factor Son of Sevenless (SOS), an activator of the GTPase p21ras. CD28 and CTLA-4 also carry proline residues that are not found in the cytoplasmic domain of ICOS. Classic RKxxPxxP and PxxPxR motifs serve as ligands for SH3 domains (Pawson et al. 2001). CD28 has two non-canonical PxxP motifs. The PRRP motif at residues 196 to 199 may bind to the tyrosine kinase, interleukin-2 inducible T-cell kinase (ITK) (Marengere et al. 1997), while a second C-terminal PYAP motif (residues 208 to 211) binds to the SH3 domain of p56lck (Holdorf et al. 1999). The Grb-2 SH3 domain can bind constitutively at low levels to the same motif, whereas the phospho-YMNM motif facilitates tandem SH2-SH3 domain binding (Kim et al. 1998;

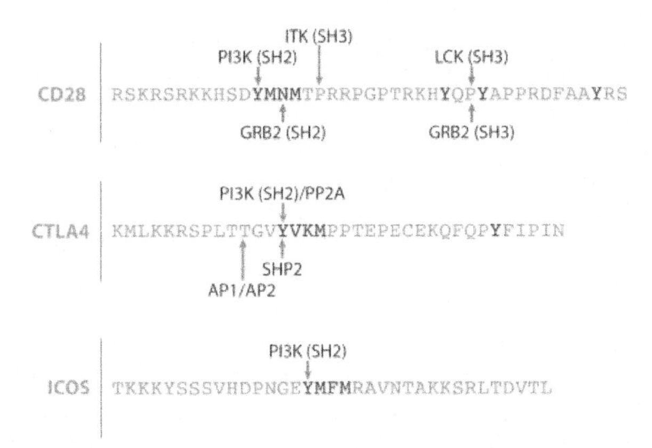

Fig. 3. Structure of the cytoplasmic domains of human CD28, ICOS, and CTLA-4. The cytoplasmic domains of CD28, ICOS, and CTLA-4 have a common YxxM motif that binds to the SH2 domain of the p85 subunit of phosphatidylinositol 3-kinase (*PI3K*). At the same time, each co-receptor has unique binding residues. CD28 has an asparagine that is needed for Grb-2 SH2 domain binding. Proline sites allow for the further binding of the SH3 domains of Grb-2 and the protein tyrosine kinase p56lck. CTLA-4 has a unique YVKM motif that binds in its non-phosphorylated form to the clathrin adapters AP-1 and AP-2. ICOS possesses a YxxM motif, but is characterized by the absence of additional proline residues

Okkenhaug and Rottapel 1998). The presence of a shared YxxM motif for PI3K binding amongst the CD28 family of co-receptors suggests they share an overlapping function, while differences in other residues point to different functions.

In this regard, PI3K binding to CD28 has been implicated in IL-2 production and rescuing from cell death or apoptosis. Mutation of the Y (i.e., loss of Grb-2 and PI3K binding) and M residues (i.e., loss of PI3K binding alone) interferes with the induction of IL-2 in T-cell hybridomas (Cai et al. 1995; Cefai et al. 1996; Kim et al. 1998; Pages et al. 1994). Further, in vivo reconstitution of Y-FMNM mutants in $CD28^{-/-}$ T cells showed a partial reduction of IL-2 production and graft-versus-host responses (Burr et al. 2001; Harada et al. 2001). The key connection appears to involve PI3K and its activation of the PDK-1/PKB/GSK-3 pathway. Serine/threonine kinases PDK1 and PKB (AKT) carry a pleckstrin homology (PH) domain which binds to phosphatidylinositol-triphosphate ($PtdInsP_3$) at the inner phase of the plasma membrane leading to the activation of these kinases. PDK1 phosphorylates various proteins such as PKC and S6 kinase, whereas the targets of PKB include pro-apoptotic molecules, transcription factors and cell-cycle regulators (Fig. 4). Glycogen synthase kinase-3 (GSK-3), a kinase that is active in resting T cells is inactivated by phosphorylation on serine 9 and 21 by PKB. Inactivation of GSK-3 can lead to prolonged residency of NFAT in the nucleus leading to enhanced IL-2 transcription. However, this pathway has not be studied in detail in the context of CD28 and T cells.

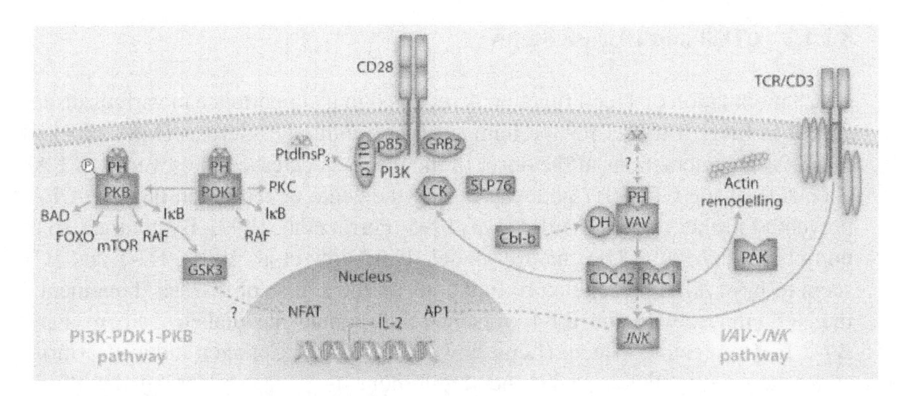

Fig. 4. CD28 pathways: PI3K–PDK-1–PKB and CD28–VAV-1. Binding of PI3K to CD28 generates phosphatidylinositol phosphates that serve as binding sites for PH domains in an array of proteins. This localizes proteins such as inducible T-cell kinase (ITK) to the plasma membrane for interaction with other proteins. One key CD28 signaling pathway is mediated by binding to PI3K. In this scenario, PI3K activation is needed for PDK1 activation leading to the phosphorylation and activation of PKB (AKT). These kinases in turn can phosphorylate multiple downstream targets that are involved in protein translation and cellular metabolism. PKB can also inactivate GSK3 by phosphorylation that can prolong the nuclear residency of NFAT needed for IL-2 production. Another pathway involves the GTP–GDP exchange factor VAV-1 that connects CD28 with the activation of Rac and Cdc42 and the subsequent activation of MEK kinase 1 (MEKK1) and JNK. Activation of Rac, Cdc42, and PAK can also induce cytoskeleton remodeling

CD28 has long been known to rescue cells from TcR-driven antigen-induced cell death (AICD) (Burr et al. 2001; Kirchhoff et al. 2000; Okkenhaug et al. 2001). Apoptosis can be induced by two pathways, via mitochondria associated proteins (i.e., the Bcl family), and death receptors i.e., Fas (CD95)/Fas ligand (FasL (CD95L)) (Fig 1, lower panel). In the context of Fas/FasL, CD28 operates via at least three mechanisms: by decreasing FasL expression, increasing expression of FLICE-inhibitory protein c-FLIP short (Kirchhoff et al. 2000) and by interfering with the formation of the death-inducing signaling complex (DISC) (Jones et al. 2002). T cells expressing active PKB were found to have reduced caspase-8, BID and caspase 3 activation due to the impaired recruitment of caspase 3 to DISC (Jones et al. 2002).

In another pathway, CD28 also plays a role in up-regulating the pro-survival factors Bcl-2 and BcL-XL. CD28-deficient mice reconstituted with a disrupted YMNM motif for PI3K binding were unable to up-regulate Bcl-XL (Okkenhaug et al. 2001). This is most probably due to the inability to recruit and activate PI3K and subsequently activate PKB. Transgenic expression of an activated form of PKB leads to constitutively elevated Bcl-XL expression (Jones et al. 2002). Lastly, the CD28-PI3K-AKT pathway plays an important role in the up-regulation of Glut4 receptor (increase in glucose transport) and glycolysis (Frauwirth et al. 2002).

8.2.1.3 CD28 and Disease States

CD28 modulation of T-cell function is reflected in its importance in various disease states. In EAE, a model for the human disease multiple sclerosis (MS), blockade of CD28/B7 interactions at the onset of the disease decreased the severity of EAE. Blockade of the CD28/B7 pathway during the acute or remission phases of EAE prevented further relapse (Miller et al. 1995; Perrin et al. 1999). By contrast, in the non-obese diabetic (NOD) mouse model (Lenschow et al. 1995), B7-1 and B7-2 seem to have different regulatory functions on the course of disease. Treatment of the NOD mouse with anti-B7-1 worsened the spontaneous diabetes, whereas anti-B7-2 had protective effects (Lenschow et al. 1995). Collagen-induced arthritis, experimental myasthenia gravis and autoimmune uveitis can be inhibited with anti-B7-2 (Salomon and Bluestone 2001).

8.2.2 Inducible T-Cell Costimulator (ICOS)

ICOS is expressed on CD4 and CD8 positive T cells following activation and remains present on effector and memory T cells (Coyle et al. 2000; Yoshinaga et al. 2000). This can be enhanced by CD28 co-ligation (Beier et al. 2000; Coyle et al. 2000; McAdam et al. 2000). Both Th1 and Th2 cells express ICOS during T-cell differentiation, though ICOS levels persists at higher levels on Th2 cells than on Th1 cells (Coyle et al. 2000; McAdam et al. 2000). ICOS is also expressed on activated NK cells, and it promotes NK cell function. Signaling via ICOS provides critical T helper function to B cells. Studies using pathway antagonists, transgenic mice, and knockout mice have revealed the important role of ICOS in B-cell differentiation, immunoglobulin class switching, germinal center formation, and memory B-cell development (Chapoval et al. 2001; Coyle et al. 2000; Hutloff et al. 1999).

Although ICOS and CD28 share an ability to induce IFN-γ, IL-4, and IL-10 production, they differ with respect to the regulation of IL-2 production. Little if any IL-2 is induced with ICOS (Arimura et al. 2002; Gonzalo et al. 2001a,b). In this context, ICOS possesses a YMFM motif that binds PI3K, but interestingly, lacks an intervening asparagine (N) that allows for Grb-2 binding (Coyle et al. 2000; Schneider et al. 1995a). The presence of this residue seems to be important since Jurkat T cells transfected with a mutant form of ICOS containing an asparagine in lieu of the phenylalanine was able to bind Grb-2 and concurrently activated the IL-2 promoter (Harada et al. 2003).

Overall, less is known about ICOS signaling than CD28 signaling. The regulation of Th2 cell function by ICOS has raised the question whether the co-receptor has a special connection with a unique Th2 signaling pathway, or whether it simply adds to the signals generated by CD28. In the latter scenario, the co-receptor would adjust the threshold of signaling by engaging ICOS ligand (B7h, GL50, B7RP-1, LICOS) expressed on APCs. This would produce D-3 lipids and as such should act to recruit PH domains to the membrane (Coyle et al. 2000). On the other hand, it lacks additional proline residues that might provide additional signals. This suggests that ICOS may be more limited in its generation of signals.

8.2.2.1 ICOS and Disease States

Given the involvement of ICOS in B-cell differentiation, immunoglobulin class switching, germinal center formation, and memory B-cell development (Chapoval et al. 2001; Coyle et al. 2000; Hutloff et al. 1999), it is not surprising that it is associated with various disease states. Mice deficient in ICOS show a major reduction in IgE and Th2 cytokine production, as well as in the development of airway hyper-reactivity, which indicates a key role for ICOS in the development of allergic respiratory inflammatory responses (Gonzalo et al. 2001a,b; Umetsu et al. 2002). EAE is a Th1 disease mediated by myelin-specific CD4 positive T cells. ICOS blockade during induction of EAE exacerbated the disease and enhanced IFN-γ production (Rottman et al. 2001).

8.3 Negative Co-stimulatory Molecules

8.3.1 CTLA-4 (CD152)

8.3.1.1 Expression and Function

CTLA-4 is the most extensively characterized inhibitory receptor. Like CD28, it binds to B7-1 (CD80) and B7-2 (CD86) although with significantly higher affinity. Unlike CD28, CTLA-4 binds as a dimer creating a lattice as seen in its crystal structure (Ikemizu et al. 2000; Stamper et al. 2001). Resting T cells express little or no surface CTLA-4, but surface levels are increased upon activation, due to both redistribution of an intracellular pool and increased synthesis. CTLA-4 has potent negative effects on a number of parameters of T-cell function (Bluestone 1995; June et al. 1994). The strong inhibitory role for CTLA-4 is underscored by the phenotype of CTLA-4-deficient mice. They develop a massive lymphoproliferative disorder characterized by polyclonal T-cell proliferation and early lethality (Tivol et al. 1995; Waterhouse et al. 1995). In this way, the co-receptor is thought to modulate the threshold of signals needed for T-cell cytokine production and proliferation (Chambers et al. 1997; Thompson and Allison 1997). With this, CTLA-4 can inhibit T-cell activation by reducing IL-2 production and IL-2 receptor expression, and by arresting T cells at the G1 phase. In addition, CTLA-4 engagement of CD80/86 on dendritic cells can induce the release of indoleamine 2,3-dioxygenase (IDO) (Boasso et al. 2004; Fallarino et al. 2003), which has regulatory effects on T cells that result from tryptophan depletion. However, the degree to which this pathway plays a significant role in CTLA-4 function is uncertain since $IDO^{-/-}$ mice fail to develop auto immunify (Mellor et al. 2003). CD4+CD25+ regulatory T cells can modulate disease in the CTLA-4$^{-/-}$ mouse (Eggena et al. 2004; Fehervari and Sakaguchi 2004; Lohr et al. 2003; Read et al. 2000).

8.3.1.2 Signaling Pathways

Reported mechanisms for CTLA-4 inhibition of T-cell responses include ectodomain competition for CD28 binding to CD80/86 (Masteller et al. 2000), disruption of CD28 localization at the immunological synapse (Pentcheva-Hoang et al. 2004), modulation of TcR signaling by phosphatases SH2-domain containing protein tyrosine phosphatase (SHP-2), and the serine-threonine phosphatase PP2A (Chuang et al. 1999, 2000; Cilio et al. 1998; Lee et al. 1998; Marengere et al.1996) and interference with the expression or composition of lipid rafts on the surface of T cells (Chikuma et al. 2003; Darlington et al. 2002; Martin et al. 2001; Rudd et al. 2002). High avidity CTLA-4 binding to CD80/86 would almost certainly compete for CD28 binding in the limited interface between T-cell and APC. In support of this, expression of tailless CTLA-4 in CTLA-4$^{-/-}$ mice can inhibit lymphocytic infiltration of organs (Masteller et al. 2000). However, the tailless mutant did not prevent lymphoproliferative disease indicating that intracellular signaling also plays a role in altering the threshold of activation. Further, CTLA-4 can inhibit T-cell responses in the absence of CD28 where competition for ligand did not occur (Fallarino et al. 1998; Lin et al. 1998). CTLA-4 function is not entirely cell autonomous as demonstrated in CTLA-4-deficient mice where reconstitution with a combination of deficient and normal T cells failed to develop a lymphoproliferative disease (Bachmann et al. 1999).

A clue to the importance of the cytoplasmic YVKM motif is evident from the swapping of the CD28 SDYMNM motif for the GVYVKM motif in CTLA-4, which can convert the co-receptor into a positive co-stimulatory receptor (Yin et al. 2003). Members of the Tec family or p59fyn can phosphorylate the motif (Miyatake et al. 1998; Schneider et al. 1998). CTLA-4 in turn has been reported to bind two phosphatases, SHP-2 and PP2A (Chuang et al. 2000; Lee et al. 1998; Marengere et al. 1996). SHP-2 binding depends on the YVKM motif (Lee et al. 1998; Marengere et al. 1996), although the interaction is indirect (Schneider and Rudd 2000). SHP-2 has generally been associated with positive signaling (Gadina et al. 1998; Hadari et al. 1998), and in certain systems, the mutation of the tyrosine in the YVKM motif has little effect on function (Baroja et al. 2000; Cinek et al. 2000; Nakaseko et al. 1999). Inhibition has also been observed without detectable SHP-2 binding. Conversely, CTLA-4 has been reported not to inhibit the function of naive CD8+ T cells, even with associated SHP-2 (Perez et al. 1997).

The PP2A binding site requires the YVKM motif (Chuang et al. 2000), and the membrane proximal lysine (K) residues (Baroja et al. 2000). PP2A has a well-established negative effect on differentiation and cell growth (Janssens and Goris 2001). PP2A also targets PKB (AKT), a key component in CD28 co-stimulation (Millward et al. 1999). This fits with the observed inhibition of CD28-PI3K/PDK1/

AKT up-regulation of glucose metabolism by CTLA-4 (Frauwirth et al. 2002). In one model, co-ligation of TcR/CTLA-4 leads to phosphorylation of PP2A that causes its dissociation from CTLA-4 and T-cell inactivation (Chuang et al. 2000; Teft et al. 2005). CD28 can also bind the phosphatase, but with different properties (Chuang et al. 2000). PP2A bound to CD28 in resting cells is enzymatically active, whereas dissociation of PP2A occurs when CD28 is tyrosine phosphorylated (Chuang et al. 2000; Teft et al. 2005). The phosphatase model involving SHP-2 or PP2A would be analogous to killer inhibitory receptors (KIRs) where the associated phosphatase dephosphorylates components needed for signaling. Supporting this, anti-CTLA-4 co-ligation can reduce linker of activation of T cells (LAT) phosphorylation (Lee et al. 1998), and various TcR signaling proteins are hyper-phosphorylated in CTLA-4-deficient mice (Marengere et al. 1996; Lee et al. 1998).

An understanding of CTLA-4 signaling must also take into account its binding to PI3K (Schneider et al. 1995a,b). This implies that CTLA-4 may generate positive as well as negative signals. In agreement, CD4+ T cells deficient in CD80 and CD86 showed an increase in JNK activity (Schneider et al. 2002). At the same time, ERK activity was inhibited indicating that CTLA-4 can differentially regulate members of the MAPK family. The relationship of this finding to CTLA-4 function remains to be established. In this manner, the inhibition of ERK alone could account for negative signaling and CTLA-4 involvement in anergy (Perez et al. 1997). On the other hand, CD4+ T cells from unprimed mice include regulatory suppressor T cells (Tregs). Future studies will be needed to determine whether Tregs differ from other T-cell subsets in their response to CTLA-4. Regardless, the activation of JNK could not account for CTLA-4 induction of TGF-β production (Schneider et al. 2002). Further downstream, CTLA-4 has effects on the regulation of cell cycle, opposing the positive effects of CD28 and inhibiting transcription factors nuclear factor kappa-B (NF-κB), NFAT, and AP-1 (Fraser et al. 1999; Olsson et al. 1999).

Major emphasis has recently been placed on the evaluation of dynamic changes in CTLA-4 expression, trafficking and degradation during T-cell activation. One of the unusual characteristics of CTLA-4 is the fact that it is primarily located in intracellular compartments (Alegre et al. 1996; Leung et al. 1995). Little of the co-receptor can be detected on the surface of cells, even during its optimal expression following T-cell activation. This tight regulation is presumably due to the need to tightly regulate the function of CTLA-4. Two pools of intracellular CTLA-4 have been described, one in the trans Golgi apparatus, and another in the endosomal and lysosomal compartments (Alegre et al. 1996; Iida et al. 2000; Leung et al. 1995; Linsley et al. 1996). TcR ligation and calcium ionophores can induce CTLA-4 release to the cell surface by an unknown mechanism (Linsley et al. 1996). Part of this regulation involves the YVKM motif since its disruption can increase the surface expression of CTLA-4 (Leung et al. 1995; Linsley et al. 1996). Recently, it was found that the adapter protein TRIM (T-cell receptor-interacting molecule) binds to CTLA-4 in the trans Golgi apparatus and promotes transport of CTLA-4 to the surface of T cells (Valk et al. 2006).

The internalization of CTLA-4 is mediated by a combination of tyrosine phosphorylation and the binding of the clathrin adapter complex AP-2 (Yi et al. 2004). Non-phosphorylated CTLA-4 can bind to AP-2 implicated in trafficking and endocytosis of CTLA-4, respectively (Bradshaw et al. 1997; Chuang et al. 1997; Schneider et al. 1999; Shiratori et al. 1997; Zhang and Allison 1997). Phosphorylation of CTLA-4 by the kinase p56lck, p59fyn, and Rlk inhibits these events (Bradshaw et al. 1997; Chuang et al. 1997; Miyatake et al. 1998; Schneider et al. 1998). Once dephosphorylated, the binding to AP-2 regulates its endocytosis (Shiratori et al. 1997; Zhang and Allison 1997). Similar processes occur to intracellular forms of CTLA-4 where AP-1 binding to CTLA-4 mediates its transport from the Golgi apparatus to the lysosomal compartment for degradation (Schneider et al. 1999).

Engagement of the TcR with antigen involves entry and aggregation of signaling proteins (i.e., LAT, PKCθ) into membrane compartments known as lipid rafts (Janes et al. 2000; Pearse and Robinson 1990; Samelson 2002; Xavier et al. 1998) and the formation of an immunological synapse (IS) between the T-cell and an APC. The IS is a characteristic structure in which an external leukocyte function-associated antigen-1 (LFA-1) ring (peripheral SMAC, pSMAC) surrounds central TcR clusters (central SMAC, cSMAC) (Dustin et al. 2001; Kupfer and Kupfer 2003). Viola and colleagues first demonstrated that CD28 coligation promotes the surface expression of lipid rafts (Viola et al. 1999), whereas Martin et al. showed that CTLA-4 can inhibit raft expression (Martin et al. 2001). Darlington et al. reported that CTLA-4 is recruited to lipid rafts during negative signaling and accumulates at the IS (Darlington et al. 2002). Further, recent work by Egen and Allison has suggested that the level of TcR signaling is directly related to the amount of CTLA-4 accumulated in the IS (Egen and Allison 2002). The level of CTLA-4 expression was increased in the IS under stronger T-cell activation and the majority of CTLA-4 in the cytoplasm was localized beneath the T-cell/APC contact site. Crystal structures of B7-ligated CTLA-4 suggest that it may form lattices within the IS (Schwartz et al. 2001; Stamper et al. 2001).

On another level, CTLA-4 can also influence the function of integrins such as LFA-1. Integrins play key roles in regulating T cell-APC conjugation and the migration of T cells (Hogg et al. 2002; Takagi and Springer 2002). A recent report by Schneider et al. have shown that CTLA-4 can generate inside-out signals that potently up-regulate LFA-1 clustering and adhesion on the surface of T cells (Schneider et al. 2005). The up-regulation of adhesion was observed under the same conditions that led to an inhibition of IL-2 production. The importance of the connection between CTLA-4 and LFA-1 adhesion was underscored by the fact that the loss of CTLA-4 expression in CTLA-4-deficient T cells resulted in an impairment of TcR/CD3 induced LFA-1 adhesion and clustering. This upregulation of adhesion was dependent on GTPase Rap-1 activation. Rap-1 is an allosteric regulatory element, switching between inactive GDP-bound and active GTP-bound conformations (Bos et al. 2003; Caron 2003). CTLA-4 activated Rap-1 at levels that were higher than those induced with anti-CD3 stimulation. The role for CTLA-4 in LFA-1 mediated adhesion provides an alternate route through which CTLA-4 may modulate T-cell immunity.

8.3.1.3 CTLA-4 and Disease States

Blockade of CTLA-4 has been employed in the therapeutic modulation of autoimmunity, transplantation and tumor immunotherapy (Hurwitz et al. 2002; Egen et al. 2002; Salomon and Bluestone 2001). CTLA-4 polymorphisms have been genetically linked to a number of human diseases (Kristiansen et al. 2000). Splice variants and mutations within the CTLA-4 gene have been identified as candidates for susceptibility to Grave's disease, autoimmune hypothyroidism and type 1 diabetes (Ueda et al. 2003). In the NOD mouse, disease susceptibility is associated with a reduced production of a splice form encoding a molecule lacking the ligand-binding domain (ligand independent CTLA-4; liCTLA-4) (Ueda et al. 2003). LiCTLA-4 is expressed in resting T cells and rapidly down-regulated during T-cell activation (Vijayakrishnan et al. 2004). Taken together, these findings suggest that genetic differences in CTLA-4 can contribute to the development of autoimmune diseases.

8.3.2 Programmed Death-1 (PD-1)

Another inhibitory member of the CD28 family is PD-1. As its name reflects, the PD-1 receptor was initially identified by subtractive hybridization studies using a T-cell hybridoma undergoing programmed cell death (Agata et al. 1996). Subsequent studies have questioned a direct role for PD-1 in cell death (Agata et al. 1996; Nishimura et al. 1996; Vibhakar et al. 1997). PD-1 is a 55 kDa transmembrane protein that shares 24% amino acid homology in the extracellular domain to CTLA-4 (Ishida et al. 1992). It binds the distinct ligands PD-L1 and PD-L2 and is expressed on CD4 CD8 double negative cells and double negative $\gamma\delta$ thymocytes (Nishimura et al. 2000). It can also be induced on activated human CD4 and CD8 positive T cells, B cells and myeloid cells. This contrasts with the more restricted expression of CD28 and CTLA-4.

Engagement of PD-1 by PD-L1 or PD-L2 inhibits TcR-mediated proliferation and cytokine production by previously activated T cells. PD-L1 and PD-L2 strongly inhibit both T -cell proliferation and cytokine production at low antigen concentrations, even in the presence of strong B7-CD28 signals. By contrast, at high antigen concentrations, PD-L1 and PD-L2 do not inhibit T-cell proliferation but markedly reduce the production of multiple cytokines (Sharpe and Freeman 2002). Interestingly, there are also studies reporting that PD-L1 and PD-L2 can provide a co-stimulatory signal to suboptimally activated T cells (Tamura et al. 2001; Dong et al. 1999; Tseng et al. 2001).

8.3.2.1 Signaling Pathways

The cytoplasmic tail of PD-1 contains two tyrosines, one that constitutes an immunoreceptor tyrosine-based inhibition motif (ITIM) (Fig. 5). The ITIM sequence is

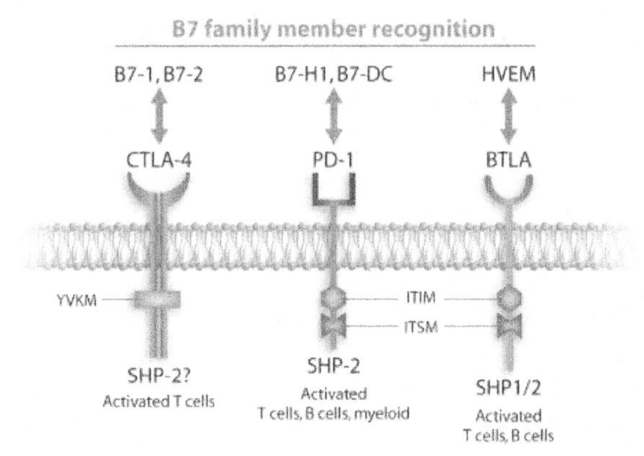

Fig. 5. Inhibitory co-receptors with binding sites for the phosphatases SHP-1 and SHP-2. Ligation-induced phosphorylation of immunoreceptor tyrosine-based inhibition motif (*ITIM*) and immunoreceptor tyrosine-based switch motif (*ITSM*) in the cytoplasmic tails of programmed death-1 (*PD-1*) and B and T lymphocyte attenuator (*BTLA*) results in the recruitment of phosphatases SHP-1 and SHP-2. Although cytotoxic T-cell antigen (*CTLA-4*) does not possess an ITIM or ITSM motif, it may indirectly bind to SHP-2 via its YVKM motif. Binding to these phosphatases leads to dephosphorylation of proximal signaling molecules

found in several classes of inhibitory receptors, including the killer inhibitory receptors found on NK cells and FcγRIIB on B cells, and functions by recruiting SH2-containing phosphatases (Vivier and Daeron 1997). The other tyrosine belongs to an immunoreceptor tyrosine-based switch motif (ITSM) (Shlapatska et al. 2001). Although both tyrosines are phosphorylated following PD-1 engagement, mutagenesis studies indicate that the tyrosine within the ITSM motif may be required for the inhibitory activity of PD-1, as opposed to the tyrosine in the ITIM that is more typically associated with inhibitory signaling (Okazaki et al. 2001). In T cells, phosphorylated ITSM recruits SHP-2 and SHP-1, whereas in B cells, only SHP-2 was recruited. In vitro studies have demonstrated that PD-L1 and PD-L2 can inhibit T-cell proliferation and cytokine production, others showed that the PD-1 ligands enhance T-cell activation. In addition to the possibility that there may be another ligand for PD-1, the ITSM motif might be relevant to the function of PD-1 ligands to deliver either positive or negative signals, dependent on the activation of the T-cell and therefore interaction with different phosphatases.

8.3.2.2 PD-1 and Disease States

The loss of negative regulation by PD-1 leads to the development of autoimmune diseases, and the genetic background of mice greatly affects the autoimmune phenotypes. PD-1-deficient mice of the C57BL/6 strain develop a late onset progressive arthritis and lupus-like glomerulonephritis with high levels of IgG3 deposition (Nishimura et al. 1999). By contrast, PD-1-deficient mice of the Balb/c strain develop an early onset dilated cardiomyopathy with 50% mortality at 20 weeks of age. The affected hearts do not have IgG and complement deposition. Affected mice develop high titers of an autoantibody that recognizes a heart-specific 33 kDa protein on the surface of cardiomyocytes, but do not have other common autoantibodies (Nishimura et al. 2001).

Experiments in PD-1-deficient mice have also revealed a role for PD-1 in regulating CD8 positive T-cell responses. 2C-TCR transgenic T cells recognize cells bearing H-2L^d, an MHC class I molecule expressed in Balb/c and B10.D2 mice. $2C^+$ autoreactive T cells are negatively selected in the thymus and only a few 2C-TCR positive cells are found in the periphery, where they do not become activated or cause disease (Nishimura et al. 1999, 2000). In the absence of the PD-1 gene, however, these cells become activated and cause a lethal graft-versus-host disease. These results indicate that potentially autoreactive T cells in the periphery can become activated in the absence of PD-1 and implicate PD-1 in regulating peripheral T and/or B-cell tolerance (Sharpe and Freeman 2002).

8.3.3 T-Cell Immunoglobulin and Mucin-Domain-Containing Molecule-3 (Tim-3)

Tim-3 is another co-receptor with an inhibitory effect on immune responses, but via a fundamentally different mechanism. It is a type 1 membrane protein consisting of a single IgV-like domain and a highly glycosylated mucin-containing domain (Monney et al. 2002). Galectin-9, a member of the galectin family that is expressed on lymphocytes and other cell types, has been identified as a ligand for Tim-3. The first Tim family member to be identified was the human hepatitis A virus cell receptor, which exhibits 39% homology to Tim-3 and is now termed Tim-1 (Feigelstock et al. 1998; Kaplan et al. 1996). Initial reports suggested that Tim-1 acts as a positive co-stimulatory molecule by stimulating cytokine production from Th2 cells without altering cytokine production from Th1 cells (Umetsu et al. 2005). The functional characterization of Tim-3 has identified its role in regulating the Th1 immune response. The Tim-3/galectin-9 pathway controls Th1 immunity by selective deletion of Tim-3 positive Th1 cells. Galectin-9 is highly expressed in the naïve immune system. This serves to inhibit the generation of Th1 responses, whereas loss of galectin-9 after activation allows the generation of Th1 cells. Although galectin-9 can induce both necrosis and apoptosis, it is believed that Tim-3/galectin-9 negatively regulate effector Th1 cells (Zhu et al. 2005).

8.3.3.1 Tim-3 and Disease States

In this context, the Tim-3/galectin pathway has also been shown to have a crucial role in the induction of peripheral tolerance. Tim-3 Ig treatment or the use of

Tim-3-deficient mice similarly prevented tolerance induction to MHC-mismatched allografts. Blockade of the pathway by treatment with anti-Tim-3 antibody or Tim-3Ig fusion proteins led to increased Th1 cell proliferation and cytokine responses, in addition to the loss of tolerance induction. Th1 cells are associated with the development of EAE, whereas the secretion of Th2 cytokines has been associated with regulation and disease recovery. The onset and progression of this autoimmune disease is dependent on a balance between the two responses. Tim-3-deficient animals also have increased severity in models of experimental autoimmune diabetes. Interestingly, Tim-3-deficient mice have increased lung inflammatory responses following allergen exposure (Coyle and Gutierrez-Ramos 2004).

8.3.4 B and T Lymphocyte Attenuator (BTLA)

BTLA is the most recently recognized member of the CD28 family. It is a type I transmembrane glycoprotein with an extracellular single IgV-like domain, a transmembrane and cytoplasmic domain. The latter contains three tyrosine residues which are part of tyrosine-based motifs (Grb-2 binding site, ITIM, ITSM) (Carreno and Collins 2003; Watanabe et al. 2003). BTLA interacts with the ligand HVEM, a member of the TNFR superfamily, and not with B7x as originally thought (Sedy et al. 2005; Watanabe et al. 2003). It is induced on T cells during activation, and remains expressed on Th1 but not Th2 cells. This suggests that BTLA may specifically down-regulate Th1-mediated inflammatory responses. In this way, the expression pattern contrasts with ICOS that remains elevated on Th2 cells, but is down-regulated on Th1 cells (Watanabe et al. 2003). BTLA also inhibits cytotoxic T lymphocyte (CTL) maturation and proliferation (Sica et al. 2003). T cells from BTLA-deficient mice have increased proliferation in response to antigen, and B cells from these mice have an increased response to anti-IgM. BTLA-deficient mice also show increased susceptibility to peptide antigen-induced EAE (Watanabe et al. 2003). BTLA-deficient mice have a more subtle phenotype than CTLA-4-deficient mice. One possibility for the lack of an obvious phenotype in BTLA-deficient mice compared to CTLA-4-deficient mice is that BTLA and PD-1 may have overlapping functions. Whether PD-1 and BTLA use similar or distinct mechanisms to block T-cell activation is less clear. Both PD-1 and BTLA have related and relatively long cytoplasmic tails that consist of an ITIM followed by a distal ITSM. BTLA also has a membrane-proximal Grb-2 binding site, but the functional importance of this motif is unknown since only mutation of the ITIM or ITSM blocked the association of SHP-1 and SHP-2 to the BTLA cytoplasmic tail (Gavrieli et al. 2003).

8.4 Conclusion

Signals provided by co-receptors act to fine-tune the response of T cells to antigen. The ability of co-receptors to modulate the level of intracellular signaling determines the degree to which the antigen-receptor complex can recognize self versus nonself. These signals can either up-regulate or inhibit the immune response, depending on the activation status of the cell and its location in the peripheral immune system. The ligands for different co-receptors are expressed on cells in different regions of the peripheral compartment. Although many co-signaling events have been characterized, there remains much to be discovered regarding the entire range of their biological functions. A greater understanding will come as new techniques and approaches are applied to the field. New imaging technologies using new fluorescent labels and sensors and the use of more sophisticated computer software for image acquisition will further deepen our understanding of the assembly and expression of these co-receptors and their intracellular mediators. Such studies should lead to a better understanding of co-receptor functions and the development of new therapeutic targets for the treatment of infectious diseases, immunodeficiency disorders, autoimmunity and cancer.

References

Acuto O, Michel F (2003) CD28-mediated co-stimulation: a quantitative support for TCR signaling. Nat Rev Immunol 3:939–951

Acuto O, Mise-Omata S, Mangino G, Michel F (2003) Molecular modifiers of T-cell antigen receptor triggering threshold: the mechanism of CD28 costimulatory receptor. Immunol Rev 192:21–31

Agata Y, Kawasaki A, Nishimura H, Ishida Y, Tsubata T, Yagita H, Honjo T (1996) Expression of the PD-1 antigen on the surface of stimulated mouse T and B lymphocytes. Int Immunol 8:765–772

Alegre ML, Noel PJ, Eisfelder BJ, Chuang E, Clark MR, Reiner SL, Thompson CB (1996) Regulation of surface and intracellular expression of CTLA-4 on mouse T cells. J Immunol 157:4762–4770

Arimura Y, Kato H, Dianzani U, Okamoto T, Kamekura S, Buonfiglio D, Miyoshi-Akiyama T, Uchiyama T (2002) A co-stimulatory molecule on activated T cells, H4/ICOS, delivers specific signals in Th cells and regulates their responses. Int Immunol 14:555–566

August A, Dupont B (1994) CD28 of T lymphocytes associates with phosphatidylinositol 3-kinase. Int Immunol 6:769–774

Bachmann MF, Köhler G, Ecabert B, Mak TW, Kopf M (1999) Cutting Edge: Lymphoproliferative disease in the absence of CTLA-4 is not cell autonomous. J Immunol 163:1128–1131

Baroja ML, Luxenberg D, Chau T, Ling V, Strathdee CA, Carreno BM, Madrenas J (2000) The inhibitory function of CTLA-4 does not require its tyrosine phosphorylation. J Immunol 164:49–55

Beier KC, Hutloff A, Dittrich AM, Heuck C, Rauch A, Buchner K. Ludewig B, Ochs HD, Mages HW, Kroczek RA (2000) Induction, binding specificity and function of human ICOS. Eur J Immunol 30:3707–3717

Bluestone J (1995) New perspectives of CD28-B7 mediated T-cell costimulation. Immunity 2, 555–559

Boasso A, Herbeuval JP, Hardy AW, Winkler C, Shearer GM (2004) Regulation of indoleamine 2,3-dioxygenase and tryptophanyl-tRNA-synthetase by CTLA-4-Fc in human CD4+ T cells. Blood 105:1574–1581

Bos JL, de Bruyn K, Enserink J, Kuiperij B, Rangarajan S, Rehmann H, Riedl J, de Rooij J, van Mansfeld F, Zwartkruis F (2003) The role of Rap 1 in integrin-mediated cell adhesion. Biochem Soc Trans 31:83–86

Bradshaw JD, Lu P, Leytze G, Rodgers J, Schieven GL, Bennett KL, Linsley PS, Kurtz SE (1997) Interaction of the cytoplasmic tail of CTLA-4 (CD152) with a clathrin-associated protein is negatively regulated by tyrosine phosphorylation. Biochemistry 36:15975–15982

Bretscher PA (1999) A two-step, two-signal model for the primary activation of precursor helper T cell. Proc Natl Acad Sci USA 96:185–190

Burr JS, Savage ND, Messah GE, Kimzey SL, Shaw AS, Arch RH, Green JM (2001) Cutting edge: distinct motifs within CD28 regulate cell proliferation and induction of Bcl-XL. J Immunol 166:5331–5335

Cai YC, Cefai D, Schneider H, Raab M, Nabavi N, Rudd CE (1995) Selective CD28pYMNM mutations implicate phosphatidylinositol 3-kinase in CD86-CD28-mediated costimulation. Immunity 3:417–426

Carreno BM, Collins M (2003) BTLA: a new inhibitory receptor with a B7-like ligand. Trends Immunol 24:524–527

Caron E (2003) Cellular functions of the Rap1 GTP-binding protein: a pattern emerges. J Cell Sci 116:435–440

Cefai D, Cai YC, Hu H, Rudd CE (1996) CD28 co-stimulatory regimes differ in their dependence on phosphatidylinositol-3 kinase: common cosignals induced by CD80 and CD86. Int Immunol 8:1609–1616

Chambers CA, Sullivan TJ, Allison JP (1997) Lymphoproliferation in CTLA-4-deficient mice is mediated by costimulation-dependent activation of CD4+ T cells. Immunity 7:885–895

Chapoval AI (2001) B7-H3: a costimulatory molecule for T-cell activation and IFN-gamma production. Nat Immunol 2:269–274

Chikuma S, Imboden JB, Bluestone JA (2003) Negative regulation of T-cell receptor-lipid raft interaction by cytotoxic T lymphocyte-associated antigen 4. J Exp Med 197:129–135

Chuang E, Alegre ML, Duckett CS, Noel PJ, Vander Heiden MG, Thompson CB (1997) Interaction of CTLA-4 with the clathrin-associated protein AP50 results in ligand-independent endocytosis that limits cell surface expression. J Immunol 159:144–151

Chuang E, Lee KM, Robbins MD, Duerr JM, Alegre ML, Hambor JE, Neveu MJ, Bluestone JA, Thompson CB (1999) Regulation of cytotoxic T lymphocyte-associated molecule-4 by src kinases. J Immunol 162:1270–1277

Chuang E, Fisher TS, Morgan RW, Robbins MD, Duerr MG, Vander Heiden JP, Gardner JE, Hambor MJ (2000) The CD28 and CTLA-4 receptors associate with the serine/threonine phosphatase PP2A. Immunity 13:313–322

Cilio CM, Daws MR, Malashicheva A, Sentman CL, Holmberg D (1998) Cytotoxic T lymphocyte antigen 4 is induced in the thymus upon in vivo activation and its blockade prevents anti-CD3-mediated depletion of thymocytes. J Exp Med 188:1239–1246

Cinek T, Sadra A, Imboden JB (2000) Cutting edge: tyrosine-independent transmission of inhibitory signals by CTLA-4. J Immunol 164:5–8

Coyle AJ, Gutierrez-Ramos JC (2004) The role of ICOS and other costimulatory molecules in allergy and asthma. Springer Semin Immun 25:349–359

Coyle AJ, Lehar S, Lloyd C, Tian J, Delaney T, Manning S, Nguyen T, Burwell T, Schneider H, Gonzalo JA, et al (2000) The CD28-related molecule ICOS is required for effective T cell-dependent immune responses. Immunity 13:95–105

Darlington PJ, Baroja ML, Chau TA, Siu E, Ling V, Carreno BM, Madrenas J (2002) Surface cytotoxic T lymphocyte-associated antigen 4 partitions within lipid rafts and relocates to the immunological synapse under conditions of inhibition of T-cell activation. J Exp Med 195:1337–1347

Dong H, Zhu G, Tamada K, Chen L (1999) B7-H1, a third member of the B7 family, co-stimulates T-cell proliferation and interleukin-10 secretion. Nat Med 5:1365–1369

Dong C, Juedes AE, Temann UA, Shresta S, Allison JP, Ruddle NH, Flavell RA (2001) ICOS co-stimulatory receptor is essential for T-cell activation and function. Nature 409:97–101

Dustin M, Allen PM, Shaw AS (2001) Environmental control of immunological synapse formation and duration. Trends Immunol 22:192–194

Egen JG, Allison JP (2002) Cytotoxic T lymphocyte antigen-4 accumulation in the immunological synapse is regulated by TCR signal strength. Immunity 16:23–35

Egen JG, Kuhns MS, Allison JP (2002) CTLA-4: new insights into its biological function and use in tumor immunotherapy. Nat Immunol 3:611–618

Eggena MP, Walker LS, Nagabhushanam V, Barron L, Chodos A, Abbas AK (2004) Cooperative roles of CTLA-4 and regulatory T cells in tolerance to an islet cell antigen. J Exp Med 199:1725–1730

Fallarino F, Fields PE, Gajewski TF (1998) B7-1 engagement of cytotoxic T lymphocyte antigen 4 inhibits T-cell activation in the absence of CD28. J Exp Med 188:205–210

Fallarino F, Grohmann U, Hwang KW, Orabona C, Vacca C, Bianchi R, Belladonna M L, Fioretti MC, Alegre ML, Puccetti P (2003) Modulation of tryptophan catabolism by regulatory T cells. Nat Immunol 4:1206–1212

Fehervari Z, Sakaguchi S (2004) CD4 (+) Tregs and immune control. J Clin Invest 114: 1209–1217

Feigelstock D, Thompson P, Mattoo P, Zhang Y, Kaplan GG (1998) The human homolog of HAVcr-1 codes for a hepatitis A virus cellular receptor. J Virol 72:6621–6628

Ferguson SE, Han S, Kelsoe G, Thompson CB (1996) CD28 is required for germinal center formation. J Immunol 156:4576–4581

Fraser JH, Rincon M, McCoy KD, Le Gros G (1999) CTLA-4 ligation attenuates AP-1, NFAT and NF-kappaB activity in activated T cells. Eur J Immunol 29:838–844

Frauwirth KA, Riley JL, Harris MH, Parry RV, Rathmell JC, Plas DR, Elstrom RL, June CH, Thompson CB (2002) The CD28 signaling pathway regulates glucose metabolism. Immunity 16:769–777

Freeman GJ, Long AJ, Iwai Y, Bourque K, Chernova T, Nishimura H, Fitz LJ, Malenkovich N, Okazaki T, Byrne MC, et al (2000) Engagement of the PD-1 immunoinhibitory receptor by a novel B7 family member leads to negative regulation of lymphocyte activation. J Exp Med 192:1027–1034

Gadina M, Stancato LM, Bacon CM, Larner AC, O'Shea JJ (1998) Involvement of SHP-2 in multiple aspects of IL-2 signaling: evidence for a positive regulatory role. J Immunol 160:4657–4661

Gavrieli M, Watanabe N, Loftin SK, Murphy TL, Murphy KM (2003) Characterization of phosphotyrosine binding motifs in the cytoplasmic domain of B and T lymphocyte attenuator required for association with protein tyrosine phosphatases SHP-1 and SHP-2. Biochem Biophys Res Commun 312:1236–1243

Gonzalo JA, Delaney T, Corcoran J, Gutierrez-Ramos JC, Coyle AJ (2001a) Cutting edge: the related molecules CD28 and inducible costimulator deliver both unique and complementary signals required for optimal T-cell activation. J Immunol 166:1–5

Gonzalo JA, Tian J, Delaney T, Corcoran J, Rottman JB, Lora J, Al-garawi A, Kroczek R, Gutierrez-Ramos JC, Coyle AJ (2001b) ICOS is critical for T helper cell-mediated lung mucosal inflammatory responses. Nat Immunol 2:597–604

Greenwald RJ, Freeman GJ, Sharpe AH (2005) The B7 family revisited. Annu Rev Immunol 23:515–548

Hadari YR, Kouhara H, Lax I, Schlessinger J (1998) Binding of Shp2 tyrosine phosphatase to FRS2 is essential for fibroblast growth factor-induced PC12 cell differentiation. Mol Cell Biol 18:3966–3973

Harada Y, Tokushima M, Matsumoto Y, Ogawa S, Otsuka M, Hayashi K, Weiss BD, June CH, Abe R (2001) Critical requirement for the membrane-proximal cytosolic tyrosine residue for CD28-mediated costimulation in vivo. J Immunol 166:3797–3803

Harada Y, Ohgai D, Watanabe R, Okano K, Koiwai O, Tanabe K, Toma H, Altman A, Abe R (2003) A single amino acid alteration in cytoplasmic domain determines IL-2 promoter activation by ligation of CD28 but not inducible costimulator (ICOS). J Exp Med 197:257–262

Hogg N, Henderson R, Leitinger B, McDowall A, Porter J, Stanley P (2002) Mechanisms contributing to the activity of integrins on leukocytes. Immunol Rev 186:164–171

Holdorf AD, Green JM, Levin SD, Denny MF, Straus DB, Link V, Changelian PS, Allen PM, Shaw AS (1999) Proline residues in CD28 and the Src homology (SH)3 domain of Lck are required for T-cell costimulation. J Exp Med 190:375–384
Hurwitz AA, Sullivan TJ, Sobel RA, Allison JP (2002) Cytotoxic T lymphocyte antigen-4 (CTLA-4) limits the expansion of encephalitogenic T cells in experimental autoimmune encephalomyelitis (EAE)-resistant BALB/c mice. Proc Natl Acad Sci USA 99:3013–3017
Hutloff A, Dittrich AM, Beier KC, Eljaschewitsch B, Kraft R, Anagnostopoulos I, Kroczek RA (1999) ICOS is an inducible T-cell co-stimulator structurally and functionally related to CD28. Nature 397:263–266
Iida T, Ohno H, Nakaseko C, Sakuma M, Takeda-Ezaki M, Arase H, Kominami E, Fujisawa T, Saito T (2000) Regulation of cell surface expression of CTLA-4 by secretion of CTLA-4-containing lysosomes upon activation of CD4+ T cells. J Immunol 165:5062–5068
Ikemizu S, Gilbert RJ, Fennelly JA, Collins AV, Harlos K, Jones EY, Stuart DI, Davis SJ (2000) Structure and dimerization of a soluble form of B7-1. Immunity 12:51–60
Ishida Y, Agata Y, Shibahara K, Honjo T (1992) Induced expression of PD-1, a novel member of the immunoglobulin gene superfamily, upon programmed cell death. EMBO J 11:3887–3895
Janes PW, Ley SC, Magee AI, Kabouridis PS (2000) The role of lipid rafts in T-cell antigen receptor (TCR) signalling. Semin Immunol 12:23–34
Janssens V, Goris J (2001) Protein phosphatase 2A: a highly regulated family of serine/threonine phosphatases implicated in cell growth and signaling. Biochem J 353:417–439
Jones RG, Elford AR, Parsons MJ, Wu L, Krawczyk CM, Yeh WC, Hakem R, Rottapel R, Woodgett JR, Ohashi PS (2002) CD28-dependent activation of protein kinase B/Akt blocks Fas-mediated apoptosis by preventing death-inducing signaling complex assembly. J Exp Med 196:335–348
June CH, Bluestone JA, Nadler LM, Thompson CB (1994) The B7 and CD28 receptor families. Immunol Today 15:321–331
Kaplan G, Totsuka A, Thompson P, Akatsuka T, Moritsugu Y, Feistone SM (1996) Identification of a surface glycoprotein on African green monkey kidney cells as a receptor for hepatitis A virus. EMBO J 15:4282–4296
Kim H-H, Tharayil M, Rudd CE (1998) Growth factor receptor-bound protein 2 SH2/SH3 domain binding to CD28 and its role in co-signaling. J Biol Chem 273:296–301
Kirchoff S, Muller WW, Li-Weber M, Krammer PH (2000) Up-regulation of c-FLIPshort and reduction of activation-induced cell death in CD28-costimulated human T cells. Eur J Immunol 30:2765–2774
Kristiansen OP, Larsen ZM, Pociot F (2000) CTLA-4 in autoimmune diseases—a general susceptibility gene to autoimmunity? Genes and Immunity 1:170–184
Kündig TM, Shahinian A, Kawai K, Mittrücker, HW, Sebzda E, Bachmann MF, Mak TW, Ohashi PS (1996) Duration of TCR stimulation determines costimulatory requirement of T cells. Immunity 5:41–52
Kupfer A, Kupfer H (2003) Imaging immune cell interactions and functions: SMACs and the immunological synapse. Semin Immunol 15:295–300
Latchman Y, Wood CR, Chernova T, Chaudhary D, Borde M, Chernova I, Iwai Y, Long AJ, Brown JA, Nunes R, et al. (2001) PD-L2 is a second ligand for PD-1 and inhibits T-cell activation. Nat Immunol 2:261–268
Lee KM, Chuang E, Griffin M, Khattri R, Hong DK, Zhang W, Straus D, Samelson LE, Thompson CB, Bluestone JA (1998) Molecular basis of T-cell inactivation by CTLA-4. Science 282:2263–2266
Lenschow DJ, Ho SC, Sattar H, Rhee L, Gray G, Nabavi N, Herold KC, Bluestone JA (1995) Differential effects of anti-B7-1 and anti-B7-2 monoclonal antibody treatment on the development of diabetes in the Nonobese Diabetic mouse. J Exp Med 181:1145–1155
Leung HT, Bradshaw J, Cleaveland JS, Linsley PS (1995) Cytotoxic T lymphocyte-associated molecule-4, a high avidity receptor for CD80 and CD86, contains an intracellular localization motif in its cytoplasmic tail. J Biol Chem 270:25107–25114

Lin H, Rathmell JC, Gray GS, Thompson CB, Leiden JM, Alegre ML (1998) Cytotoxic T lymphocyte antigen 4 (CTLA4) blockade accelerates the acute rejection of cardiac allografts in CD28-deficient mice: CTLA4 can function independently of CD28. J Exp Med 188:199–204

Ling V, Wu PW, Finnerty HF, Bean KM, Spaulding V, Fouser LA, Leonard JP, Hunter SE, Zollner R, Thomas JL, et al (2000) Cutting edge: identification of GL50, a novel B7-like protein that functionally binds to ICOS receptor. J Immunol 164:1653–1657

Linsley PS, Bradshaw J, Greene J, Peach R, Bennett KL, Mittler RS (1996) Intracellular trafficking of CTLA-4 and focal localization towards sites of TCR engagement. Immunity 4:535–543

Lohr J, Knoechel B, Jiang S, Sharpe AH, Abbas AK (2003) The inhibitory function of B7 costimulators in T-cell responses to foreign and self-antigens. Nat Immunol 4:664–669

Marengere LEM, Waterhouse P, Duncan GS, Mittrucker HW, Feng GS, Mak TW (1996) Regulation of T-cell receptor signaling by tyrosine phosphatase Syp association with CTLA-4. Science 272:1170–1173

Marengere LEM, Okkenhaug K, Clavreul A, Couez D, Gibson S, Mills GB, Mak TW, Rottapel R (1997) The SH3 domain of Itk/Emt binds to proline-rich sequences in the cytoplasmic domain of the T-cell costimulatory receptor CD28. J Immunol 159:3220–3229

Martin M, Schneider H, Azouz A, Rudd CE (2001) Cytotoxic T lymphocyte antigen 4 potently inhibits cell surface raft expression in its regulation of T-cell function. J Exp Med 194: 1675–1681

Masteller EM, Chuang E, Mullen AC, Reiner SL, Thompson CB (2000) Structural analysis of CTLA-4 function in vivo. J Immunol 164:5319–5327

McAdam AJ, Chang TT, Lumelsky AE, Greenfield EA, Boussiotis VA, Duke-Cohan JS, Chernova T, Malenkovich N, Jabs C, Kuchroo VK, et al (2000) Mouse inducible costimulatory molecule (ICOS) expression is enhanced by CD28 costimulation and regulates differentiation of CD4+ T cells. J Immunol 165:5035–5040

McAdam AJ, Greenwald RJ, Levin MA, Chernova T, Malenkovich N, Ling V, Freeman GJ, Sharpe AH (2001) ICOS is critical for CD40-mediated antibody class switching. Nature 409:102–105

Mellor AL, Baban B, Chandler P, Marshall B, Jhaver K, Hansen A, Koni PA, Iwashima M, Munn DH (2003) Cutting edge: induced indoleamine 2,3 dioxygenase expression in dendritic cell subsets suppresses T-cell clonal expansion. J Immunol 171:1652–1655

Meyers JH, Sabatos CA, Chakravarti S, Kuchroo VK (2005) The *TIM* gene family regulates autoimmune and allergic diseases. Trends Mol Med 11:362–369

Miller SD, Vanderlugt CL, Lenschow DJ, Pope JG, Karandikar NJ, Dal Canto MC, Bluestone JA (1995) Blockade of CD28/B7-1 interaction prevents epitope spreading and clinical relapses of murine EAE. Immunity 3:739–745

Millward TA, Zolnierowicz S, Hemmings BA (1999) Regulation of protein kinase cascades by protein phosphatase 2A. Trends Biochem Sci 24:186–191

Miyatake S, Nakaseko C, Umemori H, Yamamoto T, Saito T (1998). Src family tyrosine kinases associate with and phosphorylate CTLA-4 (CD152) Biochem Biophys Res Commun 249:444–448

Monney L, Sabatos CA, Gaglia JL (2002) Th1-specific cell surface protein Tim-3 regulates macrophage activation and severity of an autoimmune disease. Nature 415:536–541

Nakaseko C, Miyatake S, Iida T, Hara S, Abe R, Ohno H, Saito Y, Saito T (1999) Cytotoxic T lymphocyte antigen 4 (CTLA-4) engagement delivers an inhibitory signal through the membrane-proximal region in the absence of the tyrosine motif in the cytoplasmic tail. J Exp Med 190:765–774

Nishimura H, Honjo T (2001) PD-1: an inhibitory immunoreceptor involved in peripheral tolerance. Trends Immunol 22:265–268

Nishimura H, Agata Y, Kwasaki A, Sato M, Imamura S, Minato M, Yagita H, Nakano T, Honjo T (1996) Developmentally regulated expression of the PD-1 protein on the surface of double-negative (CD4-CD8) thymocytes. Int Immunol 8:773–780

Nishimura H, Nose M, Hiai H, Minato N, Honjo T (1999) Development of lupus-like autoimmune diseases by disruption of the PD-1 gene encoding an ITIM motif-carrying immunoreceptor. Immunity 11:141–151

Nishimura H, Honjo T, Minato N (2000) Facilitation of b selection and modification of positive selection in the thymus of PD-1-deficient mice. J Exp Med 191:891–8998

Okazaki T, Maeda A, Nishimura H, Kurosaki T, Honjo T (2001) PD-1 immunoreceptor inhibits B-cell receptor-mediated signalling by recruiting src homology 2-domain-containing tyrosine phosphatase 2 to phosphotyrosine. Proc Natl Acad Sci USA 98:13866–13871

Okkenhaug K, Rottapel R (1998) Grb-2 forms an inducible protein complex with CD28 through a Src homology 3 domain-proline interaction. J Biol Chem 273:21194–21202

Okkenhaug K, Wu L, Garza KM, LaRose J, Khoo W, Odermatt B, Mak TW, Ohashi PS, Rottapel R (2001) A point mutation in CD28 distinguishes proliferative signals from survival signals. Nat Immunol 2:325–332

Olsson C, Riebeck K, Dohlsten M, Michaëlsson E (1999) CTLA-4 ligation suppresses CD28-induced NF-κB and AP-1 activity in mouse T-cell blasts. J Biol Chem 274:14400–14405

Pages F, Ragueneau M, Rottapel R, Truneh A, Nunes J, Imbert J, Olive D (1994) Binding of phosphatidylinositol-3-OH kinase to CD28 is required for T-cell signaling. Nature 369:327–329

Pawson T, Gish GD, Nash P (2001) SH2 domains, interaction modules and cellular wiring. Trends Cell Biol 11:504–511

Pearse BM, Robinson MS (1990) Clathrin, adaptors, and sorting. Annu Rev Cell Biol 6:151–171

Pentcheva-Hoang T, Egen JG, Wojnoonski K, Allison JP (2004) B7-1 and B7-2 slectively recruit CTLA-4 and CD28 to the immunological synapse. Immunity 21:401–413

Perez VL, Van Parijs L, Bjuckians A, Zheng XX, Strom TB, Abbas AK (1997) Induction of peripheral T-cell tolerance in vivo requires CTLA-4 engagement. Immunity 6:411–417

Perrin PJ, June CH, Maldonado JH, Ratts RB, Racke MK (1999) Blockade of CD28 during in vitro activation of encephalitogenic T cells or after disease onset ameliorates experimental autoimmune encephalomyelitis. J Immunol 163:1704–1710

Prasad KVS, Cai YC, Raab M, Duckworth B Cantley L, Shoelson SE, Rudd CE (1994) T-cell antigen CD28 interacts with the lipid kinase phosphatidylinositol 3-kinase by a cytoplasmic Tyr(P)-Met-Xaa-Met motif. Proc Natl Acad Sci USA 91:2834–2838

Read S, Malmstrom V, Powrie F (2000) Cytotoxic T lymphocyte-associated antigen 4 plays an essential role in the function of CD25 (+)CD4 (+) regulatory cells that control intestinal inflammation. J Exp Med 192:295–302

Riley JL, June CH (2005) The CD28 family: a T-cell rheostat for therapeutic control of T-cell activation. Blood 105:13–21

Rottman JB, Smith T, Tonra JR, Ganley K, Bloom T, Silva R, Pierce B, Gutierrez-Ramos JC, Ozkaynak E, Coyle AJ (2001) The costimulatory molecule ICOS plays an important role in the immunopathogenesis of EAE. Nat Immunol 2:605–611

Rudd, CE (1990) CD4, CD8 and the TCR-CD3 complex: a novel class of protein-tyrosine kinase receptor. Immunol Today 11, 400–406

Rudd CE, Schneider H (2003) Unifying concepts in CD28, ICOS and CTLA-4 co-receptor signalling. Nat Rev Immunol 3:544–556

Rudd CE, Martin M, Schneider H (2002) CTLA-4 negative signaling via lipid rafts: A new perspective. Sci STKE 2002 (128) PE18

Sabatos CA, Chakravarti S, Cha E, Schubart A, Sanchez-Fueyo A, Zheng XX, Coyle AJ, Strom TB, Freeman GJ, Kuchroo VK (2003) Interaction of Tim-3 and Tim-3 ligand regulates T helper type 1 responses and induction of peripheral tolerance. Nat Immunol 4:1102–1110

Salomon B, Bluestone JA (2001) Complexities of CD28/B7: CTLA-4 costimulatory pathways in autoimmunity and transplantation. Annu Rev Immunol 19:225–252

Samelson LE (2002) Signal transduction mediated by the T-cell antigen receptor: the role of adapter proteins. Annu Rev Immunol 20:371–394

Schneider H, Rudd CE (2000) Tyrosine phosphatase SHP-2 binding to CTLA-4: absence of direct YVKM/YFIP motif recognition. Biochem Biophys Res Commun 269:279–283

Schneider H, Cai YC, Prasad KVS, Shoelson SE, Rudd CE (1995a) T-cell antigen CD28 binds to the GRB-2/SOS complex, regulators of $p21^{ras}$. Eur J Immunol 25:1044–1050

Schneider H, Prasad KVS, Shoelson SE, Rudd CE (1995b) CTLA-4 binding to the lipid kinase phosphatidylinositol 3 kinase in T cells. J Exp Med 181:351–355

Schneider H, Schwartzberg PL, Rudd CE (1998) Resting lymphocyte kinase (Rlk/Txx) phosphorylates the YVKM motif and regulates PI3K binding to T-cell antigen CTLA-4. Biochem Biophys Res Commun 252:14–19

Schneider H, Martin M, Agarraberes FA, Yin L, Rapoport I, Kirchhausen T, Rudd CE (1999) Cytolytic T lymphocyte-associated antigen-4 and the TcRζ/CD3 complex, but not CD28, interact with clathrin adaptor complexes AP-1 and AP-2. J Immunol 163:1868–1879

Schneider H, Mandelbrot DA, Greenwald RJ, Ng F, Lechler R, Sharpe AH, Rudd CE (2002) Cutting Edge: CTLA-4 (CD152) differentially regulates mitogen-activated protein kinases (extracellular signal-regulated kinase and c-Jun N-terminal kinase) in CD4+ T cells for receptor/ligand-deficient mice. J Immunol 169:3475–3479

Schneider H, Valk E, da Rocha Dias S, Wei B, Rudd CE (2005) CTLA-4 up-regulation of lymphocyte function-associated antigen 1 adhesion and clustering as an alternate basis for coreceptor function. Proc Natl Acad Sci USA 102:12861–12866

Schulze-Koops H, Lipsky PE, Davis LS (1998) Human memory T-cell differentiation into Th2-like effector cells is dependent on IL-4 and CD28 stimulation and inhibited by TCR ligation. Eur J Immunol 28:2517–2529

Schwartz, RH (1990) A cell culture model for T lymphocyte clonal anergy. Science 248:1349–1356

Schwartz JC, Zhang X, Fedorov AA, Nathenson SG, Almo SC (2001) Structural basis for co-stimulation by the human CTLA-4/B7-2 complex. Nature 410:604–608

Sedy JR, Gavrieli M, Potter KG, Hurchla MA, Coleman Lindsley R, Hildner K, Scheu S, Pfeffer K, Ware CF, Murphy TL, Murphy KM (2005) B and T lymphocyte attenuator regulates T-cell activation through interaction with herpesvirus entry mediator. Nat Immunol 6:90–98

Sharpe AH, Freeman GJ (2002) The B7-CD28 superfamily. Nat Rev Immunol 2:116–126

Shiratori T, Miyatake S, Ohno H, Nakaseko C, Isono K, Bonifacino JS, Saito T (1997) Tyrosine phosphorylation controls internalization of CTLA-4 by regulating its interaction with clathrin-associated adaptor complex AP-2. Immunity 6:583–589

Shlapatska LM, Mikhalap SV, Berdova AG, Zelensky OM, Yun TJ, Nichols KE, Clark EA, Sidorenko SP (2001) CD150 association with either the SH2-containing inositol phosphatase or the SH2-containing protein tyrosine phosphatase is regulated by the adaptor protein SH2D1A. J Immunol 166:5480–5487

Sica GL, Choi IH, Zhu G, Tamada K, Wang SD, Tamura H, Chapoval AI, Flies DB, Bajorath J, Chen L (2003) B7-H4, a molecule of the B7 family, negatively regulates T-cell immunity. Immunity 18:849–861

Stamper CC, Zhang Y, Tobin JF, Erbe DV, Ikemizu S, Davis SJ, Stahl ML, Seehra J, Somers WS, Mosyak L (2001) Crystal structure of the B7-1/CTLA-4 complex that inhibits human immune responses. Nature 410:608–611

Swallow MM, Wallin JJ, Sha WC (1999) B7h, a novel costimulatory homolog of B7-1 and B7-2, is induced by TNFalpha. Immunity 11:423–432

Tafuri A, Shahinian A, Bladt F, Yoshinaga SK, Jordana M, Wakeham A, Boucher LM, Bouchard D, Chan VS, Duncan G, et al (2001) ICOS is essential for effective T-helper-cell responses. Nature 409:31–32

Takagi J, Springer TA (2002) Integrin activation and structural rearrangement. Immunol Rev 186:141–163

Takahashi T, Tagami T, Yamazaki S, Uede T, Shimizu J, Sakaguchi N, Mak TW, Sakaguchi S (2000) Immunologic self-tolerance maintained by CD25+CD4+ regulatory T cells constitutively expressing cytotoxic T lymphocyte-associated antigen 4. J Exp Med 192:303–310

Tamura H, Dong H, Zhu G, Sica GL, Flies DB, Tamada K, Chen L (2001) B7-H1 costimulation preferentially enhances CD28-independent T-helper cell function. Blood 97:1809–1816

Teft WA, Kirchhof MG, Madrenas J (2005) A Molecular Perspective of CTLA-4 function. Annu Rev Immunol 3.1–3.33

Tesciuba AG, Subudhi S, Rother RP, Faas SJ, Frantz AM, Elliot D, Weinstock J, Matis LA, Bluestone JA, Sperling AI (2001) Inducible costimulator regulates Th2-mediated inflammation, but not Th2 differentiation, in a model of allergic airway disease. J Immunol 167:1996–2003

Thompson CB, Allison JP (1997) The emerging role of CTLA-4 as an immune attenuator. Immunity 7:445–450

Tivol EA, Borriello F, Schweitzer AN, Lynch WP, Bluestone JA, Sharpe AH (1995) Loss of CTLA-4 leads to massive lymphoproliferation and fatal multiorgan tissue destruction, revealing a critical negative regulatory role of CTLA-4. Immunity 3:541–547

Truitt KE, Hicks CM, Imboden JB (1994) Stimulation of CD28 triggers an association between CD28 and phosphatidylinositol-3-kinase in Jurkat cells. J Exp Med 179:1071–1076

Tseng SY, Otsuji M, Gorski K, Huang X, Slansky JE, Pai SI, Shalabi A, Shin T, Pardoll DM, Tsuchiya H (2001) B7-DC, a new dendritic cell molecule with potent costimulatory properties for T cells. J Exp Med 193:839–846

Ueda H, Howson JM, Esposito L, Heward J, Snook H, Chamberlain G, Rainbow DB, Hunter KM, Smith AN, Di Genova G, et al (2003) Association of the T-cell regulatory gene CTLA-4 with susceptibility to autoimmune disease. Nature 423:506–511

Umetsu DT, McIntire JJ, Akbari O, Macaubas C, DeKruyff RH (2002) Asthma: an epidemic of dysregulated immunity. Nat Immunol 3:715–720

Valk E, Leung R, Kang H, Kaneko K, Rudd CE, Schneider H (2006) T-cell receptor-interacting molecule acts as a chaperone to modulate surface expression of the CTLA-4 coreceptor. Immunity 25:1–15

Vibhakar R, Juan G, Traganos F, Darzynkiewicz Z, Finger LR (1997) Activation-induced expression of human *programmed death-1* gene in T-lymphocytes. Exp Cell Res 232:25–28

Vijayakrishnan L, Slavik JM, Illes Z, Greenwald RJ, Rainbow D, Greve B, Peterson LB, Hafler DA, Freeman GJ, Sharpe AH, et al (2004) An autoimmune disease-associated CTLA-4 splice variant lacking the B7 binding domain signals negatively in T cells. Immunity 20:563–575

Viola A, Schroeder S, Sakakibara Y, Lanzavecchia A (1999) T lymphocyte costimulation mediated by reorganization of membrane microdomains. Science 283:680–682

Vivier E, Daeron M (1997) Immunoreceptor tyrosine-based inhibition motif. Immunol Today 18:286–291

Wang S, Zhu G, Chapoval AI, Dong H, Tamada K, Ni J, Chen L (2000) Costimulation of T cells by B7-H2, a B7-like molecule that binds ICOS. Blood 96:2808–2813

Watanabe N, Gavrieli M, Sedy JR, Yang J, Fallarino F, Loftin SK, Hurchla MAS, Zimmerman N, Sim J, Zang X, et al (2003) BTLA is a lymphocyte inhibitory receptor with similarities to CTLA-4 and PD-1. Nat Immunol 4:630–638

Waterhouse P, Penninger JM, Timms E., Wakeham A, Shahinian A, Lee KP, Thompson CB, Griesser H, Mak TW (1995) Lymphoproliferative disorders with early lethality in mice deficient in Ctla-4. Science 270:985–988

Xavier R, Brennan T, Li Q, McCormack C, Seed B (1998) Membrane compartmentation is required for efficient cell activation. Immunity 8:723–732

Yi LA, Hajialiasgar, Chuang E (2004) Tyrosine-mediated inhibitory signals contribute to CTLA-4 function in vivo. Int Immunol 16:539–547

Yin L, Schneider H, Rudd CE (2003) Short cytoplasmic segment of CD28 is sufficient to convert CTLA-4 to a positive signalling receptor. J Leukoc Biol 73:178–182

Yoshinaga SK, Zhang M, Pistillo J, Horan T, Khare SD, Miner K, Sonnenberg M, Boone T, Brankow D, Dai T, et al (2000) Characterization of a new human B7-related protein: B7RP-1 is the ligand to the co-stimulatory protein ICOS. Int Immunol 12:1439–1447
Zhang Y, Allison JP (1997) Interaction of CTLA-4 with AP-50, a clathrin-coated pit adaptor protein. Proc Natl Acad Sci USA 94:9273–9278
Zhu C, Anderson AC, Schubart A, Xiong H, Imitola J, Khoury SJ, Zheng XX, Strom TB, Kuchroo VK (2005) The Tim-3 ligand galectin-9 negatively regulates T helper type 1 immunity. Nat Immunol 12:1245–1252

Index

Made in the USA
Monee, IL
31 December 2022